Thomas Lengauer (Ed.)
Bioinformatics –
From Genomes to Drugs
Vol. I: Basic Technologies

Methods and Principles
in Medicinal Chemistry

Edited by
R. Mannhold
H. Kubinyi
H. Timmerman

Editorial Board
G. Folkers, H.-D. Höltje, J. Vacca,
H. van de Waterbeemd, T. Wieland

Bioinformatics – From Genomes to Drugs
Volume I: Basic Technologies

Edited by Thomas Lengauer

WILEY-VCH

Series Editors

Prof. Dr. Raimund Mannhold
Biomedical Research Center
Molecular Drug Research Group
Heinrich-Heine-Universität
Universitätsstraße 1
D-40225 Düsseldorf
Germany

Prof. Dr. Hugo Kubinyi
BASF AG, Ludwigshaften
c/o Donnersbergstrasse g
D-67256 Weisenheim am Sand
Germany

Prof. Dr. Gerd Folkers
Department of Applied Biosciences
ETH Zürich
Winterthurer Str. 190
CH-8057 Zürich
Switzerland

Volume Editor:

Prof. Dr. Thomas Lengauer, Ph.D.
Fraunhofer Institute for Algorithms and
Scientific Computing (SCAI)
Schloss Birlinghoven
D-53754 Sankt Augustin
Germany

Library of Congress Card No.: applied for
A catalogue record for this book is available
from the British Library. Die Deutsche
Bibliothek – CIP Cataloguing-in-
Publication-Data A catalogue record for this
publication is available from Die Deutsche
Bibliothek

Printed in the Federal Republic of
Germany.
Printed on acid-free paper.

Typesetting Asco Typesetters,
Hong Kong
Printing betz-druck GmbH, Darmstadt
Bookbinding J. Schäffer GmbH & Co. KG,
Grünstadt

ISBN 3-527-29988-2

Preface

The present volume of our series "Methods and Principles in Medicinal Chemistry" focuses on a timely topic: Bioinformatics. Bioinformatics is a multidisciplinary field, which encompasses molecular biology, biochemistry and genetics on the one hand, and computer science on the other. Bioinformatics uses methods from various areas of computer science, such as algorithms, combinatorial optimization, integer linear programming, constraint programming, formal language theory, neural nets, machine learning, motif recognition, inductive logic programming, database systems, knowledge discovery and database mining. The exponential growth in biological data, generated from national and international genome projects, offers a remarkable opportunity for the application of modern computer science. The fusion of biomedicine and computer technology offers substantial benefits to all scientists involved in biomedical research in support of their general mission of improving the quality of health by increasing biological knowledge. In this context, we felt that it was time to initiate a volume on bioinformatics with a particular emphasis on aspects of designing new drugs.

The completion of the human genome sequence, published in February 2001, marks a historic event, not only in genomics, but also in biology and medicine in general. We are now able to read the text; but we understand only minor parts of it. "Making sense of the sequence" is the task of the coming years. Bioinformatics will play the leading role in this field, in understanding the regulation of gene expression, in the functional description of the gene products, the metabolic processes, disease, genetic variation and comparative biology. Correspondingly, the publication of this book is "just in time" to jump into the post-genomic era.

Basically, there are two ways of structuring the field of bioinformatics. One is intrinsically by the type of problem that is under consideration. Here, the natural way of structuring is by layers of information that are compiled, starting from the genomic data. The second is extrinsically, by the application scenario in which bioinformatics operates and by the type of molecular biology experiment that it supports. This new contribution to bioinformatics is roughly structured according to this view. The wealth of information bundled in this volume necessitated a subdivision into two parts.

The intrinsic view is the subject of Part 1: it structures bioinformatics in methodical layers. Lower layers operate directly on the genomic text that is the result of sequencing projects. Higher layers operate on higher-level information derived from this text. Accordingly, Part 1 discusses sub-problems of bioinformatics that provide components in a global bioinformatics solution. Each chapter is devoted to one relevant component: after an introductory overview, Chapters follow that are devoted to Sequence Analysis (*written by Martin Vingron*), Structure, Properties and Computer Identification of Eukaryotic Genes (*by Victor Solovyev*), Analyzing Regulatory Regions in Genomes (*by Thomas Werner*), Homology-Based Protein Modeling in Biology and Medicine (*by Roland Dunbrack*), Protein Structure Prediction and Applications in Structural Genomics, Protein Function Assignment and Drug Target Finding (*by Ralf Zimmer and Thomas Lengauer*), Protein–Ligand Docking and Drug Design (*by Matthias Rarey*) and Protein–Protein and Protein–DNA Docking (*by Mike Sternberg and Gidon Moont*). An appendix by *Thomas Lengauer*, sketching the algorithmic methods that are used in bioinformatics, concludes this first Part.

The extrinsic view is the focus of the second Part: Chapters concentrate on several important application scenarios that can only be supported effectively by combining components discussed in Part 1. These Chapters cover Integrating and Accessing Molecular Biology Resources (*by David Hansen and Thure Etzold*), Bioinformatics Support of Genome Sequencing Projects (*by Xiaoqiu Huang*), Analysis of Sequence Variations (*by Christopher Carlson et al.*), Proteome Analysis (*by Pierre-Alan Binz et al.*), Target Finding in Genomes and Proteomes (*by Stefanie Fuhrman et al.*) as well as Screening of Drug Databases (*by Martin Stahl et al.*). In a concluding Chapter, *Thomas Lengauer* highlights the Future Trends in the field of bioinformatics.

The series editors are grateful to Thomas Lengauer, who accepted the challenging task to organize this volume on bioinformatics, to convince authors to participate in the project and to finish their chapters in time, despite the fact that research runs hot these days. We are sure that the result of his coordinative work constitutes another highlight in our series on Methods and Principles in Medicinal Chemistry. In addition, we want to thank Gudrun Walter and Frank Weinreich, Wiley-VCH, for their effective collaboration.

September 2001	Raimund Mannhold	Düsseldorf
	Hugo Kubinyi	Ludwigshafen
	Henk Timmerman	Amsterdam

Contents

Part I: Basic Technologies

Part II: Applications

List of Contributors

Prof. Ron D. Appel
Swiss Institute of Bioinformatics
Proteome Informatics Group
CMU – 1, rue Michel Servet
1211 Geneva 4
Switzerland
ron.appel@isb-sib.ch

Prof. Roland L. Dunbrack, Jr
Institute for Cancer Research
Fox Chase Cancer Center
7701 Burholme Avenue
Philadelphia, PA 19111
USA
rl_dunbrack@fccc.edu

Dr. Thure Etzold
Lion Bioscience Ltd.
Sheraton House, Castle Business Park
Cambridge CB3 OAX
UK
etzold@lionbio.uk.com

Prof. Xiaoqiu Huang
Department of Computer Science
Iowa State University
226 Atanasoff Hall
Ames, IA 50011
USA
xghuang@cs.iastate.edu

Prof. Gerhard Klebe
Philipps-Universität Marburg
Institut für Pharmazeutische Chemie
Marbacher Weg 6
35032 Marburg
Germany
klebe@mailer.uni-marburg.de

Prof. Thomas Lengauer
Fraunhofer Institute for Algorithms
    and Scientific Computing
Schloss Birlinghoven
53754 Sankt Augustin
Germany
thomas.lengauer@gmd.de
(present address:
Max-Planck-Institute
    for Computer Science
Stuhlsatzenhausweg 85
66123 Saarbrücken
Germany)

Prof. Deborah Nickerson
University of Washington
Department of Molecular
    Biotechnology
Box 357730
Seattle, WA 98195
USA
debnick@u.washington.edu

Dr. Matthias Rarey
Fraunhofer Institute for Algorithms
    and Scientific Computing
Schloss Birlinghoven
53754 Sankt Augustin
Germany
matthias.rarey@gmd.de

Victor Solovyev
EOS Biotechnology
225A Gateway Boulevard
South San Francisco, CA 94080
USA
solovyev@eosbiotech.com

Dr. Roland Somogyi
Incyte Pharmaceuticals, Inc.
3174 Porter Dr.
Palo Alto, CA 94304
USA
rsomogyi@incyte.com

Dr. Michael J. E. Sternberg
Imperial Cancer Research Fund
Biomolecular Modelling Laboratory
44 Lincoln's Inn Fields
London WC2A 3PX
UK
m.sternberg@icrf.icnet.uk

Dr. Martin Vingron
German Cancer Research Centre
 (DKFZ)
Theoretical Bioinformatics Division
Im Neuenheimer Feld 280
69120 Heidelberg
Germany
m.vingron@dkfz-heidelberg.de

Dr. Thomas Werner
GSF-National Research Center for
 Environment and Health
Institute of Experimental Genetics
Ingolstädter Landstrasse 1
85764 Neuherberg
Germany
werner@gsf.de

Dr. Ralf Zimmer
Fraunhofer Institute for Algorithms
 and Scientific Computing
Schloss Birlinghoven
53754 Sankt Augustin
Germany
ralf.zimmer@gmd.de

For my son Nico

Foreword

Computational biology and *bioinformatics* are terms for an interdisciplinary field joining information technology and biology that has skyrocketed in recent years. The field is located at the interface between the two scientific and technological disciplines that can be argued to drive a significant if not the dominating part of contemporary scientific innovation. In the English language, computational biology refers mostly to the scientific part of the field, whereas bioinformatics addresses mainly the infrastructure part. In some other languages (e.g. German) bioinformatics covers both aspects of the field.

The goal of this field is to provide computer-based methods for coping with and interpreting the genomic data that are being uncovered in large volumes through the diverse genome sequencing projects and other new experimental technology in molecular biology. The field presents one of the grand challenges of our times. It has a large basic research aspect, since we cannot claim to be close to understanding biological systems on an organism or even cellular level. At the same time, the field is faced with a strong demand for immediate solutions, because the genomic data that are being uncovered encode many biological insights whose deciphering can be the basis for dramatic scientific and economical success. At the end of the pregenomic era that was characterized by the effort to sequence the human genome we are entering the postgenomic era that concentrates on harvesting the fruits hidden in the genomic text. In contrast to the pregenomic era which, from the announcement of the quest to sequence the human genome to its completion, has lasted less than 15 years, the postgenomic era can be expected to last much longer, probably extending over several generations.

While it will encompass many basic and general aspects of the field, the specific aim of the book is to point towards perspectives that bioinformatics can open towards the design of new drugs. It is this pharmaceutically oriented side of the field that provides the strongest fuel for the current widespread interest in bioinformatics.

Before this background, the book is intended as an introduction into the field of bioinformatics and is targeted to readers with a variety of backgrounds. Biologists, biochemists, pharmacologists, pharmacists and medical doctors can get from it an introduction into basic and practical issues of the computer-based part of handling and interpreting genomic data. In particular, many chapters of the book point to bioinformatics software and data resources that are available on the internet (often at no cost) and make an attempt at classifying and comparing those resources. For computer scientists and mathematicians the book contains an introduction into the biological backgrounds and the necessary information to begin to appreciate the difficulties and wonders of modeling complex biochemical and biomolecular issues in the computer.

Bioinformatics is a quickly progressing field. Both the experimental technologies and the computer-based method are in a dynamic phase of development. This book presents a snapshot of where the field stands today.

I am grateful to many people that helped make this book possible. Hugo Kubinyi first approached me with the idea for this book and since then accompanied the project with a finely balanced mixture of pressure and encouragement. Raimund Mannhold and Henk Timmerman supported the project as the other editors of this series. Above all, I thank the authors of the various chapters of the two volumes who found the time to write well thought-out chapters in a period of dramatic growth of the field that presents every participating scientist with an especially high workload. Gudrun Walter and Frank Weinreich handled the production process of the book very well. Finally I would like to express my deep gratitude to my wife Sybille and my children Sara and Nico who had to cope with my physical or mental absence while the project was ongoing.

Thomas Lengauer

Sankt Augustin, July 2001

Part I: Basic Technologies

1
From Genomes to Drugs with Bioinformatics

Thomas Lengauer

In order to set the stage for this two-volume book, this Chapter provides an introduction into the molecular basis of disease. We then continue to discuss modern biological techniques with which we have recently been empowered to screen for molecular drugs targets as well as for the drugs themselves. The Chapter finishes with an overview over the field of bioinformatics as it is covered in the two volumes of this book.

1.1
The molecular basis of disease

Diagnosing and curing diseases has always been and will continue to be an art. The reason is that man is a complex being with many facets much of which we do not and probably will never understand. Diagnosing and curing diseases has many aspects include biochemical, physiological, psychological, sociological and spiritual ones.

Molecular medicine reduces this variety to the molecular aspect. Living organisms, in general, and humans in particular, are regarded as complex networks of molecular interactions that fuel the processes of life. This "molecular circuitry" has intended modes of operation that correspond to healthy states of the organism and aberrant modes of operation that correspond to diseased states. In molecular medicine, the goal of diagnosing a disease is to identify its molecular basis, i.e. to answer the question what goes wrong in the molecular circuitry. The goal of therapy is to guide the biochemical circuitry back to a healthy state.

As we pointed out, the molecular basis of life is formed by complex biochemical processes that constantly produce and recycle molecules and do so in a highly coordinated and balanced fashion. The underlying basic principles are quite alike throughout all kingdoms of life, even though the processes are much more complex in highly developed animals and the human than in bacteria, for instance. Figure 1.1 gives an abstract view of such an

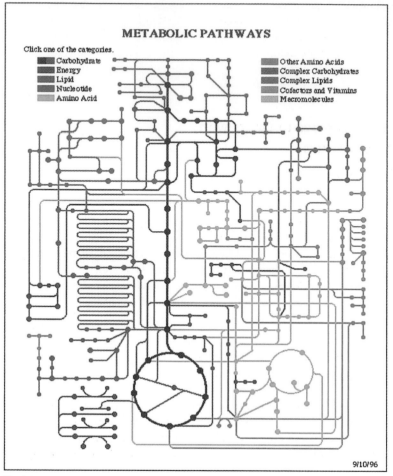

Fig. 1.1
Abstract view of part of the metabolic network of the
bacterium *E. coli*. From http://www.genome.ad.jp/kegg/
kegg.html

underlying biochemical network, the so-called *metabolic network* of a bacterial cell (the intestinal bacterium *E. coli*). The figure affords an incomplete
and highly simplified account of the actual molecular interactions, but it
nicely visualizes the view of a living cell as a biochemical circuit. The figure
has the mathematical structure of a *graph*. Each dot (*node*) stands of a small
organic molecule that is metabolized within the cell. Alcohol, glucose, and
ATP are examples for such molecules. Each line (*edge*) stands for chemical
reaction. The two nodes connected by the edge represent the substrate and
the product of the reaction. The colors in Figure 1.1 represent the role that
the respective reaction plays in metabolism. These roles include the con-

struction of molecular components that are essential for life – nucleotides (red), amino acids (orange), carbohydrates (blue), lipids (light blue), etc. – or the breakdown of molecules that are not helpful or even harmful to the cell. Other tasks of chemical reactions in a metabolic network pertain to the storage and conversion of energy. (The blue cycle in the center of the Figure is the citric acid cycle.) A third class of reactions facilitates the exchange of information in the cell or between cells. This include the control of when and in what way genes are expressed (*gene regulation*), but also such tasks as the opening and closing of molecular channels on the cell surface, and the activation or deactivation of cell processes such as replication or apoptosis (induced cell death). The reactions that facilitate communication within the cell or between cells are often collectively referred to as the *regulatory net-work*. Figure 1.1 only includes metabolic and no regulatory reactions. Of course the metabolic and the regulatory network of a cell are closely inter-twined, and many reactions can have both metabolic and regulatory aspects. In general, much more is known on metabolic than on regulatory networks, even though many relevant diseases involve regulatory rather than meta-bolic dysfunction.

The metabolic and regulatory network can be considered as composed of partial networks that we call *pathways*. Pathways can fold in on themselves, in which case we call them *cycles*. A metabolic pathway is a group of re-actions that turns a substrate into a product over several steps (pathway) or recycle a molecule by reproducing in several steps (cycle). The *glycolysis pathway* (the sequence of blue vertical lines in the center of Figure 1.1), is an example of a pathway that decomposes glucose into pyruvate. The *citric acid cycle* (the blue cycle directly below the glycolysis pathway in Figure 1.1) is an example of a cycle that produces ATP, the universal molecule for energy transport. Metabolic cycles are essential, in order that the processes of life not accumulate waste or deplete resources. (Nature is much better at recycling than man.)

There are several ways in which Figure 1.1 hides important detail of the actual metabolic pathway. In order to discuss this issue, we have extracted a metabolic cycle from Figure 1.1 (see Figure 1.2). This cycle is a component of cell replication, more precisely; it is one of the motors that drive the synthesis of thymine, a molecular component of DNA. In Figure 1.2, the

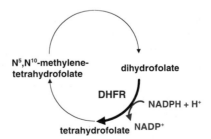

Fig. 1.2
A specific metabolic cycle

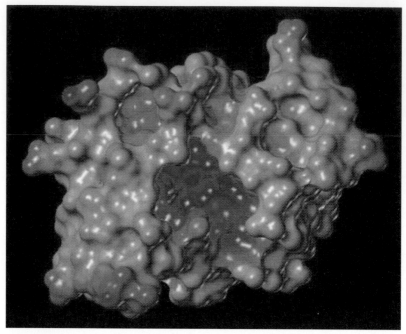

Fig. 1.3
The three dimensional structure of dihydrofolate reductase
colored by its surface potential. Positive values are
depicted in red, negative values in blue.

nodes of the metabolic cycle are labeled with the respective organic molecules, and the edges point in the direction from the substrate of the reaction
to the product. Metabolic reactions can take place spontaneously under
physiological conditions (in aqueous solution, under room temperature and
neutral pH). However, nature has equipped each reaction (each line in Figure 1.1) with a specific molecule that catalyzes that reaction. This molecule
is called an *enzyme* and, most often, it is a protein. An enzyme is a tailor-
made binding site for the transition state of the catalyzed chemical reaction.
Thus the enzyme speeds up the rate of that reaction tremendously, by rates
of as much as 10^7. Furthermore, the rate of a reaction that is catalyzed by an
enzyme can be regulated by controlling the effectivity of the enzyme or the
number of enzyme molecules that are available. Even the direction of a reaction can effectively be controlled with the help of several enzymes.

How does the enzyme do its formidable task? For an example, consider
the reaction in Figure 1.2 that turns dihydrofolate into tetrahydrofolate. This
reaction is catalyzed by an enzyme called *dihydrofolate reductase (DHFR)*.
The surface of this protein is depicted in Figure 1.3. One immediately recognizes a large and deep pocket that is colored blue (representing its negative charge). This pocket is a *binding pocket* or *binding site* of the enzyme,

and it is ideally formed in terms of geometry and chemistry, such as to bind to the substrate molecule dihydrofolate and present it in a conformation that is conducive for the desired chemical reaction to take place. In this case, this pocket is also where the reaction is catalyzed. We call this place the *active site*. (There can be other binding pockets in a protein that are far removed from the active site.)

There is another aspect of metabolic reactions that is not depicted in Figure 1.1: Many reactions involve *co-factors*. A co-factor is an organic molecule, a metal ion, or – in some cases – a protein or peptide that has to be present in order for the reaction to take place. If the co-factor is itself modified during the reaction, we call it a *co-substrate*. In the case of our example reaction, we need the co-substrate NADPH for the reaction to happen. The reaction modifies dihydrofolate to tetrahydrofolate and NADPH to NADP$^+$. Figure 1.4 shows the molecular complex of DHFR, dihydrofolate (DHF) and NADPH before the reaction happens. After the reaction has been completed, both organic molecules dissociate from DHFR and the original state of the enzyme is recovered.

Now that we have discussed some of the details of metabolic reactions let us move back to the global view of Figure 1.1. We have seen that each of the edges in that figure represents a reaction that is catalyzed by a specific pro-

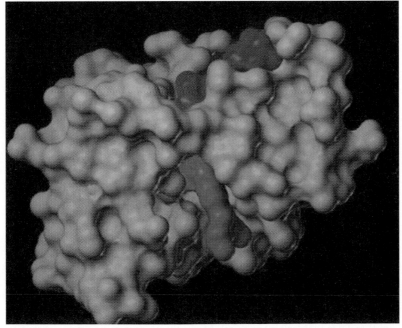

Fig. 1.4
DHFR (gray) complexed with DHF (green) and NADPH red)

tein. (However, the same protein can catalyze several reactions.) In *E. coli* there are an estimated number of 3000 enzymes, in the human there are at least twice as many. The molecular basis of a disease lies in modifications of the action of these biochemical pathways. Some reactions do not happen at their intended rate (sickle cell anemia, diabetes), resources that are needed are not present in sufficient amounts (vitamin deficiencies) or waste products accumulate in the body (Alzheimer's disease). In general imbalances induced in one part of the network spread to other parts. The aim of therapy is to replace the aberrant processes with those that restore a healthy state. The most desirable fashion, in which this could be done, would be to control the effectiveness of a whole set of enzymes, in order to regain the metabolic balance. This set probably involves many, many proteins, as we can expect many proteins to be involved in manifesting the disease. Also, each of these proteins would have to be regulated in quite a specific manner. The effectiveness of some proteins would possibly have to be increased dramatically whereas other proteins would have to be blocked entirely etc. It is obvious that this kind of therapy involves a kind of global knowledge of the workings of the cell and a refined pharmaceutical technology that is far beyond what man can do today and for some time to come.

1.2
The molecular approach to curing diseases

For this reason, the approach of today's pharmaceutical research is far more simplistic. The aim is to regulate a single protein. In some cases we aim at completely blocking an enzyme. To this end, we can provide a drug molecule that effectively competes with the natural substrate of the enzyme. The drug molecule, the so-called *inhibitor*, has to be made up such that it binds more strongly to the protein than the substrate. Then, the binding pockets of most enzyme molecules will contain drug molecules and cannot catalyze the desired reaction in the substrate. In some cases, the drug molecule even binds very tightly (covalently) to the enzyme (suicide inhibitor). This bond persists for the remaining lifetime of the protein molecule. Eventually, the deactivated protein molecule is broken down by the cell and a new identical enzyme molecule takes its place. Aspirin is an example of a suicide inhibitor. The effect of the drug persists until the drug molecules themselves are removed from the cell by its metabolic processes, and no new drug molecules are administered to replace them. Thus, one can control the effect of the drug by the time and dose it is administered.

There are several potent inhibitors of DHFR. One of them is *methotrexate*. Figure 1.5 shows methotrexate (color) both unbound (left) and bound (right) to DHFR (black and white). Methotrexate has been administered as an effective cytostatic cancer drug for over two decades.

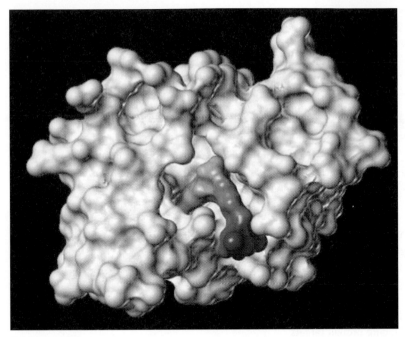

Fig. 1.5
Methotrexate (colored by surface potential, see Figure 3)
and bound DHFR (gray)

There are many other ways of influencing the activity of a protein by providing a drug that binds to it. Drugs interact with all kinds of proteins:

- with receptor molecules that are located in the cell membrane and fulfill regulatory tasks
- with ion channels and transporter systems (again proteins residing in the cell membrane) that monitor the flux of molecules into and out of the cell

The mode of interaction between drug and protein does not always have the effect to block the protein, but we are generally looking for drugs that bind tightly to the protein.

Most drugs that are on the market today modify the enzymatic or regulatory action of a protein by strongly binding to it as described above. Among these drugs are long-standing, widespread and highly popular medications and more modern drugs against diseases such as AIDS, depression, or cancer. Even the life-style drugs that came into use in the past few years, such as Viagra and Xenical, belong to the class of protein inhibitors.

In this view, the quest for a molecular therapy of a disease decomposes into two parts:

- *Question 1: Which protein should we target?* As we have seen, there are many thousands candidate proteins in the human. We are looking for one of them that, by binding the drug molecule to it, provides the most effective remedy of the disease. This protein is called the *target protein.*
- *Question 2: Which drug molecules should be used to bind to the target protein?* Here, the molecular variety is even larger. Large pharmaceutical companies have compound archives with millions of compounds at their disposal. Every new target protein raises the questions, which of all of these compounds would be the best drug candidate. Nature is using billions of molecules. With the new technology of combinatorial chemistry, where compounds can be synthesized systematically from a limited set of building blocks, this number of *potential* drug candidates is also becoming accessible to the laboratory.

We will now give a short summary of the history of research on both of these questions.

1.3
Finding protein targets

Question 1 could not really be asked realistically until a few years ago. Historically, few target proteins were known at the time that the respective drug has been discovered. The reason is that new drugs were developed by modifying known drugs, based on some intuitive notion of molecular similarity. Each modification was immediately tested in the laboratory either in vitro or in vivo. Thus, the effectiveness of the drug could be assessed without even considering the target protein. To this day, all drugs that are on the marketplace world-wide target to an estimated set of 500 proteins [1]. Thus the search for target proteins is definitely the dominant bottleneck of today's pharmaceutical research.

Today, new experimental methods of molecular biology, the first versions of which have been developed just a few years ago, afford us with a fundamentally new way – the first systematic way – of looking for protein targets. We exemplify this progress at a specific DNA chip technology [2]. However, the general picture extends to many other experimental methods under development.

Figure 1.6 shows a DNA-chip that provides us with a differential census of the proteins manufactured by a yeast cell in two different cell states, one governed by the presence of glucose (green) and one by the absence of glucose (red). In effect the red picture is that of a starving yeast cell, whereas the green picture show the "healthy" state. Each bright green dot stands for a protein that is manufactured (expressed) in high numbers in the "healthy" state. Each bright red dot stands for a protein that is expressed

Fig. 1.6
A DNA chip (from http://cmgm.stanford.edu/pbrown/
explore/)

highly in the starving cell. If the protein occurs frequently in both the healthy and the starving state, the corresponding dot is bright yellow (resulting from an additive mixture of the colors green and red). Dark dots stand for proteins that are not frequent, the tint of the color again signifies whether the protein occurs more often in the healthy cell (green), equally often in both cells (yellow) or more often in the diseased cell (red).

At this point it is of secondary importance, exactly which experimental procedures generate the picture of Figure 1.6. What is important is, how much information is attached to colored dots in the picture. Here, we can make the following general statements.

1. The identity of the protein is determined by the coordinates of the colored dot. We will assume, for simplicity, that dots at different locations also represent different proteins. (In reality, multiple dots that represent

the same protein are introduced, on purpose, for the sake of calibration.) The exact arrangement of the dots is determined before the chip is manufactured. This involves identifying a number of proteins to be represented on the chip and laying them out on the chip surface. This layout is governed by boundary conditions and preferences of the experimental procedures and is not important for the interpretation of the information.
2. Only rudimentary information is attached to each dot. At best, the experiment reveals the complete sequence of the gene or protein. Sometimes, only short segments of the relevant sequence are available.

Given this general picture, the new technologies of molecular biology can be classified according to two criteria, as shown by the next subsections.

1.3.1
Genomics vs proteomics

In genomics, not the proteins themselves are monitored but rather we screen the expressed genes whose translation ultimately yields the respective proteins. In proteomics, the synthesized proteins themselves are monitored. The chip in Figure 1.6 is a DNA-chip, i.e., it contains information on the expressed genes and, thus, only indirectly on the final protein products. The advantage of the genomics approach is that genes are more accessible experimentally and easier to handle than proteins. For this reason, genomics is ahead of proteomics, today. However, there also are disadvantages to genomics. First the expression level of a gene need not be closely correlated with the concentration of the respective protein in the cell. But the latter figure may be more important to us if we want to elicit a causal connection between protein expression and disease processes. Even more important, proteins are modified post-translationally (i.e. after they are manufactured). These modifications involve glycosylation (attaching complex sugar molecules to the protein surface), and phosphorylation (attaching phosphates to the protein surface), for instance, and they lead to many versions of protein molecules with the same amino-acid sequence. Genomics cannot monitor these modifications, which are essential for many diseases. Therefore, it can be expected that, as the experimental technology matures, proteomics will gain importance over genomics.

1.3.2
Extent of information available on the genes/proteins

Technologies vary widely in this respect. The chip in Figure 1.6 is generated by a technology that identifies the gene sequence. We are missing information on the structure and the function of the protein, its molecular interac-

tion partners and its location inside the metabolic or regulatory network of
the organism. All of this information is missing the majority of the genes
on the chip.

There are many variations on the DNA-chip theme. There are technologies based on so-called *expressed sequence tags (ESTs)* that tend to provide
more inaccurate information on expression levels, and various sorts of
microarray techniques (see Chapter 5 of Volume 2). Proteomics uses different kinds of gels, and experiments involving mass spectrometry (see Chapter 4 of Volume 2). As is the case with genomics, proteomics technologies
tend to generate information on the sequences of the involved proteins and
on their molecular weight, and possibly information on post-translational
modifications such as glycosylation and phosphorylation. Again, all higher-
order information is missing. It is infeasible to generate this information
exclusively in the wet lab – we need bioinformatics to make educated
guesses here. Furthermore basically all facets of bioinformatics that start
with an assembled sequence can be of help. This includes the comparative
analysis of genes and proteins, protein structure prediction, the analysis and
prediction of molecular interactions involving proteins as well as bioinformatics for analyzing metabolic and regulatory networks. This is why all of
bioinformatics is relevant for the purpose of this book. Thus, part 1 of the
book summarizes the state of the art of the relevant basic problems in bioinformatics.

If, with the help of bioinformatics, we can retrieve this information then
we have a chance of composing a detailed picture of the disease process
that can guide us to the identification of possible target proteins for the
development of an effective drug. Note that the experimental technology
described above is universally applicable. The chip in Figure 1.6 contains all
genes of the (fully sequenced) organism *S. cerevisiae* (yeast). The cell transition analyzed here is the diauxic shift – the change of metabolism upon
removal of glucose. But we could exchange this with almost any other cell
condition of any tissue of any conceivable organism. The number of spots
that can be put on a single chip goes into the hundred thousands. This is
enough for putting all human genes on a single chip. Also, we do not have
to restrict ourselves to disease conditions; all kinds of environmental conditions (heat, cold, low or high pH, chemical stress, drug treatment, diverse
stimuli etc.) or intrinsic conditions (presence or absence of certain genes,
mutations etc.) can be the subject of study.

This paradigm of searching for target proteins in genomics data is considered to be so promising that pharmaceutical industry has been investing
heavily in it in over the past years. Contracts between pharmaceutical companies and new biotech industry that is providing the data and clues to new
targets have involved hundreds of million dollars. In turn, each new target
protein can afford a completely new approach to disease therapy and a potentially highly lucrative worldwide market share. Yet, the paradigm still has
to prove its value. These commercial aspects of the research have also

strongly affected the scientific discipline of bioinformatics. We will comment more on this in the second volume of this book.

Providing adequate bioinformatics for finding new target proteins is a formidable challenge that is the focus of much of this book. But once we have a target protein our job is not done.

1.4
Developing drugs

If the target protein has been selected, we are looking for a molecule that binds tightly to a binding site of the protein. Nature often uses macromolecules, such as proteins or peptides to inhibit other proteins. However, proteins do not make good drugs: They are easily broken down the digestive system, they can elicit immune reaction, and they cannot be stored for a long time. Thus, after a short excursion into drug design based on proteins, pharmaceutical research has gone back to looking for small drug molecules. Here, one idea is to use a peptide as the template for an appropriate drug (peptidomimetics).

Due to the lack of fundamental knowledge of the biological processes involved, the search for drugs was governed by chance, until recently. However, as long as chemists have thought in terms of chemical formulas, pharmaceutical research has attempted to optimize drug molecules based on chemical intuition and on the concept of molecular similarity. The basis for this approach is the lock-and-key principle formulated by Emil Fischer [3] about a hundred years ago. Figure 1.4 nicely illustrates that principle: In order to bind tightly the two binding molecules have to be complementary to each other both sterically and chemically (colors). The drug molecule fits into the binding pocket of the protein like a key inside a lock. The lock-and-key principle has been the dominating paradigm for drug research ever since its invention. It has been refined to include the phenomenon of induced fit, by which the binding pocket of the protein undergoes subtle steric changes in order to adapt to the geometry of the drug molecule.

For most of the century, the structure of protein binding pockets has not been available to the medicinal chemist, so drug design was based on the idea that molecules whose surface is similar in shape and chemical features should bind to the target protein with comparable strength. Thus drug design was based on comparing drug molecules, either intuitively or, more recently, systematically with the computer. As 3D protein structures became available, the so-called *rational* or *structure-based* approach to drug development was invented, which exploited this information to develop effective drugs. Rational drug design is a highly interactive process with the computer originally mostly visualizing the protein structure and allowing queries on its chemical features. The medicinal chemist modified drug molecules inside the binding pocket of the protein at the computer screen. As rational drug design began to involve more systematic optimization procedures interest rose in *molecular docking*, i.e. the prediction of the structure and binding affinity of the molecular complex involving a structurally resolved protein and its binding partner, in the computer. Synthesizing and testing a drug in the laboratory used to be comparatively expensive. Thus it was of interest to have the computer suggest a small set of highly promising drug candidates. However, to this day rational drug design has been hampered by the fact that the structure of the target protein will not be known for many pharmaceutical projects for some time to come. For instance, many diseases involve target proteins that reside in the cell membrane, and we cannot expect the 3D structure of such proteins to become known soon.

With the advent of *high-throughput screening*, the binding affinity of as many as several hundred thousands drug candidates to the target protein can now be assayed within a day. Furthermore, *combinatorial chemistry* allows for the systematic synthesis of molecules that are composed of preselected molecular groups that are linked with preselected chemical reactions. The number of molecules that is accessible in such a combinatorial library, in principle, can exceed many billions. Thus we need the computer to suggest promising sublibraries that contain a large number of compounds that bind tightly to the protein.

As in the case of target finding, the new experimental technologies in drug design require new computer methods for screening and interpreting the voluminous data assembled by the experiment. These methods are seldom considered part of bioinformatics, since the biological object, namely the target protein is not the focus of the investigation. Rather, people speak of *cheminformatics*, the computer aspect of medicinal chemistry. Whatever the name, it is our conviction that both aspects of the process that guides us from the genome to the drug have to be considered together, and we will do so in this book.

1.5
A bioinformatics landscape

In the Sections above, we have described the application scenario that is the viewpoint from which we are interested in bioinformatics. In this Section, we attempt to chart out the field of bioinformatics in terms of its scientific subproblems and the application challenges that it faces.

Basically, there are two ways of structuring the field of bioinformatics. One is *intrinsically*, by the type of problem that is under consideration. Here, the natural way of structuring is by layers of information that are compiled starting from the genomic data and working our way towards various levels of the phenotype. The second is *extrinsically*, by the medical or pharmaceutical application scenario to which bioinformatics contributes and by the type of biological experimentation that it supports. This book is roughly structured according to this distinction. The intrinsic view is the subject of Volume 1. This volume discusses subproblems of bioinformatics that provide components in a global bioinformatics solution. Each Chapter is devoted to one relevant component. The extrinsic view is the focus of Volume 2. This volume has Chapters which each concentrate on some important application scenarios that can only be supported effectively by combining components discussed in Volume 1.

Both views are summarized in an integrated fashion in Figure 1.7, where the bioinformatics circle is depicted reaching from the genome sequence to the macroscopic phenotype. We will be guided by this figure as we now chart out the bioinformatics landscape in more detail.

1.5.1
The intrinsic view

The intrinsic view may be the more basic research-oriented one. This view on bioinformatics tends to produce reasonably well-defined and timeless scientific problems that often have the character of grand challenges. As described above, we will structure this view of bioinformatics in layers of derived information that we add on top of the primary information provided by the genomic sequence.

1.5.1.1 Layer −1: Sequencing support

This part of bioinformatics addresses problems that occur even before the genomic sequence is available. The object is to interpret experimental data that are generated by sequencing efforts. Problems to be addressed here are *base calling*, i.e. the interpretation of the signals output by sequencers in terms of nucleic acid sequences, *physical mapping*, i.e. providing a rough map of relevant marker loci along the genome, and *fragment assembly*, i.e. the process of piecing together short segments of sequenced DNA to form a contiguous sequence of the genome or chromosome. Layer −1 plays a special role, since it is both a subproblem and an application scenario. The application-oriented character of this problem originates from its close relationship with the applied sequencing procedure. The two competing approaches to sequencing the human genomes, namely whole-genome-shotgun sequencing, as performed by Celera Genomics [4] and BAC-assembly as performed by the publically funded human genome project [5], which lead to quite different approaches for assembly, have illustrated this point very visibly. We chose to discuss this layer in Volume 2 (Chapter 2 by Xiaoqiu Huang).

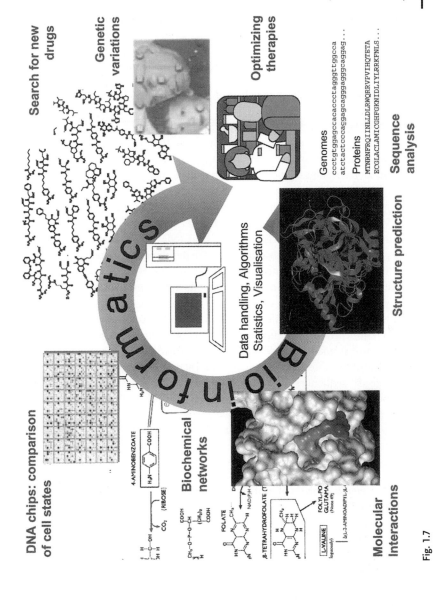

DNA chips: comparison of cell states

Search for new drugs

Genetic variations

Optimizing therapies

Genomes
ccctggagccacaccctaggtttggcca
atctactcccagagcaggggaggcaggag...

Proteins
MTNRNFRQIINLLDLRWQRRVPVIHQTETA
ECGLACLAMICGHFGKNIDLIYLRRKFNLS...

Sequence analysis

Biochemical networks

Data handling, Algorithms
Statistics, Visualisation

Bioinformatics

Structure prediction

Molecular Interactions

4-AMINOBENZOATE

CO₂ (RIBOSE)

FOLATE

β-TETRAHYDROFOLATE

FOLYLPO GLUTAMA (None 4P)

L-VALINE

Fig. 1.7
A schematic overview of bioinformatics

After exercising layer −1 of bioinformatics successfully the raw genomic sequence is available.

1.5.1.2 Layer 1: Analysis of nucleic acid sequences (5 o'clock position in Figure 1.7)

This layer is concerned with annotating the raw genomic sequence with information that can be derived directly from the sequence. Problems addressed in this layer are *gene finding* (Chapter 3 of Volume 1, Victor Solovyev), i.e. the identification of those – as it happens, rare – stretches of genomic DNA that code for proteins, and the analysis of the much more prevalent and quite a bit more mysterious *non-coding regions* in the genome (Chapter 4 of Volume 1, Thomas Werner). The former problem is our entry to understanding the proteome. It is so hard, that today experimental procedures that go directly after protein sequences are developed to help along the way. The latter problem currently concentrates on the regions upstream of sequences that encode proteins. These regions are believed to contain patterns that govern the regulation of the expression of the genes, i.e. their translation to proteins. Both problems can be considered basic research problems and scientific grand challenges. Numerous other annotations are desirable, such as the annotation of genomic sequences with homologous sequences in the same or different species, the distinction between orthologous[1] and paralogous[2] sequences. A comparison between many sequences at a time can yield a *multiple alignment* or, if we are interested in the lineages of the participating sequences, an *evolutionary tree*.

The processes of layer 1 provide the data for layer 2.

1.5.1.3 Layer 2: Analysis of protein sequences (5 o'clock position in Figure 1.7)

This layer annotates protein sequences. Here, the most voluminous source of information is the comparison with other homologous sequences, either on the protein or on the nucleic acid level. The results desired are relationships between different proteins that allow for making conclusions about

[1] Orthologous sequences are homologous sequences in *different* species that have common origin and whose distinctions come about by gradual evolutionary modifications from the common ancestor. Orthologous sequences tend to have the same function in different species.

[2] Paralogous sequences are homologous sequences that exist within a species. Paralogous sequences have a common origin but involve gene duplication events to arise. Often the gene duplication has the purpose of using the sequence to implement a new function. Thus paralogous sequences often perform different functions.

protein function, cellular localization and the like. The methods in layer 2 are basically the same ones as in layer 1. Chapter 2 of Volume 1 (Martin Vingron) concentrates on the bioinformatics methods for layers 1 and 2.

1.5.1.4 Layer 3: Analysis and prediction of molecular structure (6 o'clock position in Figure 1.7)

This layer generates three-dimensional models of the biomolecules under consideration. In the introductory part of the Chapter we have seen that this structure is the key to the function of a molecule. Thus, modeling molecular structure is a very important problem of bioinformatics. There are several kinds of molecules to be considered.

1.5.1.4.1 DNA

The structure of DNA is quite well preserved throughout nature. It is the double helix. There are small differences, the so-called fine structure of DNA, which are modeled in the computer with methods like energy minimization and molecular dynamics. These methods are also used for modeling other biomolecules, notably, proteins. Therefore modeling the structure of DNA is not covered in this book.

1.5.1.4.2 RNA

This molecule is structurally more flexible than DNA. The database of resolved three-dimensional RNA structures is much smaller than that for proteins; it contains only a few hundred molecules. Modeling three-dimensional structures of RNA is basically an unsolved problem, however, there are effective methods for modeling two-dimensional RNA structures. We do not discuss modeling RNA structures in this book and refer the reader to [6].

1.5.1.4.3 Proteins

Proteins display a wide variety of structures. The protein structure prediction problem has been a challenge for a long time. Many successes have been accomplished but the general problem is still far from being solved. What can be achieved is to model (target) proteins given a structurally resolved protein that acts as a template. However this process is only successful if the target and the template protein exhibit a sufficient amount of similarity in sequence. Chapter 5 of Volume 1 (Roland Dunbrack, Jr) describes the methods used in this approach to protein modeling. If the sequence similarity is low, then one cannot generate an accurate full-atom model of the protein, in general. But often one can still find out a lot of significant aspects of the structure of the protein, such as the overall architecture, the coordinates of the protein backbone, or even more accurate models of relevant binding pockets. Chapter 6 of Volume 1 (Ralf Zimmer, Thomas Lengauer) reports on the state of the art in this field.

1.5.1.4.4 Lipids, carbohydrates

Lipids form membranes inside and around the cell. Carbohydrates form complex tree-like molecules that become attached to the surface of proteins and cellular membranes. In both cases, the three-dimensional molecular structure is not unique, but the molecular assemblies are highly flexible. Thus, analyzing the molecular structure involves the inspection of a process in time. Molecular dynamics is the only available computer-based method for doing so. Compared with protein structures there are relatively few results on lipids and carbohydrates. The book does not detail this topic.

1.5.1.5 Layer 4: Molecular interactions (7 o'clock position in Figure 1.7)

If the molecular structure is known we can attempt to analyze the interactions between molecules. Among the many possible pairs of interaction partners two are both relevant and approachable by computer.

1.5.1.5.1 Protein–ligand interactions

Here, one molecular partner is a protein; and the other is a small, often flexible, organic molecule. This is the problem discussed in detail in the introductory parts of this Chapter. The problem is a basic research problem and, at the same time, is of prime importance for drug design. The problem has two aspects. One is to determine the correct geometry of the molecular complex. The second is to provide an accurate estimate of the differential free energy of binding. Whereas much progress has been made on the first aspect, the second aspect remains a grand challenge. Chapter 7 of Volume 1 (Matthias Rarey) discusses this problem.

1.5.1.5.2 Protein–protein interactions

Here two proteins bind to each other. This problem is important if we want to understand molecular interaction inside living organisms; it plays less of a role for drug design. The problem is different from protein–ligand docking in several respects. In protein–ligand docking, the binding mode is mostly determined by strong enthalpic forces between the protein and the ligand. In addition, the contributions of desolvation (replacing the water molecules inside the pocket by the ligand) are essential. The notion of molecular surface is not as much relevant, especially, since the ligand can be highly flexible and does not have a unique surface. In contrast, geometric complementarity of both proteins is a dominating issue in protein–protein docking, where both binding partners meet over a much larger contact surface area. Issues of desolvation can be essential here. Induced fit, i.e. subtle structural changes on the protein surface to accommodate binding are important in both problems, but seem to be more essential in protein–protein docking. Chapter 8 of Volume 1 (Michael Sternberg) discusses this problem. Estimating the free energy of binding is a major bottleneck in all docking problems.

1.5.1.5.3 Other interactions

Of course, all kinds of molecular interactions are important, be it protein–DNA (essential for understanding gene regulation) or reactions involving RNA, lipids or carbohydrates. Protein–DNA docking seems even harder than protein–protein docking. Because of the rigidity of DNA stronger conformational changes seem to be necessary on the protein side. Also, water plays an important role in protein–DNA docking. Chapter 8 discusses protein–DNA docking. The other problems are not discussed in this book.

1.5.1.6 Layer 5: Metabolic and regulatory networks (9 o'clock position in Figure 1.7)

This layer takes the step from analyzing single interactions to analyzing networks of interactions. It has a database aspect, which just collects the voluminous data and makes them generally accessible, and an algorithmic aspect that performs simulations on the resulting networks. Both aspects are in a preliminary stage. The most development has taken place in metabolic databases. Here, we have quite a detailed picture of parts of the metabolism for several organisms. This book does not detail the simulation of metabolic networks.

1.6
The extrinsic view

This view structures bioinformatics according to different application scenarios and types of experimental data that are under scrutiny.

1.6.1
Basic contributions: molecular biology databases and genome comparison

The first completely sequenced bacterial genome, that of the bacterium, *H. influenzae*, was available in 1994. Since then, a few dozen bacterial genomes have been sequenced. Sequencing a bacterium has become state of the art, and we can expect many more genomes to appear in the near future. The first eucaryote (*S. cerevisiae*, yeast [7]) has been sequenced in 1996, and the genome of the first multicellular organism (*C. elegans*, nematode, [8]) became available in 1998. The genome of *D. melanogaster* (fruitfly [9]) became available in February 2000, and two drafts of the human genome sequence have been published in February 2001 [4, 5]. As the number and variety of completely sequenced genomes rises, their comparison can yield important scientific insights. What is it that makes the bacterium *E. coli* a symbiotic beneficial parasite, whereas the bacterium *H. influenzae* that generates ear infections basically has toxic effects in the relevant parts of the human body? Huynen et al. [10] have carried out a differential analysis that looks for homologies among proteins in both bacteria and concentrates on those (about 200) proteins that occur only in *H. influenzae*. Using homology searches over the database of available protein sequences from all kinds of organisms, Huynen et al. [10] found that many of these 200 proteins are similar to proteins that act toxic in other organisms. Thus, in some sense, the set of candidate proteins that are responsible for the adverse effects of *H. influenzae* has been narrowed down from the about 1800 protein in the bacterium to 200. This is an example of the kinds of results genome comparison can achieve.

In order to carry out such tasks, we need to combine a wide variety of methods and data. Chapter 1 of Volume 2 (Thure Etzold et al.) describes a successful and widely used approach for linking the vastly growing variety of biomolecular databases.

1.6.2
Scenario 1: Gene and protein expression data (10 o'clock position in Figure 1.7)

The measurement and interpretation of expression levels of genes and proteins is at the center of today's attention in genomics, proteomics, and bioinformatics. The goal is to create a detailed differential functional picture of the cell's inner workings, as it is determined by the cell state at hand. Cell states can be influenced and determined by external environmental factors (temperature, pH, chemical stress factors), and by internal factors such as the cell belonging to a special tissue or be in a certain disease state.

Historically the first version of expression of data is *expressed sequence tags (ESTs)*. These are short stretches of cDNA (complementary DNA), i.e. DNA that has been translated back from messenger RNA (mRNA) expressed in the cell. The messenger RNA is in a mature state, i.e. introns have been spliced out before the cDNA was taken from the mRNA. ESTs provide a census of the genes expressed in a cell that is subject to certain internal and external conditions. The level of expression of a certain gene is roughly correlated with the frequency with which this gene is hit by sequenced ESTs. Today, there are large collections of ESTs available for many organisms and cells subject to many different conditions.

The DNA chips described above (see Figure 1.6) provide a different variant of the kind of data provided by ESTs. The advantage of DNA chips is that they can provide more accurate expression data on expression levels. Whereas, EST counts can be low and generally provide little resolution. One way of reading out expressions levels from DNA chips is via fluorescent marking. The high correlation between the light intensity and the expression level provides more resolution than EST counts.

Expression levels on DNA or protein gels are generally deduced from the size of spots on the gel. 2D gels and mass spectrometry can be used as fast tools for identifying expressed proteins in various states of the cell. These experimental data and their interpretation with bioinformatics are subject to Chapter 4 (Roland Appel et al., proteomics) and Chapter 5 (Roland Somogyi, genomics) of Volume 2.

1.6.3
Scenario 2: Drug screening (1 o'clock position in Figure 1.7)

Once the target protein is identified, the search for the drug starts. That search is again composed of a mixture of experimental and computer-based procedures. In a first phase, one is looking for a *lead compound*. The only requirement on the lead compound is that it binds tightly to the binding site of the target protein. In a second step of *lead optimization*, the lead compound is modified to be non-toxic, have few side effects and be bio-accessible, i.e. be easily delivered to the site in the body, at which its effect is desired.

Basically, there are two approaches to developing a new drug. One is to create one from scratch. This has been attempted on the basis of the knowledge of the three-dimensional structure of the binding site of the protein. In effect, the drug molecule can be designed to be complementary to the binding site in steric and physicochemical terms. This approach is called *ab initio* drug design (not to be confused with *ab initio* protein structure prediction or *ab initio* methods in theoretical chemistry). *Ab initio* drug design has been hampered by the fact that often the developed molecules were hard to synthesize. Furthermore lead optimization turned out to be hard for many lead compounds developed with *ab initio* design techniques.

The second approach is to screen through a large database of known molecules and check their binding affinity to the target protein. The advantage of this approach is that, mostly, the compounds that have been accumulated in a database have been investigated before – though mostly in a different context. Their bioaccessibility and toxicity may have been studied and they probably have been synthesized. Screening compounds from a library can been done in the laboratory or in the computer. In the latter case, the process is called *virtual screening*. Compound databases become large, so a mixture of the two processes may be advisable. Here, a set of compounds is preselected from a large library with virtual screening. These compounds are then tested in the laboratory. Chapter 6 of volume 2 by Matthias Rarey, Martin Stahl, and Gerhard Klebe gives more detail on virtual drug screening.

1.6.4
Scenario 3: Genetic variability (2 o'clock position in Figure 1.7)

Recently, new experimental methods have been developed that uncover variations of genomic text for organisms of the same species. The sequencing data discriminate between different external or internal states of the organism or provide insight into intra-species genetic variability. Genetic variability is just starting to be investigated on a genomic scale both experimentally and with bioinformatics. Chapter 3 of Volume 2 (Deborah Nickerson) introduces this new and exciting field of molecular biology and bioinformatics.

Genetic variability does not only occur in the host (patient) but also in the guest (infectious agent). For many antibacterial and several antiviral therapies (e.g. AIDS), the guest population reacts to the administering of drugs, and resistant strains develop. In this context, drug selection is a difficult problem. Given appropriate genotypic and phenotypic data, bioinformaticcs can eventually help in optimizing therapies in this context (3 o'clock position in Figure 1.7).

The following Chapters of both volumes provide the necessary detail for an in-depth introduction into bioinformatics. Chapter 7 of volume 2 will make an attempt at wrapping up where we stand today and balance hopes and expectations with limitations of this new and rapidly developing field.

References

1 DREWS, J., Die verspielte Zukunft, **1998**, Basel, Switzerland: Birkhäuser Verlag.

2 DeRisi, J. L., V. R. Iyer, and P. O. Brown, Exploring the metabolic and genetic control of gene expression on a genomic scale, *Science*, **1997**. 278(5338): p. 680–6.

3 Fischer, E., Berichte der Deutschen Chemischen Gesellschaft, **1894**. 27: p. 2985–2993.

4 Venter, J. C., et al., The Sequence of the Human Genome, *Science*, **2001**. 291(5507): p. 1304–1351.

5 Lander, E. S., et al., Initial sequencing and analysis of the human genome, *Nature*, **2001**. 409(6822): p. 860–921.

6 Zuker, M., Calculating nucleic acid secondary structure, *Curr Opin Struct Biol*, **2000**. 10(3): p. 303–10.

7 Goffeau, A., et al., Life with 6000 genes, *Science*, **1996**. 274(5287): p. 546, 563–7.

8 Genome sequence of the nematode C. elegans: a platform for investigating biology. The C. elegans Sequencing Consortium, *Science*, **1998**. 282(5396): p. 2012–8.

9 Adams, M. D., et al., The genome sequence of Drosophila melanogaster, *Science*, **2000**. 287(5461): p. 2185–95.

10 Huynen, M. A., Y. Diaz-Lazcoz, and P. Bork, Differential genome display, *Trends Genet*, **1997**. 13(10): p. 389–90.

2
Sequence Analysis

Martin Vingron

2.1
Introduction

Although a theoretical method to predict the three-dimensional structure of a protein from its sequence alone is not in sight yet, researchers have uncovered a multitude of connections between the primary sequence on one hand and various functional features of proteins on the other hand. Among the success stories are the recognition of transmembrane proteins or the classification of proteins into classes of similar function based on sequence similarity. The latter achievement uses the observation that proteins that are similar in sequence are likely to share similar functional features. This is also giving rise to the enormous utility of similarity searches in sequence data bases. This Chapter will deal with the kinds of analysis and predictions based on primary sequence alone and the algorithms used in this field.

Roughly speaking, methods fall into two classes, based on whether they analyze the individual protein sequence in isolation or whether implications are drawn from a comparison with or among many sequences. Individual sequences are analyzed primarily based on characteristics of amino acids as derived either from physical chemistry or statistically. Such characteristics are hydrophobicity scales or statistical preferences for a particular secondary structure. Section 2.2 will cover these methods.

Naturally the type of information that can be extracted from a sequence in isolation is very much restricted compared to what can be gained from analyzing many sequences. The key observation allowing us to exploit the information that is collected in our databases is the close relationship between homology, i.e., common ancestry and sequence similarity. In searching for sequence similarity we strive for the recognition of homology because homology lets us suggest common functions. The basis for the study of sequence similarity is the comparison of two sequences which will be dealt with in Section 2.3.

Sequence comparisons are effected in large numbers when searching sequence databases for sequences that are similar to a query sequence. Algorithms for this purpose need to be fast even at the expense of sensitivity. Section 2.4 discusses the widely used heuristic approaches to database searching. However, the algorithms we are designing for the purpose of quantifying sequence similarity can only be as good as our understanding of evolutionary processes and thus they are far from perfect. Therefore, results of algorithms need to be subjected to a critical test using statistics. Methods for the assessment of the statistical significance of a finding are introduced in Section 2.5.

Genes do not come in pairs but rather in large families. Consequently, the need arises to align more than two sequences at a time which is done by multiple alignment programs. Computationally a very hard problem, it has found considerable attention from the side of algorithms development. Section 2.6 presents the basic approaches to multiple sequence alignment. Section 2.7 will build on the knowledge of a multiple alignment and introduce how to exploit the information contained in several related sequences for the purpose of identifying additional related sequences in a database. The last section will cover methods for identifying domains in one or among several sequences. For the annotation of sequences and for prediction of function this aspect has become increasingly important during recent years.

2.2
Analysis of individual sequences

This Chapter focuses on analyses that can be performed based solely on the primary sequence of a protein. Several rationales can be applied. Physico-chemical characteristics of individual amino acids are one basis for predictions of gross structural features. For example, particular repetitive patterns may suggest a coiled-coil structure while in general secondary structure can be predicted based on an a statistical analysis of the primary sequence. The definition of signals recognized by the cellular transport machinery allow the prediction of subcellular location. Although somewhat unsystematic such observations can provide valuable hints as to the structure and/or function of a protein.

Amino acids side chains differ in their physico-chemical features. For example, some like to be exposed to water, i.e., they are hydrophilic, while the hydrophobic amino acids tend to avoid exposure to water. Charge, size, or flexibility in the backbone are only some of the other examples of amino acid parameters. These parameters are usually measured on a numerical scale such that for every parameter there exists a table assigning a number to each amino acid. For the case of hydrophobicity two such scales have become famous. The first is due to Hopp and Woods [1] while the other is due to Kyte and Doolittle [2]. A large collection of amino acid parameters

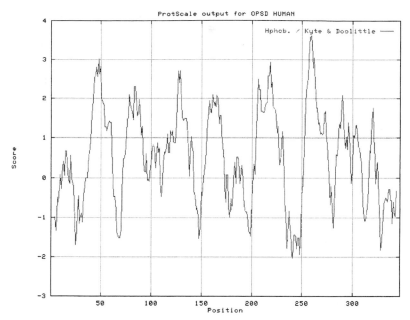

Fig. 1
Hydrophobicity plot for the transmembrane protein opsin.
Note the peaks in hydrophobicity which roughly denote
the putative membrane spanning regions.

have been collected by Argos [3] who found that subgroups of them are
correlated with each other such that the real information content of this
large number of parameters is in fact lower than it seems. He selected the
following parameters as a non-redundant set: hydrophobicity, turn preference [4], residue bulk [5], refractivity index [6], and antiparallel strand preference [7].

Standard sequence analysis software today offers programs that plot various parameters for a given protein (Figure 2.1). Serious software packages
tend to provide the user with a selection of informative and non-redundant
parameters similar to the one given. Some other packages pretend to offer
new insights by plotting large numbers of parameters. In practice, the various parameters are used to plot a curve along the amino acid chain. Values
are averaged within a sliding window to smoothen the curve. The selection
of the window width is, of course, arbitrary but values between 7 and 15
would generally seem appropriate. Within a window a weighting scheme
may be applied which would typically assign more importance to residues
in the middle of the window than to the ones on the edges.

Features of the individual amino acids also play a key role in protein secondary structure formation. Based on experimental data, scales have been

assigned to the amino acids describing their preference for assuming, e.g., a helical structure [8]. Consequently, early secondary structure prediction methods have assigned preferences to the amino acids according to which secondary structure they tend to assume. For example, Glutamate is frequently found in alpha helices while Valine has a preference for beta strands and Proline is known to be strongly avoided in helices. Modern secondary structure prediction methods are more involved though (see below).

The functional features of proteins that are grasped by such parameters are manifold. Hydrophobic amino acids tend to occur in the interior of globular proteins, while at the surface of a protein one will preferentially find hydrophilic residues. One application of the latter is the prediction of antigenic epitopes because these are assumed to coincide with patches of hydrophilic residues. In transmembrane proteins, the regions of the chain that span the membrane tend to be strongly hydrophobic and the recognition of several such regions is an integral part of methods aiming at the recognition of transmembrane proteins based on their sequence alone [9, 10]. Recognition of transmembrane regions has been found to be remarkably successful, also leading to the acceptance of the output of these programs as annotation in sequence databases.

Certain periodicities in the occurrence of hydrophilic and hydrophobic residues may indicate particular secondary structures. Exposed helices tend to display a hydrophilic residue every 3–4 positions, while the other residues are rather hydrophobic. This is well visible in a "helical wheel representation" where the amino acid sequence is printed on a circle with the appropriate 3.6 amino acids per 360°. Likewise, a strand whose one side is exposed may display a hydrophilic residue at every other position. Based on these observations, Eisenberg devised the method of hydrophobic moments [11]. Many structural proteins interact with each other through intertwined helices forming a so-called coiled-coil structure. The physical constraints on the interface between these helices is reflected in a certain periodic arrangement of hydrophilic and hydrophobic amino acids. Generally, the sequence of an alpha helix that participates in a coiled-coil region will display a periodicity with a repeated unit of length 7 amino acids. Denote those 7 positions by *a* through *g*, then position *e* and *g* tend to be charged or polar while *a* and *d* are hydrophobic [12]. Prediction methods for coiled-coil regions are making use of these preferences [13] and even attempt to distinguish between two-stranded and three-stranded coiled-coils [14].

The primary sequence of a protein also contains the information whether a protein is secreted or which cellular compartment it is destined for. In particular, secretory proteins contain a N-terminal signal peptide for the recognition of which programs have been developed [15]. Other signals like, e.g., nuclear localization signals are notoriously hard to describe. Many posttranslational modifications, on the other hand, are linked to particular amino acid patterns as described in the Prosite database.

2.2.1
Secondary structure prediction

Linus Pauling already suggested that amino acid chains could assume regular local structures, namely alpha helices and beta strands [16]. In between these secondary structure elements there are turns or loops [17]. There is a long tradition of attempts to predict local secondary structure based on sequence. Secondary structure prediction generally aims at correlating the frequencies of occurrences of short amino acid words with particular secondary structures. The data set for this statistic must be derived from known protein structures with the secondary structures assigned to the primary sequence which is in itself not a trivial task [18]. Based on such a learning set, statistical methods are applied that range from information theory (the GOR method [19]) to neural networks [20] and linear discriminant analysis [21].

The basic approach essentially due to Chou and Fasman [22] uses a log-odds ratio comparing the frequency with which a particular amino acid assumes, say, a helical structure over the overall frequency of an arbitrary amino acid to occur in an alpha helix. The logarithm of this ratio is a measure indicating when this residue preferentially assumes a helical structure (positive value) as opposed to avoiding a helix (negative value). These preference values can be learnt from a large data set. For a prediction the preference is used like one of the parameters from above. Values are averaged within a sliding window and the resulting curves for helix, strand and loop prediction are compared, looking for the highest one in every part of the sequence.

The GOR method phrases a very similar approach in an information theoretic framework, computing not only preferences for individual residues but aiming at the delineation of preferences for short stretches of amino acids. Since the given data set will generally not supply sufficient data for estimation of the log-odds for every k-tuple certain approximations have to made [23]. At the same time it has become clear that even this approach is unlikely to give perfect predictions because, in known crystal structures, one and the same 5-mer of residues will be found in different secondary structures [24]. Other approaches like the one due to Solovyev and Salamov [25] assemble different characteristics for a short stretch (singlet and doublet secondary structure preferences, hydrophobic moment) of amino acids and apply linear discriminant analysis in order to derive a predictor for the secondary structure of a region.

While the information contained in one primary sequence alone seems to be insufficient to predict a protein's secondary structure, multiply aligned sequences (see below) offer a means to push the limits of the prediction. Originally due to Zvelebil et al. [26] this line of attack on the problem is the basis of the PHD method by Rost and Sander [27]. The PHD method takes as input a multiple alignment of a set of homologous protein sequences and uses several neural networks to effect the prediction. The first neural network maps the distribution vectors of amino acid occurrences in a window of adjacent alignment columns onto one of the states helix, strand, and loop. Outputs of this neural net are then reconciled by a second neural net which computes a consensus prediction. A last level of consensus-taking combines several predictions from neural nets that were trained on different data or with different techniques. The learning set for this procedure is taken from the HSSP database [28] of multiply aligned sequences that are annotated with their common secondary structure.

Development of prediction methods can, of course, only become fruitful when there are standards available to test and compare methods. It has been the achievement of Kabsch and Sander [29] to put forward a validation scheme for secondary structure prediction. This is based on an automatic assignment of secondary structures for a given crystal structure followed by an assessment of the degree of accuracy of helix, sheet and loop predictions. As long as one restricts oneself to the problem of secondary structure prediction for a single sequence the limit in predicting accuracy today seems to lie around 63% [30]. PHD claims an accuracy up to 72% due to the inclusion of multiply aligned sequences. In practical applications, however, the concrete data set will decide about the success of secondary structure prediction. Even when a substantial number of protein sequences can be aligned prior to predicting secondary structure, the information that they contribute depends on the degree of divergence between them. Some methods (like GOR and PHD, among others) supply the user with an estimate of how reliable a prediction in a particular area is which is of course helpful in practice. The success of overlaying the output from several secondary structure prediction programs is hard to predict because it is not clear whether the individual methods are sufficiently different to actually produce new information through such an approach. Recent methods by Cuff et al. [31] and by Selbig et al. [32] have automated and evaluated this approach.

The expasy server [33] offers references to many Web sites for protein sequence analysis and secondary structure prediction.

2.3
Pairwise sequence comparison

The methods of the last chapter analyzed a sequence by its own virtues. We now turn to the comparison of two sequences. The rationales behind the comparison of sequences may be manifold. Above all, the theory of evolution tells us that gene sequences may have derived from common ancestral sequences. Thus it is of interest to trace the evolutionary history of mutations and other evolutionary changes. Comparison of biological sequences in this context is understood as comparison based on the criteria of evolution. For example, the number of mutations, insertions, and deletions of bases necessary to transform one DNA sequence into another one is a measure reflecting evolutionary relatedness. On the other hand, a comparison may be more pragmatic in that it is not aimed at a detailed reconstruction of the evolutionary course of events but rather aims at pinpointing regions of common origin which may in turn coincide with regions of similar structure or similar function. Physical characteristics of amino acids play a more important role in this viewpoint than they do when studying evolution.

2.3.1
Dot plots

Dot plots are probably the oldest way of comparing sequences [34]. A dot plot is a visual representation of the similarities between two sequenes. Each axis of a rectangular array represents one of the two sequences to be compared. A window length is fixed, together with a criterion when two sequence windows are deemed to be similar. Whenever one window in one sequence resembles another window in the other sequence, a dot or short diagonal is drawn at the corresponding position of the array. Thus, when two sequences share similarity over their entire length a diagonal line will extend from one corner of the dot plot to the diagonally opposite corner. If two sequences only share patches of similarity this will be revealed by diagonal stretches.

Figure 2.2 shows an example of a dot plot. There, the coding DNA sequences of the alpha chain of human hemoglobin and of the beta chain of human hemoglobin are compared to each other. For this computation the window length was set to 31. The program adds up the matches within a window and the grayscale value of the diagonal is set according to the quality of the match at that position. One can clearly discern a diagonal trace along the entire length of the two sequences. Note the jumps where this trace changes to another diagonal of the array. These jumps correspond to position where one sequence has more (or fewer) letters than the other one. The figure was produced using the program "dotter" [35].

Dot plots are a powerful method of comparing two sequences. They do not predispose the analysis in any way such that they constitute the ideal first-pass analysis method. Based on the dot plot the user can decide whether he deals with a case of global, i.e., beginning-to-end similarity, or local similarity. "Local similarity" denotes the existence of similar regions between two sequences that are embedded in the overall sequences which lack similarity. Sequences may contain regions of self-similarity which are frequently termed internal repeats. A dot plot comparison of the sequence itself will reveal internal repeats by displaying several parallel diagonals.

Instead of simply deciding when two windows are similar a quality function may be defined. In the simplest case, this could be the number of

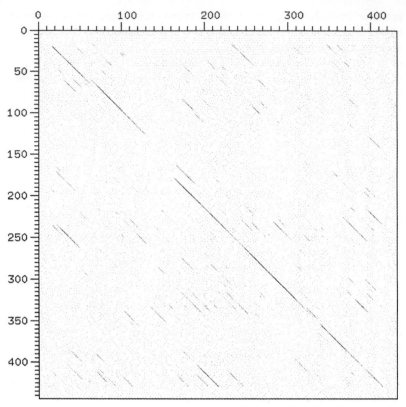

Fig. 2
Dot plot comparing two hemoglobin sequences.

matches in the window. For amino acid sequences the physical relatedness between amino acids may give rise to a quantification of the similarity of two windows. For example, when a similarity matrix on the amino acids (like the Dayhoff matrix, see below) is used one might sum up these values along the window. However, when this similarity matrix contains unequal values for exact matches this leads to exactly matching windows of different quality. The dot plot method of Argos [36] is an intricate design that reflects the physical relatedness of amino acids. The program dotter [37] is an X-windows based program that allows to display dot plots for DNA, for proteins, and for comparison of DNA to protein.

2.3.2
Sequence alignment

A sequence alignment [38] is a scheme of writing one sequence on top of another where the residues in one position are deemed to have a common evolutionary origin. If the same letter occurs in both sequences then this position has been conserved in evolution (or, coincidentally, mutations from another ancestral residue has given rise to the same letter twice). If the letters differ it is assumed that the two derive from an ancestral letter, which could be one of the two or neither. Homologous sequences may have different length, though, which is generally explained through insertions or deletions in sequences. Thus, a letter or a stretch of letters may be paired up with dashes in the other sequence to signify such an insertion or deletion. Since an insertion in one sequence can always be seen as a deletion in the other one sometimes uses the term 'indel', or simply 'gap'.

In such a simple evolutionarily motivated scheme, an alignment mediates the definition of a distance for two sequences. One generally assigns 0 to a match, some positive number to a mismatch and a larger positive number to an indel. By adding these values along an alignment one obtains a score for this alignment. A distance function for two sequences can be defined by looking for the alignment which yields the minimum score.

Naively, the alignment that realizes the minimal distance between two sequences could be identified by testing all possible alignments. This number, however, is prohibitively large but luckily, using dynamic programming, the minimization can be effected without explicitly enumerating all possible alignments of two sequences. To describe this algorithm [39], denote the two sequences by $s = s_1, \ldots, s_n$ and $t = t_1, \ldots, t_m$. The key to the dynamic programming algorithm is the realization that for the construction of an optimal alignment between two stretches of sequence $s_1, \ldots, s_i$ and $t_1, \ldots, t_j$ if suffices to inspect the following three alternatives:

(i) the optimal alignment of $s_1, \ldots, s_{i-1}$ with $t_1, \ldots, t_{j-1}$, extended by the match between s_i and t_j;
(ii) the optimal alignment of $s_1, \ldots, s_{i-1}$ with $t_1, \ldots, t_j$, extended by matching s_i with a gap character '−';
(iii) the optimal alignment of $s_1, \ldots, s_i$ with $t_1, \ldots, t_{j-1}$, extended by matching a gap character '−' with t_j.

Each of these cases also defines a score for the resulting alignment. This score is made up of the score of the alignment of the so far unaligned sequences that is used plus the cost of extending this alignment. In case: (i) this cost is determined by whether or not the two letters are identical and in cases (ii) and (iii) the cost of extension is the penalty assigned to a gap. The winning alternative will be the one with the best score (Figure 2.3).

To implement this computation one fills in a matrix the axes of which are annotated with the two sequences s and t. It is helpful to use north, south, west, and east to denote the sides of the matrix. Let the first sequence extend from west to east on the north side of the matrix. The second sequence extends from north to south on the west side of the matrix. We want to fill the matrix starting in the north-western corner, working our way southward

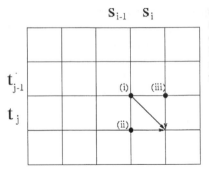

S_{i-1} S_i

t_{j-1}

t_j

(i) (iii)

(ii)

Fig. 3
Schematic representation of the edit matrix
comparing two sequences. The arrows
indicate how an alignment may end
according to the three cases described in
the text.

row by row, filling each row from west to east. To start one initializes the
northern and western margin of the matrix, typically with gap penalty values. After this initialization the above rules can be applied. A cell (i, j) that
is already filled contains the score of the optimal alignment of the sequence
$s_1, \ldots, s_i$ with $t_1, \ldots, t_j$. The score of each such cell can be determined by
inspecting the cell immediately north-west of it (case (i)), the cell west (case
(ii)), and the one north (case (iii)) of it and deciding for the best scoring
option. When the procedure reaches the south-eastern corner that last cell
contains the score of the best alignment. The alignment itself can be recovered as one backtracks from this cell to the beginning, each time selecting the path that had given rise to the best option.

The idea of assigning a score to an alignment and then minimizing or
maximizing over all alignments is at the heart of all biological sequence
alignment. However, many more considerations have influenced the definition of the scores and made sequence alignment applicable to a wide
range of biological settings. Firstly, note that one may either define a distance or a similarity function to an alignment. The difference lies in the
interpretation of the values. A distance function will define positive values
for mismatches or gaps and then aim at minimizing this distance. A similarity function will give high values to matches and low values to gaps and
then maximize the resulting score. The basic structure of the algorithm is
the same for both cases. In 1981, Smith and Waterman [40] showed that for
global alignment, i.e., when a score is computed over the entire length of
both sequences, the two concepts are in fact equivalent. Thus, it is now
customary to choose the setting that gives more freedom in appropriately
modeling the biological question that one is interested in.

In the similarity framework one can easily distinguish among the different possible mismatches and also among different kinds of matches. For
example, a match between two Tryptophanes is usually seen to be more
important than a match between two Alanines. Likewise, the pairing of two
hydrophobic amino acids like Leucine and Isoleucine is preferable to the
pairing of a hydrophobic with a hydrophilic residue. Scores are used to de-

scribe these similarities and are ususally represented in the form of a symmetric 20 by 20 matrix, assigning a similarity score to each pair of amino acids. Although easy to understand from the physical characteristics of the amino acids, the values in such a matrix are usually derived based on an evolutionary model that allows one to estimate whether particular substitutions are preferred or avoided. This approach has been pioneered by M. Dayhoff [41] who computed a series of amino acid similarity matrices. Each matrix in this series corresponds to a particular evolutionary distance among sequences. This distance is measured in an arbitrary unit called 1 PAM, for 1 Accepted Point Mutation (in 100 positions). The matrices carry names like PAM120 or PAM250 and are supposed to be characteristic for evolutionary distances of 120 or 250 PAM, respectively. Other more recent series of matrices are the BLOSUM matrices [42] or Gonnet's [43] matrices. For every matrix one needs to find appropriate penalties for gaps.

The treatment of gaps deserves special care. The famous algorithm by Needleman and Wunsch [44] did not impose any restrictions on the penalty assigned to a gap of a certain length. For reasons of computational speed this was later specified to assigning a cost function linear in the number of deleted (inserted) residues [45]. This amounts to penalizing every single indel. However, since a single indel tends to be penalized such that it is considerably inferior to a mismatch, this choice resulted in longer gaps being quite expensive and thus unrealistically rare. As a remedy, one mostly uses a gap penalty function which charges a *gap open penalty* for every gap that is introduced and penalizes the length with a *gap extension penalty* which is charged for every inserted or deleted letter in that gap. Clearly, this results in an affine linear function in the gap length, frequently written as $g(k) = a + b * k$ [46].

With the variant of the dynamic programming algorithm first published by Gotoh [47] it became possible to compute optimal alignments with affine linear gap penalties in time proportional to the product of the lengths of the two sequences to be aligned. This afforded a speed-up by one order of magnitude compared to a naive algorithm using the more general gap function. A further breakthrough in alignment algorithms development was provided by an algorithm that could compute an optimal alignment using computer memory only proportional to the length of one sequence instead of their product. This algorithm by Myers and Miller [48] is based on work by Hirschberg [49].

Depending on the biological setting several kinds of alignment are in use. When sequences are expected to share similarity extending from the beginning of the sequences to their ends they are aligned globally. This means that each residue of either sequence is part either of a residue pair or a gap. In particular it implies that gaps at the ends are charged like any other gap. This, however, is a particularly unrealistic feature of a global alignment. While sequences may very well share similarity over their entire length (see the example dot plot of two hemoglobin chains in Figure 2.2), their respective N- and C-termini usually are difficult to match up and differences in length at the ends are more of a rule than an exception. Consequently, one prefers to leave gaps at the ends of the sequences un-penalized. This variant is easy to implement in the dynamic programming algorithm. Two modifications are required. Firstly the initialization of the matrix needs to reflect the gap cost of 0 in the margin of the matrix. Secondly, upon backtracking, one does not necessarily start in the corner of the matrix but much rather searches the margins for the maximum from which to start. Variants of this algorithm that penalize only particular end-gaps are easy to derive and can be used, e.g., to fit one sequence into another or to overlap the end of one sequence with the start of another.

In many cases, however, sequences share only a limited region of similarity. This may be a common domain or simply a short region of recognizable similarity. This case is dealt with by so-called local alignment in an algorithm due to Smith and Waterman [50]. Local alignment aims at identifying the best pair of regions, one from each sequence, such that the optimal (global) alignment of these two regions is the best possible. This relies on a scoring scheme that maximizes a similarity score because otherwise an empty alignment would always yield the smallest distance. Naively, the algorithm to compute a local alignment would need to inspect every pair of regions and apply a global alignment algorithm to it. The decisive idea of Smith and Waterman was to offer the maximization in each cell of the matrix a fourth alternative: a zero to signify the beginning of a new alignment. After filling the dynamic programming matrix according to this scheme, backtracking starts from the cell in the matrix that contains the largest value.

Upon comparing a dot plot and a local alignment one might notice regions of similarity visible in the dot plot but missing in the alignment. While in many cases there exist gap penalty settings that would include all interesting matching regions in the alignment, generally it requires the comparison with the dot plot to notice possible misses. This problem is remedied by an algorithm due to Waterman and Eggert [51] which computes suboptimal, local, and non-overlapping alignments. It starts with the application of the Smith-Waterman algorithm, i.e., a dynamic programming matrix is filled and backtracking from the matrix cell with the largest entry yields the best local alignment. Then the algorithm proceeds to delineate a second-best local alignment. Note that this cannot be obtained by backtracking from the second-best matrix cell. Such an approach would yield an alignment largely overlapping the first one and thus containing little new information. Instead, those cells in the dynamic programming matrix are set to zero from where backtracking would lead into the prior alignment. This can be seen as "resetting" the dynamic programming matrix after having deleted the first alignment. Then the second best alignment is identified by looking for the maximal cell in the new matrix and starting backtracking from there. Iteration of this procedure yields one alternative, non-overlapping alignment after the other in order of descending quality. Application of this algorithm avoids possibly missing matching regions because even under strong gap penalties the procedure will eventually show all matching regions.

There is an interesting interplay between parameters, in particular the gap penalty, and the algorithmic variant used. Consider a pair of sequences whose similar regions can in principle be strung together into an alignment (as opposed to sequences containing repeats which cannot all be seen in a single alignment). Under a weak gap penalty the Smith–Waterman algorithm has a chance to identify this entire alignment. On the other hand, not knowing about the similarity between the sequences ahead of time, a weak gap penalty might also yield all kinds of spurious aligned regions. The Waterman–Eggert algorithm is a valid alternative. The gap penalty can be chosen fairly stringently. The first (i.e., the Smith–Waterman) alignment will then identify only the best matching region out of all the similar regions. By iterating the procedure, though, this algorithm will successively identify the other similar regions as well. For a detailed discussion of these issues see Vingron and Waterman [52].

2.4
Database searching I: single sequence heuristic algorithms

This section takes a first look at the problem of identifying those sequences in a sequence database that are similar to a given sequence. This task arises, e.g., when a gene has been newly sequenced and one wants to determine whether a related sequence already exists in a database. Generally, two settings can be distinguished. The starting point for the search may either be a single sequence with the goal of identifying its relatives, or a family of sequences with the goal of identifying further members of that family. Searching a data base needs to be fast and sensitive but the two objectives contradict each other. Fast methods have been developed primarily for searching with a single sequence and this shall be the topic of this section.

When searching a database with a newly determined DNA or amino acid sequence – the so-called query sequence – the user will typically lack knowledge of whether an expected similarity might span the entire query or just part of it. Likewise, he will be ignorant of whether the match will extend along the full length of some database sequence or only part of it. Therefore, one needs to look for a local alignment between the query and any sequence in the database. This immediately suggests the application of the Smith–Waterman algorithm to each database sequence. One should take care, though, to apply a fairly stringent gap penalty such that the algorithm focuses on the regions that really match. After sorting the resulting scores the top scoring database sequences are the candidates one is interested in.

Several implementations of this procedure are available, most prominently the SSEARCH program from the FASTA package [53]. There exist implementations of the Smith–Waterman algorithm that are tuned for speed like one using special processor instructions [54] and, among others, one by Barton [55]. Depending on implementation, computer, and database size, a search with such a program will take on the order of one minute.

The motivation behind the development of other database search programs has been to emulate the Smith–Waterman algorithm's ability to discern related sequences while at the same time performing the job in much less time. To this end, one usually makes the assumption that any good alignment as one wishes to identify contains, in particular, some stretch of ungapped similarity. Furthermore, this stretch will tend to contain a certain number of identically matching residues and not only conservative replacements. Based on these assumptions most heuristic programs rely on identifying a well-matching core and then extending it or combining several of these. With hindsight, the different developments in this area can further be classified according to a traditional distinction in computer science according to which one either preprocesses the query or the text (i.e., the database). Preprocessing means that the string is represented in different form that allows for faster answer to particular questions, e.g., whether it contains a certain subword.

The FASTA program (part of a package [56] that usually goes by the same name) sets a size k for k-tuple subwords. The program then looks for diagonals in the comparison matrix between query and search sequence along which many k-tuples match. This can be done very quickly based on a preprocessed list of k-tuples contained in the query sequence. The set of k-tuples can be identified with an array whose length corresponds to the number of possible tuples of size k. This array is linked to the indices where the particular k-tuples occur in the query sequence. Note that a matching k-tuple at index i in the query and at index j in the database sequence can be attributed to a diagonal by subtracting the one index from the other. Therefore, when inspecting a new sequence for similarity, one walks along this sequence inspecting each k-tuple. For each of them one looks up the indices where it occurs in the query, computes the index-difference to identify the diagonal and increases a counter for this diagonal. After inspecting the search sequence in this way a diagonal with a high count is likely to contain a well-matching region. In terms of the execution time, this procedure is only linear in the length of the database sequence and can easily be iterated for a whole database. Of course this rough outline needs to be adapted to focus on regions where the match density is high and link nearby, good diagonals into alignments.

The other widely used program to search a database is called BLAST [57, 58]. BLAST follows a similar scheme in that it relies on a core similarity, although with less emphasis on the occurrence of exact matches. This program also aims at identifying core similarities for later extension. The core similarity is defined by a window with a certain match density on DNA or with an amino acid similarity score above some threshold for proteins. Independent of the exact definition of the core similarity, BLAST rests on the precomputation of all strings which are in the given sense similar to any position in the query. The resulting list may be on the order of thousand or more words long, each of which if detected in a database gives rise to a core similarity. In Blast nomenclature this set of strings is called the neighborhood of the query. The code to generate this neighborhood is in fact exceedingly fast.

Given the neighborhood, a finite automaton is used to detect occurrences in the database of any string from the neighborhood. This automaton is a program, constructed on the fly and specifically for the particular word neighborhood that has been computed for a query. Upon reading through a database of sequences, the automaton is given an additional letter at a time and decides whether the string that ends in this letter is part of the neighborhood. If so, BLAST attempts to extend the similarity around the neighborhood and if this is successful reports a match.

Like FASTA, BLAST has also been adapted to connect good diagonals and report local alignments with gaps. BLAST converts the database file into its own format to allow for faster reading. This makes it somewhat unwieldy to use in a local installation unless someone takes care of the installation. FASTA, on the other hand, is slower but easier to use. There exist excellent web servers that offer these programs, in particular at the National Center for Biotechnology Information (NCBI [59]) and at the European Bioinformatics Institute (EBI [60]) where BLAST or FASTA can be used on up-to-date DNA and protein databases.

According to the above mentioned distinction among search methods into those that preprocess the pattern and those that preprocess the text, there also is the option of transforming a DNA or amino acid database such that it becomes easier to search. This route was taken, e.g., by a group from IBM developing the FLASH [61] program. They devised an intricate though supposedly very space consuming technique of transforming the database into an index for storing the offsets of gapped k-tuples. The work of Heumann and Mewes [62] has focused on further developing the well known data structure of a suffix tree. A suffix tree allows for quick lookup of where subwords of arbitrary length occur in a database. The problem with a suffix tree is that its size is several times the magnitude of the original data which for a sequence database means that it is unlikely to fit into main memory. Heumann and Mewes developed a data structure that allows for efficient mapping to disk such that the size problems can be circumvented. Likewise, the QUASAR program by Burkhard et al. [63] preprocesses the database into a so-called suffix array, similar to a suffix tree yet simple to keep on disk.

Progress in computational speed using either specially designed or massively parallel hardware has led to the availability of extremely fast versions of the Smith–Waterman algorithm. The EBI, among other institutions, offers a service where this algorithm is executed on a massively parallel computer resulting in search times of a few seconds. Companies like Compugen [64] or Paracel [65] have developed special hardware to do this job.

With the availability of expressed sequence tags (ESTs) it has become very important to match DNA sequence with protein sequence in such a way that a possible translation can be maintained throughout the alignment. Both the FASTA and the BLAST package contain programs for this and related tasks. When coding DNA is compared to proteins, gaps are inserted in such a way as to maintain a reading frame. Likewise, a protein sequence can be searched versus a DNA sequence database. The search of DNA vs. DNA with an emphasis on matching regions that allow for a contiguous translation is not so well supported. Although a dynamic programming algorithm for this task is feasible the existing implementation in BLAST compares all reading frames.

2.5
Alignment and search statistics

Alignment score is the product of an optimization, mostly a maximization procedure. As such it tends to be a large number sometimes suggesting biological relatedness where there is none. In pairwise comparisons the user still has a chance to study an alignment by eye in order to judge its validity but upon searching an entire database automatic methods are necessary to attribute a statistical significance to an alignment score.

In the early days of sequence alignment, the statistical significance of the score of a given pairwise alignment was assessed using the following procedure. The letters of the sequences are permuted randomly and a new alignment score is calculated. This is repeated roughly 100 times and mean and standard deviation of this sample are calculated. The significance of the given alignment score is reported in 'number of standard deviations above the mean', also called the Z-score. Studying large numbers of random alignments is in principle correct. However, the significance of the alignment should then be reported as the fraction of random alignments that score better than the given alignment. The procedure described assumes that these scores were normally distributed. Since the random variable under study – the score of an optimal alignment – is the maximum over a large number of values this is not a reasonable assumption. In fact, when trying to fit a normal distribution to the data the lack of fit quickly becomes obvious. The second argument against this way of calculating significance is a pragmatic one. The procedure needs to be repeated for every alignment under study because the effect of the sequence length cannot be accounted for.

Based on the work of several researchers [66, 67] it has meanwhile become apparent that alignment score as well as scores from database searches obey a so-called extreme value distribution. This is not surprising given that extreme value distributions typically describe random variables which are the result of a maximization. In sequence alignment, there are analytical results confirming the asymptotic convergence to an extreme value distribution for the case of local alignment without gaps, i.e., the score of the best-matching contiguous diagonal in a comparison [68]. This is also a valid approximation to the type of matching effected in the database search program BLAST. Thus this approach has become widely used and in fact has contributed significantly to the popularity of database search programs because significance measures have made the results of the search much easier to interpret.

The statistical significance of an event like observing a sequence alignment of a certain quality is the probability to observe a better value as a result of chance alone. Thus one needs to model chance alignments, which is precisely what the statistician means by deriving the distribution of a random variable. The probability that a chance result would exceed an actually obtained threshold S is 1 minus the value of the cumulative distribution function evaluated at that threshold. (If S is the score of a given sequence segment this probability is the famed P-value of the segment. It is to be contrasted with the E-value which gives the expected number of segments in the sequence that exceed the score. For interesting outcomes both of these values are very small and almost identical.) In sequence alignment, the respective cumulative distribution function is generally expressed as [69]:

$$\exp(mnKe^{\lambda S}).$$

Here, m and n are the lengths of the sequences compared and K and λ are parameters which need to be computed (where possible) or derived by simulation. K and λ depend on the scoring matrix used (e.g., the PAM120 matrix) and the distribution of residues. Hence, for any scoring system these parameters are computed beforehand and the statistical significance of an alignment score S is then computed by evaluating the formula with the length of the two sequences compared.

The most prominent case where the parameters K and λ can be defined analytically is local alignment without gaps. Algorithmically this amounts to computing a Smith–Waterman alignment under very high gap penalties such that the resulting alignment will simply not contain any gaps. Since this notion of alignment also guides the heuristic used by the BLAST database search program the resulting statistical estimates are primarily used in database searching. In this application, one of the lengths is the length of the input sequence and the other length is on the order of the length of the concatenated sequences from the database that is being searched.

When gaps are allowed the situation is more complex because an approximation of the distribution function of alignment score by an extreme value distribution as above is not always valid. Generally speaking, it is allowed only for sufficiently strong gap penalties where alignments remain compact as opposed to spanning the entire sequences. Under sufficiently strong gap penalties, though, it has been demonstrated that the approximation is indeed valid just like for infinite gap penalties [70]. However, it is not possible any more to compute analytically the values of the parameters K and λ. As a remedy, one applies simulations where many alignments of randomly generated sequences are computed and the parameters are determined based on fitting the empirical distribution function with an extreme value distribution [71]. Like in the case above, this procedure allows to determine parameters beforehand and compute significance by putting the lengths of the sequences into the formula.

The question remains how to determine whether approximation by an extreme value distribution is admissible for a certain scoring scheme and gap penalty setting one is using. This can be tested on randomly generated (or, simply, unrelated) sequences by computing a global alignment between sequences under that particular parameter setting. If the result has negative sign (averaged over many trials or on very long random sequences) then the approximation is admissible. This is based on a theorem due to Arratia and Waterman [72] and subsequent simulation results reported by Waterman and Vingron [73]. In particular, a gap open penalty of 12 with extension penalty of 2 or 3 for the case of the PAM250 matrix, as well as any stronger combination allow for approximation by the extreme value distribution.

In database searching the fitting need not be done on randomly generated sequences. Under the assumption that the large majority of sequences in a database are not related to the query, the bulk of the scores generated upon searching can be used for fitting. This approach is taken by W. Pearson in the FASTA package. It has the advantage that the implicit random model is more realistic since it is taken directly from the data actually searched. Along a similar line of thought, Spang and Vingron [74] tested significance calculations in data base searching by evaluating a large number of search results. Their study showed that one should not simply use the sum of the lengths of all the sequences in the database as the length-parameter in the formula for the extreme value distribution. This would overestimate the length that actually governs the statistics. Instead, a considerably shorter effective length can determined for a particular database using simulations. This effect is probably due to the fact that alignments cannot start in one sequence and end in the next one which makes the number of feasible starting points for random alignments smaller than the actual length of the database.

2.6
Multiple sequence alignment

For many genes a database search will reveal a whole number of homologous sequences. One then wishes to learn about the evolution and the sequence conservation in such a group. This question surpasses what can reasonably be achieved by the sequence comparison methods described in Section 2.3. Pairwise comparisons do not readily show positions that are conserved among a whole set of sequences and tend to miss subtle similarities that become visible when observed simultaneously among many sequences. Thus, one wants to simultaneously compare several sequences.

A multiple alignment arranges a set of sequences in a scheme where positions believed to be homologous are written in a common column. Like in a pairwise alignment, when a sequence does not possess an amino acid in a particular position this is denoted by a dash. There also are conventions similar to the ones for pairwise alignment regarding the scoring of a multiple alignment. The so-called sum-of-pairs (SOP) [75] score adds the scores of all the induced pairwise aligments contained in a multiple alignment. For a linear gap penalty this amounts to scoring each column of the alignment by the sum of the amino acid pair scores or gap penalties in this column. Although it would be biologically meaningful, the distinctions between global, local, and other forms of alignment are rarely made in a multiple alignment. The reason for this will become apparent below, when we describe the computational difficulties in computing multiple alignments.

Note that the full set of optimal pairwise alignments among a given set of sequences will generally overdetermine the multiple alignment. If one wishes to assemble a multiple alignment from pairwise alignments one has to avoid "closing loops", i.e., one can put together pairwise alignments as long as no new pairwise alignment is included to a sequence which is already part of the multiple alignment. In particular, pairwise alignments can be merged when they align one sequence to all others, when a linear order of the given sequences is maintained, or when the sequence pairs for which pairwise alignments are given form a tree. While all these schemes allow for the ready definition of algorithms that output multiply aligned sequences, they do not include any information stemming from the simultaneous analysis of several sequences.

An alternative approach is to generalize the dynamic programming optimization procedure applied for pairwise alignment to the delineation of a multiple alignment that maximizes, e.g., the SOP score. The algorithm used [76] is a straight-forward generalization of the global alignment algorithm. This is easy to see, in particular, for the case of the column-oriented SOP scoring function avoiding affine gap penalty in favor of the simpler linear one. With this scoring, the arrangement of gaps and letters in a column can be represented by a Boolean vector indicating which sequences contain a gap in a particular column. Given the letters that are being compared, one needs to evaluate the scores for all these arrangements. However conceptually simple this algorithm may be, its computational complexity is rather forbidding. For n sequences it is proportional to 2^n times the product of the lengths of all sequences.

In practice this algorithm can be run only for a modest number of sequences being compared. There exists software to compare three sequences with this algorithm that additionally implements a space-saving technique [77]. For more than three sequences, algorithms have been developed that aim at reducing the search space while still optimizing the given scoring function. The most prominent program of this kind is MSA2 [78, 79]. An alternative approach is used by DCA [80, 81] which implements a divide-and-conquer philosophy. The search space is repeatedly subdivided by identifying anchor points through which the alignment is highly likely to pass.

None of these approaches, however, would work independent of the number of sequences to be aligned. The most common remedy is reducing the multiple alignment problem to an iterated application of the pairwise alignment algorithm. However, in doing so, one also aims at drawing on the increased amount of information contained in a set of sequences. Instead of simply merging pairwise alignments of sequences, the notion of a profile [82] has been introduced in order to grasp the conservation patterns within subgroups of sequences. A profile is essentially a representation of an already computed multiple alignment of a subgroup. This alignment is "frozen" for the remaining computation. Other sequences or other profiles can be compared to a given profile based on a generalized scoring scheme defined for this purpose. The advantage of scoring a sequence versus a profile over scoring individual sequences lies in the fact that the scoring schemes for profile matching reflect the conservation patterns among the already aligned sequences.

Given a profile and a single sequence, the two can be aligned using the basic dynamic programming algorithm together with the accompanying scoring scheme. The result will be an alignment between sequence and profile that can readily be converted into a multiple alignment now comprising the sequences underlying the profile plus the new one. Likewise, two profiles can be aligned with each other resulting in a multiple alignment containing all sequences from both profiles. With these tools various multiple alignment strategies can be implemented. Most commonly, a hierachical tree is generated for the given sequences which is then used as a guide for iterative profile construction and alignment. This alignment strategy was introduced in papers by Taylor [83], Corpet [84], and Higgins [85]. Higgins' program Clustal [86] has meanwhile become the de facto standard for multiple sequence alignment [87]. Other programs in practical use are the MSA2 program, DCA, and Dialign. Dialign [88, 89] is different in that it aims at the delineation of regions of similarity among the given sequences. Chapter 8 will discuss approaches to multiple sequence comparison where common patterns are sought rather than overall alignments.

Since iterative profile alignment tends to be guided by a hierarchical tree, this step of the computation is also influencing the final result. Usually the hierarchical tree is computed based on pairwise comparisons and their resulting alignment scores. Subsequently this score matrix is used as input to a clustering procedure like single linkage clustering or UPGMA [90]. However, it is well understood that in an evolutionary sense such a hierarchical clustering does not necessarily result in a biologically valid tree. Thus, when allowing this tree to determine the multiple alignment there is the danger of pointing further evolutionary analysis of this alignment into the wrong direction. Consequently, the question has arisen of a common formulation of evolutionary reconstruction and multiple sequence alignment. The cleanest although biologically somewhat simplistic model attempts to reconstruct ancestral sequences to attribute to the inner nodes of a tree [91]. Such reconstructed sequences at the same time determine the multiple alignment among the sequences. In this 'generalized tree alignment' one aims at minimizing the sum of the edge-lengths of this tree, where the length of an edge is determined by the alignment distance between the sequences at its incident nodes. As to be expected, the computational complexity of this problem again makes its solution unpractical. The practical efforts in this direction go back to the work of Sankoff [92]. Hein [93] and Schwikowski and Vingron [94] produced software [95, 96] relying on these ideas.

2.7
Multiple alignments and database searching

Information about which residues are conserved and thus important for a particular family is crucial not only for the purpose of multiply aligning a set of sequences. Also in the context of identifying related sequences in a database this information is very valuable. A multitude of methods has been developed that aim at identifying sequences in a database which are related to a given family. The first one was the notion of a profile that was described above and was actually introduced in the context of data base searching. Like in multiple alignment, profiles help in emphasizing conserved regions in a database search. Thus, a sequence that matches the query profile in a conserved region will receive a higher score than a database sequence matching only in a divergent part of an alignment. This feature is of enormous help in distinguishing truly related sequences.

Algorithmically, profile searching simply uses the dynamic programming alignment algorithm for aligning a sequence to a profile on each sequence in the database. Of course, this is computationally quite demanding and much slower than the heuristic database search algorithms like BLAST or FASTA. Typically, the multiple alignment underlying the profile will describe a conserved domain which one expects to find within a database sequence. Therefore, in this context it is important that end gaps should not be penalized. Furthermore, gap penalties for profile matching frequently vary along the profile in order to reflect the existence of gaps within the underlying multiple alignment. Through this mechanism, one attempts to allow new gaps preferentially in regions where gaps have been observed already. However, different suggestions exist as to the choice and derivation method for these gap penalties [97].

In 1994, Haussler and co-workers [98] introduced Hidden Markov Models (HMMs) for the purpose of identifying family members in a database. HMMs are a class of probabilistic models well suited for describing the relevant parameters in matching a given multiple alignment against a database. For HMMs there exist automatic learning algorithms that adapt the parameters of the HMM for best identification of family members. Thus, they offer in particular a solution to the question of gap penalty settings along a profile. Sequences are matched to HMMs in much the same way as they are aligned to a profile although the interpretation of the procedure is different. The HMM is thought of as producing sequences by going through different states. Aligning a sequence to an HMM amounts to delineating the series of model-states that is most likely to have produced the sequence.

Based on this interpretation one can make an interesting distinction between the optimal alignment of a sequence with a HMM and the computation of the probability that a model has produced a particular sequence. The optimal alignment is computed with an algorithm exactly analogous to the dynamic programming algorithm and maximizing the probability that a series of states has given rise to a particular sequence. In contrast hereto, in absence of knowledge of the correct path of states, the probability that a model has given rise to a particular sequence should rather be computed as the sum over the different sets of states that could have produced the sequence. This interpretation leads to a summation over all paths instead of the choice of the best one. These issues are discussed by Durbin et al. [99] (Section 2.5.4) although, practically, there is little known about the difference in performance between the two approaches. Bucher and Karplus [100] introduced generalized profiles and showed that the two concepts are equally powerful in their abilities to model sequence families and detect related sequences.

The fact that a profile or HMM can pick out new sequences also related to the given family suggests that these should be used to update the profile or HMM used as search pattern. This idea leads to iterative search algorithms where the database is searched repeatedly, each time updating the query pattern with some or all of the newly identified sequences. Psi-Blast [101] is a very successful implementation of this idea. It starts with a single sequence and after the first search builds profiles from conserved regions among the query and newly identified sequences. Without allowing for gaps (to speed up the search) these new profiles are used to repeat the search. Generally, Psi-Blast quickly converges after updating these profiles again and generally is very successful in delineating all the conserved regions a sequence may share with other sequences in a database. In the realm of HMMs, SAM is a very careful implementation of the idea of iterated searches [102, 103].

It is the generally held view that searching a database with a profile or HMM will produce extreme-value distributed random scores, just like single-sequence database searching. The quality of the fit to the extreme-value distribution may depend on the particular given alignment, though. This has been substantiated with mathematical arguments only for the case of ungapped profile matching. Nevertheless, this basic understanding of the statistical behavior of database matching methods is a crucial element in the iterative search programs. Without clear and reliable cutoff values one could not decide which sequences to integrate into the next search pattern and would run the danger of including false positives, thus blurring the information in the pattern.

Both single sequence search methods and profile/HMM-based ones have been thoroughly validated during recent years [104]. Databases of structurally derived families like, e.g., SCOP [105, 106] have made it possible to search a sequence database with a query and exactly determine the number of false positives and false negatives. For every search one determines how many sequences one misses (false negatives) in dependence of the number of false positive matches. If the sequence statistics is accurate the number of false positives correlates well with the E-value which is the number of false positives expected by chance. Although not always quite decided, this way of validating search methods allows to make objective comparisons and to determine how much quality one actually gains with slower methods over faster, less accurate one.

2.8
Protein families and protein domains

The dual question to the one that assigns related sequences from a database to a given query sequence or family is the question that tries to assign to a query sequence the family that it is a member of or the domains that it contains. One simple yet very effective resource for this purpose is the Prosite database [107, 108] which contains amino acid patterns that are descriptive for particular domains, families, or functions. These patterns allow to specify alternative residues in particular positions or variable length spacers between positions. Matching a sequence against a Prosite entry amounts simply to looking for the particular pattern of characters in the given sequence. For many sequences that are contained in a protein sequence database this is not even necessary any more because they have already been annotated with the Prosite patterns they contain. The expasy server [109] offers the possibility to screen a sequence versus the entire collection of Prosite patterns.

For certain protein sequence domains character patterns may be rather poor descriptors, i.e., they may cause many false positive and false negative matches. The Pfam database remedies this situation by supplying pre-computed Hidden Markov Models for protein domains. A query sequence can be matched against this library of HMMs in order to identify known domains in the query sequence. Here, too, match statistics plays a crucial role in order to determine the significantly matching domains. A server that allows to scan a sequence versus all Pfam domains can be found at the Sanger center [110]. Software has also been developed to recognize the Pfam HMMs in either coding DNA or in genomic DNA. In the latter case the program combines the HMM matching with the distinction between coding and non-coding DNA. Recently, a database of domains, InterPro [111], has emerged that unifies sources like Prosite and Pfam.

Many attempts have been made at developing algorithms that will auto-matically determine domains shared by several input sequences. In a way this is a generalization of the multiple alignment problem in the direction of local alignment, however in this context it seems to be considerably more difficult than in the pairwise case. There exist two programs that identify Prosite-like patterns contained in many sequences. These are Pratt [112, 113] and TEIRESIAS [114, 115]. The program Blockmaker [116, 117] iden-tifies non-gapped sequence blocks that are characteristic of a given set of sequences. This program has also been used to construct a large collection [118] of Blocks that can be searched like Prosite or Pfam. Krause and Vingron [119] used an iterated search procedure SYSTERS [120] to delin-eate protein families and supply consensus sequences of these families to be searched with a DNA or protein query sequence. The most sophisticated programs for the detection of common sequence motifs in a given set use probabilistic modeling and/or machine learning approaches. In particular, the mathematical technique of the Gibbs sampler has lent its name also to a motif-finding program, the Gibbs Motif Sampler [121, 122]. Bailey and Elkan [123] designed the MEME [124] program which relies on an expecta-tion maximization algorithm.

2.9
Conclusion

The problems and methods introduced above have been instrumental in the advance in our understanding of genome function, organization, and structure. While some years ago human experts would check every program output, nowadays sequence analysis routines are being applied in an automatic fashion creating annotation that is included in various database. This holds true for similarity relationships among sequences and extends all the way to the prediction of genomic structure or to function prediction based on similarity. Although the quality of the tools has increased dramatically, the possibility of error and in particular its perpetuation by further automatic methods exists. Thus, it is apparent that the availability of these high-throughput computational analysis tools is a blessing and a problem at the same time.

References

1 Hopp, T. P. and Woods, K. R. (1981) Prediction of protein antigenic determinants from amino acid sequences, *Proc. Natl. Acad. Sci. USA* 78:3824–3828.

2 Kyte, J. and Doolittle, R. F. (1982) A simple method for displaying the hydropathic character of a protein, *J. Mol. Biol.* 157:105–132.

3 Argos, P. (1987) A Sensitive Procedure to Compare Amino Acid Sequences, *J. Mol. Biol.* 193:385–396.

4 Palau, J., Argos, P., Puigdomenech, P. (1982) Protein Secondary Structure – Studies on the Limits of Prediction Accuracy, *Int. J. Pept. Protein Res.* 19:394–401.

5 Jones, D. D. (1975) Amino-Acid Properties and Side-Chain Orientation in Proteins – Cross-Correlation Approach, *J. Theor. Biol.* 50:167–183.

6 Jones, D. D. (1975) Amino-Acid Properties and Side-Chain Orientation in Proteins – Cross-Correlation Approach, *J. Theor. Biol.* 50:167–183.

7 Lifson, S., Sander, C. (1979) Antiparallel and parallel Beta-Strands differ in Amino-Acid Residue Preferences, *Nature* 261:109–111.

8 O'Neil, K. T., DeGrado, W. F. (1990) A thermodynamic scale for the helix-forming tendencies of the commonly occurring amino acids, *Science* 250:646–651.

9 Hofmann, K., Stoffel, W. (1993) TMbase: A database of membrane-spanning protein segments, *Biol. Chem. Hoppe-Seyler* 347:166.

10 Persson, B., Argos, P. (1994) Prediction of transmembrane segments in proteins utilising multiple sequence alignments, *J. Mol. Biol.* 237:182–192.

11 Eisenberg, D., Weiss, R. M., and Terwilliger, T. C. (1984) The hydrophobic moment detects periodicity in protein hydrophobicity, *Proc. Natl. Acad. Sci. USA* 81:140–144.

12 Cohen, C., Parry, D. A. D. (1986) α-helical coiled-coils: A widespread motif in proteins, *Trends Biochem. Sci.* 11:245–248.

13 Lupas, A., Van Dyke, M., Stock, J. (1991) Predicting coiled coils from protein sequences, *Science* 252:1162–1164.

14 WOLF, E., KIM, P. S., BERGER, B. **(1997)** MultiCoil: a program for predicting two- and three stranded coiled coils, *Protein Sci.* 6:1179–1189.

15 NIELSEN, H., ENGELBRECHTE, J., BRUNAK, S., VON HEIJNE, G. **(1997)** Identification of prokaryotic and eukaryotic signal peptides and prediction of their cleavage sites, *Protein Engineering* 10:1–6.

16 PAULING, L. and COREY, R. B. **(1951)** The structure of proteins: Two hydrogen-bonded helical configurations of the polypeptide chain, *Proc. Natl. Acad. Sci. USA.* 37:205–211.

17 BRANDEN, C. and TOOZE, J. **(1999)** *Introduction to Protein Structure, 2nd ed.*, Garland Publishing, New York.

18 COLLC'H, N., ETCHEBEST, C., THOREAU, E., HENRISSAT, B., MORNON, J. P. **(1993)** Comparison of three algorithms for the assignment of secondary structure in proteins: the advantages of a consensus assignment, *Protein Eng.* 6:377–382.

19 GARNIER, J., GIBRAT, J.-F., and ROBSON, B. **(1996)** GOR Method for Predicting Protein Secondary Structure from Amino Acid Sequence, *Meth. Enz.* 266:540–553.

20 ROST, B. and SANDER, C. **(1994)** Combining Evolutionary Information and Neural Networks to Predict Protein Secondary Structure, *PROTEINS: Structure, Function, and Genetics* 19:55–72.

21 SOLOVYEV, V. V. and SALAMOV, A. A. **(1994)** Predicting α-helix and β-strand segments of globular proteins, *Comp. Appl. Biosci.* 10:661–669.

22 CHOU, P. Y. and FASMAN, G. D. **(1978)** Empirical predictions of protein conformations, *Ann. Rev. Biochem.* 47:251–276.

23 GARNIER, J., GIBRAT, J.-F., and ROBSON, B. **(1996)** GOR Method for Predicting Protein Secondary Structure from Amino Acid Sequence, *Meth. Enz.* 266:540–553.

24 ARGOS, P. **(1987)** Analysis of sequence-similar pentapeptides in unrelated protein tertiary structures, *J. Mol. Biol.* 20:331–348.

25 SOLOVYEV, V. V. and SALAMOV, A. A. **(1994)** Predicting α-helix and β-strand segments of globular proteins, *Comp. Appl. Biosci.* 10:661–669.

26 ZVELEBIL, M. J., BARTON, G. J., TAYLOR, W. R., STERNBERG, M. J. E. **(1987)** Prediction of Protein Secondary Structure and Active Sites using the Alignment of Homologous Sequences, *J. Mol. Biol.* 195:957–961.

27 ROST, B. and SANDER, C. **(1994)** Combining Evolutionary Information and Neural Networks to Predict Protein Secondary Structure, *PROTEINS: Structure, Function, and Genetics* 19:55–72.

28 SCHNEIDER, R. and SANDER, C. **(1991)** Database of homology-derived structures and the structural meaning of sequence alignment, *Proteins* 9:56–68.

29 KABSCH, W., SANDER, C. **(1983)** How good are predictions of protein secondary structure? *FEBS Letters* 155:179–182.

30 GARNIER, J., GIBRAT, J.-F., and ROBSON, B. **(1996)** GOR Method for Predicting Protein Secondary Structure from Amino Acid Sequence, *Meth. Enz.* 266:540–553.

31 CUFF, J. A., CLAMP, M. E., SIDDIQUI, A. S., FINLAY, M., BARTON, G. J. **(1998)** JPred: a consensus secondary structure prediction server, *Bioinformatics* 14:892–893.

32 SELBIG, J., MEVISSEN, T., LENGAUER, T. **(1999)** Decision tree-based formation of consensus protein secondary structure prediction, *Bioinformatics* 15:1039–1046.

33 http://www.expasy.ch/

34 MAIZEL, J. V., LENK, R. P. **(1981)** Enhanced graphic matrix analysis of nucleic acid and protein sequences, *Proc. Natl. Acad. Sci. USA* 78:7665–7669.

35 SONNHAMMER, E. L., DURBIN, R. **(1995)** A dot-matrix program with dynamic threshold control suited for genomic DNA and protein sequence analysis, *Gene* 167:GC1–10.

36 ARGOS, P. **(1987)** A Sensitive Procedure to Compare Amino Acid Sequences, *J. Mol. Biol.* 193:385–396.

37 SONNHAMMER, E. L., DURBIN, R. **(1995)** A dot-matrix program with dynamic threshold control suited for genomic DNA and protein sequence analysis, *Gene* 167:GC1–10.

38 WATERMAN, M. S. **(1995)** *Introduction to Computational Molecular Biology*, Chapman & Hall (London).

39 SANKOFF, D. **(1972)** Matching sequences under deletion/insertion constraints, *Proc. Natl. Acad. Sci. USA* 69:4–6.

40 SMITH, T. F., WATERMAN, M. S. **(1981)** Identification of common molecular subsequences, *J. Mol. Biol.* 147:195–197.

41 DAYHOFF, M. O., BARKER, W. C., HUNT, L. T. **(1978)** *Establishing homologies in protein sequences*, In: Atlas of Protrein Sequences and Structure 5.

42 HENIKOFF, S. and HENIKOFF, J. G. **(1992)** Amino acid substitution matrices from protein blocks, *Proc. Natl. Acad. Sci. USA* 89:10915–10919.

43 BENNER, S. A., COHEN, M. A., GONNET, G. H. **(1994)** Amino acid substitution during functionally constrained divergent evolution of protein sequences, *Protein Engineering* 7:1323–1332.

44 NEEDLEMAN, S. B. and WUNSCH, C. D. **(1970)** A general method applicable to the search for similarities in the amino-acid sequence of two proteins, *J. Mol. Biol.* 48:443–453.

45 SANKOFF, D. **(1972)** Matching sequences under deletion/insertion constraints, *Proc. Natl. Acad. Sci. USA* 69:4–6.

46 WATERMAN, M. S. **(1984)** Efficient sequence alignment algorithms, *J. Theor. Biol.* 108:333–337.

47 GOTOH, O. **(1982)** An improved algorithm for matching biological sequences, *J. Mol. Biol.* 162:705–708.

48 MYERS, E. W., MILLER, W. **(1988)** Optimal alignments in linear space, *Comp. Appl. Biosciences* 4:11–17.

49 HIRSCHBERG **(1977)** Algorithms for the longest common subsequence problem, *J. ACM* 24:664–675.

50 SMITH, T. F., WATERMAN, M. S. **(1981)** Identification of common molecular subsequences, *J. Mol. Biol.* 147:195–197.

51 WATERMAN, M. S., EGGERT, M. **(1987)** A new algorithm for best subsequence alignments with application to tRNA-rRNA comparisons, *J. Mol. Biol.* 197:723–728.

52 VINGRON, M., WATERMAN, M. S. **(1994)** Sequence Alignment and Penalty Choice: Review of Concepts, Case Studies, and Implications, *J. Mol. Biol.* 235:1–12.

53 PEARSON, W. R., LIPMAN, D. J. **(1988)** Improved tools for biological sequence comparison, *Proc. Natl. Acad. Sci. USA* 85:2444–2448.

54 WOZNIAK, A. **(1997)** Using video-oriented instructions to speed up sequence comparison, *Comp. Appl. Biosci.* 13:145–150.

55 BARTON, G. J. **(1993)** An efficient algorithm to locate all locally optimal alignments between two sequences allowing for gaps, *Comp. Appl. Biosci.* 9:729–734.

56 PEARSON, W. R., LIPMAN, D. J. **(1988)** Improved tools for biological sequence comparison, *Proc. Natl. Acad. Sci. USA* 85:2444–2448.

57 ALTSCHUL, S. F., GISH, W., MILLER, W., MYERS, E. W., LIPMAN, D. J. **(1990)** Basic local alignment search tool, *J. Mol. Biol.* 215:403–410.

58 ALTSCHUL, S. F., MADDEN, T. L., SCHAFFER, A. A., ZHANG, J., MILLER, W., LIPMAN, D. J. **(1997)** Gapped BLAST and PSI-BLAST: a new generation of protein database search programs, *Nucl. Acids Res.* 25:3389–3402.

59 http://www.ncbi.nlm.nih.gov/

60 http://www.ebi.ac.uk/

61 CALIFANO, A., RIGOUTSOS, I. **(1993)** FLASH: A fast look-up algorithm for string homology. In: Proceedings of the 1st International Conference on Intelligent Systems in Molecular Biology, 56–64.

62 HEUMANN, K., MEWES, H.-W. **(1996)** The Hashed Position Tree (HPT): a suffix tree variant for large data sets stored on slow mass storage devices. In: Proceedings of the 3rd South American Workshop on String Processing, Carleton University Press, Ottawa 101–114.

63 BURKHARD, S., CRAUSER, A., FERRAGINA, P., LENHOF, H.-P., RIVALS, E., VINGRON, M. **(1999)** q-gram based database searching using a suffix array (QUASAR). In: Proceedings of the 3rd International Conference on Computational Molecular Biology, 77–83.

64 http://www.compugen.co.il/

65 http://www.paracel.com/

66 SMITH, T. F., WATERMAN, M. S., BURKS, C. **(1985)** The statistical distribution of nucleic acid similarities, *Nucl. Acids Res.* 13:645–656.

67 KARLIN, S., ALTSCHUL, S. F. **(1990)** Methods for assessing the statistical significance of molecular sequence features by using general scoring schemes, *Proc. Natl. Acad. Sci. USA* 87:2264–2268.

68 DEMBO, A., KARLIN, S., ZEITOUNI, O. **(1994)** Critical phenomena for sequence matching with scoring, *Ann. Prob.* 22:2022–2039.

69 KARLIN, S., ALTSCHUL, S. F. **(1990)** Methods for assessing the statistical significance of molecular sequence features by using general scoring schemes, *Proc. Natl. Acad. Sci. USA* 87:2264–2268.

70 VINGRON, M., WATERMAN, M. S. **(1994)** Sequence Alignment and Penalty Choice: Review of Concepts, Case Studies, and Implications, *J. Mol. Biol.* 235:1–12.

71 WATERMAN, M. S., VINGRON, M. **(1994)** Rapid and accurate estimates of statistical significance for database searching, *Proc. Natl. Acad. Sci. USA* 91:4625–4628.

72 ARRATIA, R., WATERMAN, M. S. **(1994)** A phase transition for the score in matching random sequences allowing deletions, *Ann. Appl. Prob.* 4:200–225.

73 WATERMAN, M. S., VINGRON, M. **(1994)** Sequence comparison significance and Poisson approximation, *Statistical Science* 9:401–418.

74 SPANG, R., VINGRON, M. **(1998)** Statistics of large scale sequence searching, *Bioinformatics* 14:279–284.

75 ALTSCHUL, S. F., LIPMAN, D. J. **(1989)** Trees, stars, and multiple biological sequence alignment, *SIAM J. Appl. Math.* 49:179–209.

76 WATERMAN, M. S. **(1984)** Efficient sequence alignment algorithms, *J. Theor. Biol.* 108:333–337.

77 HUANG, X. **(1993)** Alignment of three sequences in quadratic space, *Applied Computing Review* 1:7–11.

78 GUPTA, S. K., KECECIOGLU, J. D., SCHAFFER, A. A. **(1995)** Improving the practical space and time efficiency of the shortest-path approach to sum-of-pairs multiple sequence alignment, *J. Comp. Biol.* 2:459–472.

79 http://www.ncbi.nlm.nih.gov/CBBresearch/Schaffer/msa.html

80 STOYE, J. **(1998)** Multiple Sequence Alignment with the Divide-and-Conquer Method, *Gene* 211:GC45–56.

81 http://bibiserv.techfak.uni-bielefeld.de/dca/

82 GRIBSKOV, M., MCLACHLAN, A. D., EISENBERG, D. **(1987)** Profile Analysis: Detection of distantly related proteins, *Proc. Natl. Acad. Sci. USA* 84:455–4358.

83 TAYLOR, W. R. **(1987)** Multiple sequence alignment by a pairwise algorithm, *Comp. Appl. Biosci.* 3:81–87.

84 CORPET, F. **(1988)** Multiple sequence alignment with hierarchical clustering, *Nucl. Acids Res.* 16:10881–10890.

85 HIGGINS, D. G., BLEASBY, A. J., FUCHS, R. **(1992)** CLUSTAL V: improved software for multiple sequence alignment, *Comp. Appl. Biosci.* 8:189–191.

86 http://www.ebi.ac.uk/clustalw/

87 JEANMOUGIN, F., THOMPSON, J. D., GOUY, M., HIGGINS, D. G. **(1998)** Trends Biochem, *Sci.* 23:403–405.

88 MORGENSTERN, B. **(1999)** DIALIGN 2: Improvement of the segment-to-segment approach to multiple sequence alignment, *Bioinformatics* 15:211–218.

89 http://bibiserv.techfak.uni-bielefeld.de/dialign/

90 SWOFFORD, D. L., OLSEN, G. J. **(1990)** Phylogeny Reconstruction. In: Molecular Systematics. Eds. Hillis, D. M., Moritz, C., Sinauer Associates, Sunderlang, Massachusetts. 411–501.

91 SANKOFF, D. **(1973)** Minimal mutation trees of sequences, *SIAM J. Appl. Math.* 28:35–42.

92 Sankoff, D., Cedergren, R. J., Lapalme, G. **(1976)** Frequency of insertion-deletion, transversion, and transition in evolution of 5S ribosomal RNA, *J. Mol. Evol.* 7:133–149.

93 Hein, J. **(1990)** Unified approach to alignment and phylogenies, *Meth. Enz.* 183:626–645.

94 Schwikowski, B., Vingron, M. **(1997)** The deferred path heuristic for the generalized tree alignment problem, *J. Comp. Biol.* 4:415–431.

95 http://www-bioweb.pasteur.fr/seqanal/interfaces/treealign-simple.html

96 http://www.dkfz.de/tbi/services/3w/start

97 Taylor, W. R. **(1996)** A non-local gap-penalty for profile alignment, *Bull. Math. Biol.* 58:1–18.

98 Krogh, A., Brown, M., Mian, I. S., Sjölander, K., Haussler, D. **(1994)** Hidden Markov Models in Computational Biology: Applications to Protein Design, *J. Mol. Biol.* 235:1501–1531.

99 Durbin, R., Eddy, S., Krogh, A., Mitchison, G. **(1998)** *Biological Sequence Analysis*, Cambridge University Press, Cambridge, UK.

100 Bucher, P., Karplus, K., Moeri, N., Hofmann, K. **(1996)** A flexible motif search technique based on generalized profiles, *Comput. Chem.* 20:3–23.

101 Altschul, S. F., Madden, T. L., Schaffer, A. A., Zhang, J., Miller, W., Lipman, D. J. **(1997)** Gapped BLAST and PSI-BLAST: a new generation of protein database search programs, *Nucl. Acids Res.* 25:3389–3402.

102 Karplus, K., Barrett, C., Hughey, R. **(1998)** Hidden Markov models for detecting remote protein homologies, *Bioinformatics* 14:846–856.

103 http://www.cse.ucsc.edu/research/compbio/sam.html

104 Brenner, S. E., Chothia, C., Hubbard, T. J. **(1998)** Assessing sequence comparison methods with reliable structurally identified distant evolutionary relationships, *Proc. Natl. Acad. Sci. USA* 95:6073–6078.

105 Murzin, A. G., Brenner, S. E., Hubbard, T., Chothia, C. **(1995)** SCOP: a structural classification of proteins database for the investigation of sequences and structures, *J. Mol. Biol.* 247:536–540.

106 http://scop.mrc-lmb.cam.ac.uk/scop/

107 Hofmann, K., Bucher, P., Falquet, L., Bairoch, A. **(1999)** The PROSITE database, its status in 1999, *Nucl. Acids Res.* 27:215–219.

108 http://www.expasy.ch/prosite/

109 http://www.expasy.ch/

110 http://www.sanger.ac.uk/

111 http://www.ebi.ac.uk/interpro

112 Jonassen, I. **(1997)** Efficient discovery of conserved patterns using a pattern graph, *Comp. Appl. Biosci.* 13:509–522.

113 http://www.ii.uib.no/~inge/Pratt.html

114 Rigoutscos, I., Floratos, A. **(1998)** Combinatorial pattern discovery in biological sequences: The TEIRESIAS algorithm, *Bioinformatics* 14:55–67.

115 http://www.research.ibm.com/bioinformatics/

116 HENIKOFF, S., HENIKOFF, J. G., ALFORD, W. J., PIETROKOVSKI, S. **(1995)** Automated construction and graphical presentation of protein blocks from unaligned sequences, *Gene* 163:GC17–26.

117 http://www.blocks.fhcrc.org/blockmkr/make_blocks.html

118 http://www.block.fhcrc.org/

119 KRAUSE, A., STOYE, J., VINGRON, M. **(2000)** The SYSTERS protein sequence clusters set, *Nucl. Acids Res.* 28:270–272.

120 http://www.dkfz.de/tbi/services/cluster/systersform

121 LAWRENCE, C. E., ALTSCHUL, S. F., BOGUSKI, M. S., LIU, J. S., NEUWALD, A. F., WOOTTON, J. C. **(1993)** Detecting subtle sequence signals: a Gibbs sampling strategy for multiple alignment, *Science* 262:208–214.

122 http://bayesweb.wadsworth.org/gibbs/gibbs.html

123 BAILEY, T. L., ELKAN, C. **(1995)** The value of prior knowledge in discovering motifs with MEME, *Intelligent Systems for Molecular Biology* 3:21–29.

124 http://meme.sdsc.edu/

3
Structure, Properties and Computer Identification of Eukaryotic Genes

Victor Solovyev

3.1
Structural characteristics of eukaryotic genes

The gene is a fragment of nucleic sequence that carries the information representing a particular polypeptide or RNA molecule. In eukaryotcs, genes lie in a linear array on chromosomes, which consist of a long molecule of duplex DNA and chromatin proteins (mostly histones that form a structure called a nucleosome). The complex of DNA and proteins (chromatin) can maintain genes in an inactive state by restricting access to RNA polymerase and its accessory factors. To activate a gene, the chromatin encompassing that gene and its control region must be modified to permit transcription [1]. The principal steps in gene expression of protein-coding genes are transcription and post-transcriptional processing of messenger RNA precursors including 5′-capping, 3′-polyadenilation and splicing. The order in which these events occur is not entirely clear: some splicing events could take place during transcription [2]. The processing events produce the mature mRNA in the nucleus and then it is transported to the cytoplasm for translation. The mature mRNA includes sequences called exons that encode the protein product according to the rules of the genetic code. However, the gene sequence often includes non-coding regions: introns that are removed from the primary transcript during RNA splicing and 5′- and 3′- untranslated regions. Model and stages of gene expression of a typical protein-coding gene are presented in Figure 3.1.

Knowledge about structural gene characteristics is accumulated in GenBank and EMBL nucleotide sequence databases, where one gene can be described in dozens of entries assigned to partially sequenced gene regions, alternative splicing forms or mRNAs. Processing GenBank [3] data, a gene-centered database InfoGene was created [4, 5], which contains description of known genes and their basic functional signals and structural components. All major organisms are presented in the separate divisions. The Human InfoGene division contains about 21 000 genes (including 16 000 partially sequenced genes), 53 435 coding regions, 83 488 exons and about

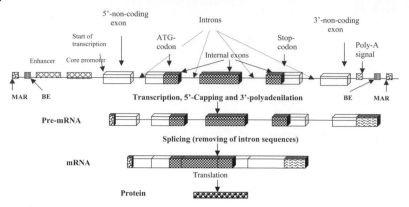

Fig. 3.1
Model of eukaryotic gene structure and gene expression
stages. BEs are boundary elements that bind proteins to
prevent the enhancer effects on outside genes. MARs are
matrix attachment regions.

58 000 donor and acceptor splice sites. Table 3.1 shows the major structural
characteristics of Human, Mouse, *D. melanogaster*, *C. elegans*, *S. cerevisiae*
and *A. thaliana* genes deposited in GenBank, release 119.

We do not observe any significant difference in the size of protein coding
mRNAs in different types of organisms, but the gene sizes are often larger
in vertebrates and especially in primates. We can notice that the human
coding exons are significantly shorter that the sizes of the respective genes.
The average size of an exon is about 200 bp, that is close to the DNA length
associated with the nucleosome particle. Coding and non-coding exons can
be as short as several bases and as long as dozens or thousands of bases.
Usually protein coding exons occupy just a few percent of the gene size.
Different kind of repeats cover 41% of sequenced human DNA, and coding
exons account only for 2–3% of the genomic sequence.

The structural characteristics of eukaryotic genes considered above create
two major problems in computational gene identification.

1. Low quality recognizers will generate a lot of false positive predictions,
   the number of which is comparable with the true exon number (Figure
   3.2);
2. Recognition of small exons (1–20 bp) can not be done using any com-
   position-based coding measure that is often successful for prokaryotes.

We need to develop gene prediction approaches that are based significantly
on the recognition of functional signals encoded in the gene sequence.

The main information about exon location is encoded in splice site sequences. In the next section we will consider essential characteristics of
splice sites. The other functional signals as promoter, poly-A, start and stop

Tab. 3.1

Structural characteristics of genes in eukaryotic model organisms. The numbers reflect gene characteristics described in GenBank (Release 119, 2000), which might deviate from the average parameters for organisms. Gene numbers were calculated for DNA loci only. Many long genes have partially sequence introns, therefore the actual average gene size is bigger. The range (or maximal) and average values are shown. For donor and acceptor sites the percentage of annotated canonical (GT-AG) splice pairs is provided.

	Homo sapiens	Mus musculus	Drosophila melanogaster	C. elegans	S. cerevisiae	Arabidopsis thaliana
CDS/partial	53435/29404	24527/13060	20314/1510	20634/526	12635/1016	31194/1461
Exons/partial	83488/21342	24508/7913	66960/19343	122951/38293	13572/13127	145942/42844
Genes/partial	20791/16141	7428/5573	17435/1154	19658/1263	12513/1098	28346/1023
Alternative splicing	2167, 10.4%	749, 10%	1785, 10%	1194, 6.1%	598, 4.7%	227, 0.1%
No introns genes	1552, 7.4%	748, 10%	3583, 20%	669, 3%	11070, 88.5%	5776, 19.7%
Number of exons	117, 5.7	64, 4.72	50, 3.88	52, 6.1	3, 1.03	78, 5.1
Exon length	1-1088, 201.6	1-6642, 207.1	6-10785, 419.5	1-14975, 22125	1-7471, 1500.0	3-75916, 192.0
Intron length	259776, 2203.5	42573, 818.3	205244, 613.7	19397, 244.0	7317, 300	118637, 174.4
Gene length	401910, 9033	150523, 3963	155515, 2854	45315, 2624	14733, 1462	170191, 2027
Donor sites	58707, 98.0%	6225, 96.9%	49592, 98.0%	102872, 99.5%	471, 93.0%	117658, 99.2%
Acceptor sites	58112, 98.53%	5627, 97.5%	49602, 97.9%	102933, 99.7%	475, 95.6%	121917, 96.9%

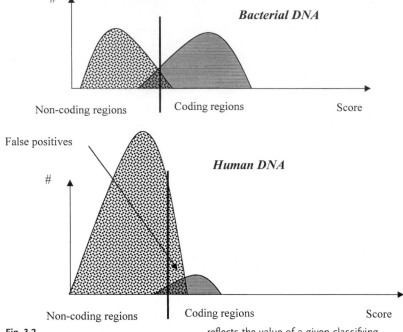

Fig. 3.2

The huge amount of non-coding DNA in the human genome can significantly increase the number of false positive coding exon predictions. The score x-axis reflects the value of a given classifying function to assign a sequence region to coding or non-coding classes. The y-axis presents the number of sequence fragments with a particular score.

of translation will be discussed later with description of their recognition functions used in many gene prediction programs.

3.2
Classification of splice sites in mammalian genomes

Since the discovery of split genes it was observed that practically all introns contain two very conservative dinucleotides. The donor site has GT exactly at the intron's 5′-boundary and the acceptor site has AG exactly at its 3′-boundary [6, 7]. We call splice sites of this type canonical. Introns flanked by the standard GT-AG pairs excised from pre-mRNA by the spliceosome including U1, U2, U4/U6 and U5 snRNPs [7]. Recently, a few examples of a new type of splice pair, a AT-AC, has been discovered. It is processed by a related, but different splicing machinery [8, 9]. AT-AC introns are excised by a novel type of spliceosome composed of snRNPs U11, U12, U4atac/U6atac, and U5 [10, 11, 12]. Several other cases of non-canonical splice sites with

GC-AG, GG-AG, GT-TG, GT-CG or CT-AG dinucleotides at the splice junctions have been reported [8, 13].

The most comprehensive investigation of different types of annotated splice sites has been performed by Burset, Seledtsov and Solovyev [14]. They have extracted 43 337 pairs of exon-intron boundaries and their sequences from the InfoGene database [15] covering all annotated genes in mammalian genomic sequences. Annotation errors present a real problem in obtaining accurate information on eukaryotic gene functional signals from nucleotide sequence databases, such as GenBank or EMBL [3, 16]. This is especially crucial for the analysis of non-standard splice sites. For instance, carefully checking 50 such examples revealed 21 cases of clear EMBL annotation errors [17]. To verify the annotated splice sites Burset et al. [14] used the alignment of spliced exons with known mammalian ESTs and high throughput genomic sequences.

Of 43 337 pairs of donor and acceptor splice sites (splice pairs) 22 489 were supported by EST sequences. 98.71% of those contain canonical dinucleotides GT and AG for donor and acceptor sites. 0.56% hold non-canonical GC-AG splice site pairs. The reminder 0.73% occurs in a lot of small groups (with maximum size of 0.05%). 53.6% of canonical and just 27.3% of non-canonical splice pairs were supported by ESTs. Based on these figures it was supposed that at least half of annotated non-canonical sites presents annotation errors, as was shown in some previous works [8, 17].

In addition to the conserved dinucleotides AG-GT, canonical splice sites demonstrate well defined, conserved donor site consensus: AG|GTRAGT[1], and acceptor site consensus: YYTTYYYYYNC|AGG [18]. For the much smaller AT-AC group different conserved positions have been noticed: |ATATCCTTT for donor site and YAC| for acceptor site [12, 19, 20]. These differences reflect some specific interactions with the components of splicing machineries and they can be used to judge if a particular splice site group belongs to the GT-AG or AT-AC splice system. Weight matrices for the GT-AG, GC-AG pairs constructed based on EST-supported splice pairs are presented in Table 3.2 and consensus sequences for the AT-AC pair are depicted in Figure 3.3.

The occurrence of canonical dinucleotides upstream or downstream in each EST-supported non-canonical splice site group was carefully analyzed. For example, the telethonin gene has only one intron annotated in positions 639–885. This junction is completely supported by ESTs and the annotated splice sites are GG-CA. Analyzing the telethonin sequence uncovered an occurrence of a canonical splice pair GT-AG just one position downstream from the annotated site (Figure 3.4). Taking into account that the non-canonical splice junctions occur very rarely, we can suspect that the canon-

---

[1] Here, we use the common IUPAC
     ambiguity codes: e.g., R = A, G, Y = C, T,
     S = C, G, N = A, C, G, T

Tab. 3.2
Sequence composition of major group of splice sites.
GT-AG group (**98.70%**).
Frequency of nucleotides in donor splice site. 1 and 2 are positions of conserved GT dinucleotide.

%	−3	−2	−1	1	2	3	4	5	6
A	34.0	60.4	9.2	0.0	0.0	52.6	71.3	7.1	16.0
C	36.3	12.9	3.3	0.0	0.0	2.8	7.6	5.5	16.5
G	18.3	12.5	80.3	100	0.0	41.9	11.8	81.4	20.9
U	11.4	14.2	7.3	0.0	100	2.5	9.3	5.9	46.2

Frequency of nucleotides in acceptor splice site. −1 and −2 are positions of conserved AG dinucleotide.

%	−14	−13	−12	−11	−10	−9	−8	−7	−6	−5	−4	−3	−2	−1	1
A	9.0	8.4	7.5	6.8	7.6	8.0	9.7	9.2	7.6	7.8	23.7	4.2	100	0.0	23.9
C	31.0	31.0	30.7	29.3	32.6	33.0	37.3	38.5	41.0	35.2	30.9	70.8	0.0	0.0	13.8
G	12.5	11.5	10.6	10.4	11.0	11.3	11.3	8.5	6.6	6.4	21.2	0.3	0.0	100	52.0
U	42.3	44.0	47.0	49.4	47.1	46.3	40.8	42.9	44.5	50.4	24.0	24.6	0.0	0.0	10.4

GC-AG group (0.56%)
Frequency of nucleotides in donor splice site. 1 and 2 are positions of conserved GT dinucleotide.

%	−3	−2	−1	1	2	3	4	5	6
A	40.5	88.9	1.6	0.0	0.0	87.3	84.1	1.6	7.9
C	42.1	0.8	0.8	0.0	100	0.0	3.2	0.8	11.9
G	15.9	1.6	97.6	100	0.0	12.7	6.3	96.8	9.5
U	1.6	8.7	0.0	0.0	0.0	0.0	6.3	0.8	70.6

Frequency of nucleotides in acceptor splice site. −1 and −2 are positions of conserved AG dinucleotide.

%	−14	−13	−12	−11	−10	−9	−8	−7	−6	−5	−4	−3	−2	−1	1
A	11.1	12.7	3.2	4.8	12.7	8.7	16.7	16.7	12.7	9.5	26.2	6.3	100	0.0	21.4
C	36.5	30.9	19.1	23.0	34.9	39.7	34.9	40.5	40.5	36.5	33.3	68.2	0.0	0.0	7.9
G	9.5	10.3	15.1	12.7	8.7	9.5	16.7	4.8	2.4	6.3	13.5	0.0	0.0	100	62.7
U	38.9	41.3	58.7	55.6	42.1	40.5	30.9	37.3	44.4	47.6	27.0	25.4	0.0	0.0	7.9

Donor: $S_{90}|ATA_{100}T_{100}C_{100}C_{100}T_{100}T_{90}T_{70}$

Acceptor: $T_{70}G_{50}C_{70}NC_{60}AC|A_{60}T_{60}$

Fig. 3.3
Consensus sequences for the AT-AC pair of the alternative splicing machinery.

	Donor	Acceptor
Annotated:	...**CCCGAGGAGG**|ggtgagtgtg.........cctctcccca|**GCTGCTCCCT**...	
Possible:	...**CCCGAGGAGGG**|gtgagtgtg.........cctctccccag|**CTGCTCCCT**...	

Annotated junction: ...**CCCGAGGAGG\|GCTGCTCCCT**...

(EST=AI802984)

Possible junction: ...**CCCGAGGAGGG\|CTGCTCCCT**...

Fig. 3.4
Possible errors in EST-supported splice pairs. Example of an annotated non-canonical junction from the *H. sapiens* telethonin gene, intron 1 (Genbank accession #: AJ010063). This junction is supported by ESTs, which also supports a canonical splice junction. The annotated non-canonical junction and the putative canonical junction produce the same final spliced sequence.

ical splice sites are very likely the real ones. This suggested to explain these observations of shifted canonical dinucleotides by annotation errors involving inserting/deleting one nucleotide, which is actually absent/present in the real genomic sequence. This hypothesis was tested by comparing human gene sequences deposited to GenBank earlier with the sequence of the same region obtained in high throughput genomic sequencing projects (HTGs). Several examples of clear annotation and sequencing errors identified by the comparison are presented in Figure 3.5. 156 out of 171 human non-canonical and EST-supported splice site sequences had a clear match in the human HTG. They can be classified after corrections as: 79 GC-AG pairs (of which 1 was an error that corrected to GC-AG), 61 errors that were corrected to GT-AG canonical pairs, 6 AT-AC pairs (of which 2 were errors that corrected to AT-AC), 1 case that was produced from non-existent intron, 7 cases that were found in HTG that were deposited to GenBank, and finally there were only 2 cases left of supported non-canonical splice sites.

It was concluded that 99.24% of the splice site pairs should be GT-AG, 0.69% GC-AG, 0.05% AT-AC and finally only 0.02% could consist of other types of non-canonical splice sites (Table 3.3). Therefore, gene finding approaches using just standard GT-AG splice sites can potentially predict accurately 97% genes (assuming 4–5 exons per gene on average). If the

Sequences of homeodomain protein, HOXA9EC (AF010258)

	Donor	Acceptor
Genbank:	**CGATCCCAAT**|aa-tgtctcct	cccgcagaat|**AACCCAGCAG**
High throughput:	**CGATCCCA**|gtaagtgtctcct	cccgcag|**AT-AACCCAGCAG**

Sequences of poly(A) binding protein II, PABP2 (AF026029)

	Donor	Acceptor
Genbank:	**TCCAGGCAAT**|gctgagtaac	tttcctgata|**GCTGGCCCGG**
High throughput:	**TCCAGGCAATG**|gtgagtaac	tttcctgatag|**CTGGCCCGG**

Fig. 3.5
Errors found by comparing GenBank and human high throughput sequences for several annotated non-canonical splice sites.

Tab. 3.3
Frequencies of canonical and non-canonical splice sites in
human genes.

GT-AG	99.24%
GC-AG	0.69%
AT-AC	0.05%
Other non-canonical	0.02%

method takes into consideration GC-AG splice pairs, it can increase the
recognition quality to 99%. The GC-AG splice group has several interesting
features. The first is its relatively high frequency (0.56% of all EST-supported splicing pairs belong to this type and 0.69% is the final estimated
frequency of this group). The frequency matrix of this site shows significantly higher degree of conservation in relation to the canonical donor
matrix. It provides a possibility to implement this information in gene prediction programs without generating many false-positive predictions. The
Fgenesh HMM-based gene prediction program [21] has been modified to
predict genes with canonical as well as with GC-donor splice sites. This
version of Fgenesh (http://www.softberry.com/gf/gf.html) can identify non-
standard GC-exons with approximately the same level of false-positive predictions as the original program.

22 199 verified examples of canonical splice pairs are presented in the
SpliceDB database, which is publicly available through the WWW (http://
genomic.sanger.ac.uk/SpliceDB.html) [14]. It also includes 1615 annotated
and 292 EST-supported and shift-verified non-canonical pairs.

3.3
Methods for the recognition of functional signals

This Section is intended to describe several approaches for the recognition
of functional signals in genes and some features of these signals used
in gene identification. A traditional way to find functional sites is based
on using consensus sequences or weight matrices reflecting conservative
nucleotides of a signal.

3.3.1
Search for nonrandom similarity with consensus sequences

A statistical method for estimating the significance of the similarity between
a consensus of a functional signal and a similar sequence fragment is
briefly described below [22, 23].

Let us assume that we are searching for a site in a sequence of length N with a random arrangement of nucleotides A, T(U), G and C, where the frequencies of these nucleotides are P_A, $P_{T(U)}$, P_G and P_C, respectively. If we accept that $P_1 = P_A$, $P_2 = P_G$, $P_3 = P_T$, $P_4 = P_C$, then the frequencies of the nucleotides of the other classes P_j ($j = 5, \ldots, 15$) representing the non-empty subsets of the set $\{A, C, G, T\}$ of nucleotides are determined as sums of frequencies of nucleotides of all the types of the j-th class. P_{15} represents the set of all 4 nucleotides, depicted by the IUPAC code N.

3.3.1.1 Single block site

Let us consider a functional site of length L. Let N_l be the number of positions in the consensus sequence whose bases belong to the l-th class and let $N_1 + N_2 + \cdots + N_{15} = L$. Assume that the consensus has M conserved positions M_l ($l = 1, \ldots, 14$), where M is the number of conserved nucleotides of the l-th class ($M_1 + M_2 + \cdots + M_{14} = M$). Then k ($k = 0, 1, \ldots$) mismatches between the site and the segment of length L belonging to the sequence under consideration are allowed only at the L-M variable positions. R_l ($l = 1, \ldots, 14$) are the numbers of mismatches between the consensus and the DNA segment of the l-th class of nucleotides. The number of mismatches should meet the following conditions: $0 \leq R_1 \leq \min(k, N_1 - M_1)$, $0 \leq R_2 < \min(k - R_1, N_{15} - M_{15}), \ldots, 0 \leq R_{15} < \min(k - R_1 - R_2 - \cdots - R_{13}, N_{14} - M_{14})$.

Therefore, the probability $P(L, k)$ of detecting the segment (L, k) of length L with mismatches in k variable positions, what we will call an (L, k) site, is:

$$P(L, k) = \sum_{R_l=0}^{\min(k, N_1 - M_1)} \cdots \sum_{R_{15}=0}^{\min(k - R_1 - R_2 \ldots R_{14}, N_{14} - M_{15})} C_{N_1 - M_1}^{R_1} P_1^{N_1 - P_1}(1 - P_1)^{R_1}$$

$$\cdots C_{N_{15} - M_{15}}^{R_{15}} P_{15}^{N_{15} - R_{15}}(1 - P_{15})^{R_{15}} \tag{1}$$

In this case the expected number $\bar{T}(L, k)$ of structures (L, k) in a random sequence of length N is:

$$\bar{T}(L, k) = P(L, k) \times F_L$$

Here F_L is the number of possible site positions in the sequence: $F_L = N - L + 1$.

The probability of having precisely T structures (L, k) in the sequence may be estimated using the binomial distribution:

$$P(T) = C_{F_L}^T P^T(L, k)[1 - P(L, k)]^{F_L - T}. \tag{2}$$

The probability of detecting T structures with k or fewer mismatches is:

$$P(T) = \sum_{z=0}^{k} C_{F_L}^{T} P^{T}(L, z)[1 - P(L, z)]^{F_L - T}. \tag{3}$$

Now we can derive the upper boundary of the confidence interval T_0 (with the significance level q) for the expected number of structures in a random sequence:

$$\sum_{t=0}^{T_0-1} P(t) < q \quad \text{and} \quad \sum_{t=0}^{T_0} P(t) \geq q. \tag{4}$$

If the number of (L, k) sites detected in the sequence meets the condition $T \geq T_0$, they can be considered as potential functional signals with the significance level q.

3.3.1.2 Composite (two-blocks) site

Let us consider a composite consensus from two blocks of lengths L_1 and L_2 at a distance D ($D_1 \leq D \leq D_2$, i.e., D_1 and D_2 are, respectively, the minimum and maximum allowed distances between the blocks). Let N_{1l} and N_{2l} be the number of nucleotides of the l-th class in the first and second blocks, respectively ($l = 1, \ldots, 15$). It is clear that $N_{j1} + N_{j2} + \cdots + N_{j15} = L_j$ ($j = 1, 2$).

Let the first and second blocks have M_{1l} and M_{2l} conserved positions of the nucleotides of l-th class. Then the probability $P(L_j, k_j)$ of finding in a random sequence the segment (L_j, k_j) of size L_j differing in k_j non-conserved positions from the j-th block of the site is calculated using Eq. (2) with the substitutions L, k, N_l and M_l by L_j, k_j, N_{jl} and M_{jl} ($l = 1, \ldots, 14$; $j = 1, 2$), respectively. The probability of simultaneous and independent occurrence of (L_1, k_1) and (L_2, k_2) in the random sequence is:

$$P(L_1, k_1, ; L_2, k_2) = P(L_1, k_1) \times P(L_2, k_2). \tag{5}$$

The number of possible ways of arranging the segments (L_1, k_1) and (L_2, k_2) in the random sequence of length N is:

$$F(L_1, L_2, D_1, D_2) = (D_2 - D_1 + 1)\left[N - L_1 - L_2 - \frac{D_1 + D_2}{2} + 1\right]. \tag{6}$$

Then, the expected number of sequences $(L_1, k_1, ; L_2, k_2, D_1, D_2)$ is:

$$(L_1, k_1, L_2, k_2, D_1, D_2) = F(L_1, L_2, D_1, D_2) \times P(L_1, k_1, ; L_2, k_2) \tag{7}$$

The probability $P(T)$ of detecting T sequences $(L_1, k_1, ; L_2, k_2, D_1, D_2)$ and the upper boundary of the confidence interval T_0 can be computed analogously to Eq. (2–4).

This statistics has been implemented in the program NSITE (http://www.softberry.com/gf.html) [24], that identifies nonrandom similarity between fragments of a given sequence and consensuses of regulatory motifs from the *TRANSFAC* database [25].

3.3.2
Position-specific sensors

Weight matrices are often applied for functional signal description [26, 27, 28]. They are usually more accurate than the consensus technique and often incorporated as a signal scoring method in gene finding algorithms. A weight matrix can be considered as a simple model based on a set of position-specific probability distributions $\{p_s^i\}$, that provide probabilities of observing a particular type of nucleotide s in a particular position i of the functional signal sequence S. The probability of generating a sequence X $(x_1, \ldots, x_k)$ under this model is:

$$P(X/S) = \prod_{i=1}^{k} p_{x_i}^i, \tag{8}$$

where nucleotides of the signal are generated independently. A corresponding model can also be constructed for non-site (N) sequences: $\{\pi_s^i\}$. A good discriminative score based on these models is the log-likelihood ratio:

$$\text{LLR}(X) = \log P(X/S)/P(X/N) \tag{9}$$

This score can be computed as an average sum of weights of observed nucleotides in a given sequence fragment using a corresponding weight matrix $w(i, s) = \{\log(p_s^i/\pi_s^i)\}$:

$$\text{LLR}(X) = \frac{1}{k} \sum_{i=1}^{k} w(i, x_i). \tag{10}$$

The more the $\text{LLR}(X)$ (log likelihood ratio) exceeds 0, the better chances this sequence has to represent a real functional signal.

There are some other weight functions that are used to search for functional signals, for example, weights can be received by optimization procedures such as perceptrons or neural networks [29, 30]. Also, different position-specific probability distributions $\{p_s^i\}$ can be considered. One typical generalization is to use position-specific probability distributions $\{p_s^i\}$ of k-base oligonucleotides (instead of mononucleotides), another one is to exploit Markov chain models, where the probability to generate a particular nucleotide x_i of the signal sequence depends on $k_0 - 1$ previous bases (i.e.

depends on the $k_0 - 1$ base oligonucleotide ending at the position $i - 1$. Then the probability of generating the signal sequence X is:

$$P(X/S) = p_0 \prod_{i=k_0}^{k} p_{s_{i-1}, x_i}^{i-1, i} \tag{11}$$

where $p_{s_{i-1}, x_i}^{i-1, i}$ is the conditional probability of generating nucleotide x_i in position i given oligonucleotide s_{i-1} ending at position $i - 1$, p_0 is the probability of generating oligonucleotide $x_1 \ldots x_{k_0-1}$. For example, a simple weight matrix represents the independent mononucleotide model (or 0-order Markov chain), where $k_0 = 1$, $p_0 = 1$ and $p_{x_{i-1}, x_i}^{i-1, i} = p_{x_i}^i$. When we use dinucleotides (1st order Markov chain) $k_0 = 2$, $p_0 = p_{x_1}^1$, and $p_{x_{i-1}, x_i}^{i-1, i}$ is the conditional probability of generating nucleotide x_i in position i given nucleotide x_{i-1} at position $i - 1$. The conditional probability can be estimated from the ratio of the observed frequency of the k_0-base oligonucleotide ($k_0 > 1$) ending at position i divided by the frequency of the $k_0 - 1$-base oligonucleotide ending at position $i - 1$ in a set of aligned sequences of some functional signal.

$$p_{s_{i-1}, x_i}^{i-1, i} = f(s_{i-1}, x_i)/f(s_{i-1})$$

By the same method we can construct a model for non-site sequences for computing $P(X/N)$, where often the 0-order Markov chain with genomic base frequencies (or even equal frequencies (0.25)) is used. A log likelihood ratio (10) with Markov chains was used to select CpG island regions [31] as well in as a description of functional signals in gene finding programs such as Genscan [32], Fgenesh [21, 33] and GeneFinder [34].

Useful discriminative measure taking into account some *a priori* knowledge can be based on computing Bayesian probabilities as components of position specific distributions $\{p_s^i\}$:

$$P(S/o_s^i) = P(o_s^i/S)P(S)/(P(o_s^i/S)P(S) + P(o_s^i/N)P(N)), \tag{12}$$

where $P(o_s^i/S)$ and $P(o_s^i/N)$ can be estimated as position-specific frequencies of oligonucleotides o_s^i in the set of aligned sites and non-sites; $P(s)$ and $P(N)$ are the a priori probabilities of site and non-site sequences, respectively. s is the type of oligonucleotide starting (or ending) in the i-th position [35]. If one assumes independence of oligonucleotides in different positions the probability of a sequence X belonging to the signal is given by:

$$P(S/X) = \prod_{i=1}^{k} P(S/o_m^i).$$

Another empirical discriminator called "Preference" uses the average positional probability of belong to a signal:

$$Pr(S/X) = 1/k \sum_{i=1}^{k} P(S/o_m^i).$$ (13)

This measure which is used in constructing discriminant functions for the
Fgenes gene finding program [36] can be more stable than the above one on
short sequences and has a simple interpretation: if $Pr(S/X) > 0.5$, then our
sequence is more likely belong to signal than to non-signal sequences.

3.3.3
Content-specific measures

Some functional signal sequences have distinctive general oligonucleotide
composition. For example, many eukaryotic promoters are found in GC-rich
chromosome fragments. We can characterize these regions applying scoring functions that are similar to the ones discussed above, but using probability distributions and their estimations by oligonucleotide frequencies,
which are computed on the whole sequence of functional signal and are not
position specific. For example, the Markov chain based probability (11) of
generating the signal sequence X will be:

$$P(X/S) = p_0 \prod_{i=k_0}^{k} p_{s_{i-1}, x_i}$$ (14)

3.3.4
Frame-specific measures for recognition of protein coding regions

An important problem is to compute the probability of generating a protein
coding sequence X. A coding sequence is a sequence of triplets (codons)
read continuously from a fixed starting point. Three different reading
frames with different codons are possible for any nucleotide sequence (or
six if the complementary chain is also considered). It was noted that nucleotides are distributed quite unevenly among different codon positions,
therefore the probability of observing a specific oligonucleotide in coding
sequences depends on its position relative to the reading frame (three possible variants) as well as on adjacent nucleotides [37, 38, 39, 40]. Asymmetry
in base composition between codon positions arises due to uneven usage of
amino acids and synonymous codons in addition to the particular structure
of the genetic code [41]. Comprehensive assessment of various protein
coding measures was done by Fickett and Tung [42]. They estimate the
quality of more than 20 measures and showed that the most powerful

measures, such as the 'in phase hexanucleotide composition', codon or amino acids usage give about 81% accuracy as coding region recognition functions on 54 base windows. Some powerful recognisers of coding gene regions based on neural network approaches have been constructed [43, 44]. Their sensitive coding region sensor, based on codon and dicodon statistics in six frames and a neural network output function, reaches 99% accuracy on predicting 180 nt ORFs (open reading frame), but it was analyzed only for genes expressed in liver and tested on 1000 examples. The prediction accuracy over 54 bp coding windows was about 85% [44]. Using a neural network to combine the information from 7 sensors that describe large amounts and diverse types of information, Uberbacher and Mural created the "coding recognition module", which identifies 90% of the coding exons of length 100 nt or greater and about 50% of those less than 100 nt long [43]. Solovyev and Lawrence [45] used as the recognition function a modification of a Bayesian prediction scheme Eq. (12) that demonstrated good results on prokaryotic coding regions prediction [40]. The accuracy of classification using oligonucleotides of 8 bp length and function Eq. (13) is 90% for 54 nt and 95% for sequences 108 nt long [45], which is better than the accuracy for the combined six most powerful measures by LDA function in the Fickett and Tung [42] investigation.

In Markov chain approaches the frame-dependent probabilities p_{s_{i-1}, x_i}^{f} ($f = \{1, 2, 3\}$) can be used to model coding regions:

$$P(X/S) = p_0 \prod_{i=k_0}^{k} p_{s_{i-1}, x_i}^{f}, \qquad (15)$$

where f is equal 1, 2 or 3 for oligonucleotides ending at codon position 1, 2 or 3, respectively.

3.3.5
Accuracy measures

For estimation of the performance of an algorithm or a recognition function we will use several quality measures [42, 46, 47]. True positives (T_P) is the number of correctly predicted and false positives (F_P) is the number of falsely predicted authentic sites. True negatives (T_N) is the number of correctly predicted and false negative (F_N) is the number of falsely predicted non-sites. Sensitivity (S_n) measures the fraction of the true examples that are correctly predicted: $S_n = T_P/T_P + F_N$. Specificity (S_p) measures the fraction of the predicted examples that are correct: $S_p = T_P/T_P + F_P$. Only consideration of both S_n and S_p values makes sense when we aim at providing accuracy information. If we want to concentrate on a single value for accuracy estimation the average of the correctly predicted number of sites and non-sites: $AC = 0.5\ (TP + TN)$ is a suitable measure. However, this latter measure does not take into account possible difference in sizes of site and non-sites sets. A better single measure (correlation coefficient) takes into account the relation between correctly predictive positives and negatives as well as false positives and negatives [48]:

$$CC = (T_p T_n - F_p F_n)/\sqrt{(T_p + F_p)(T_n + F_n)(T_p + F_n)(T_n + F_p)}.$$

3.3.6
Application of linear discriminant analysis

Many eukaryotic functional signals have very short conservative regions and we should use their additional specific features (other than oligonucleotide frequencies) to increase the accuracy of signal identification. Different features of a functional signal may have different significance for the recognition and might not be independent. Classical linear discriminant analysis provides a method to combine such features in a discriminant function. In general, a discriminant function, when applied to a pattern, yields an output that is an estimate of the class membership of this pattern. The discriminative technique provides a minimization of the error rate of classification [49]. Let assume that each given sequence can be described by vector X of p characteristics $(x_1, x_2, \ldots, x_p)$, which could be measured. The procedure of linear discriminant analysis is to find a linear combination of the measures (called the linear discriminant function or LDF), that provides maximum discrimination between sites sequences (class 1) and non-site examples (class 2).

The LDF:

$$Z = \sum_{i=1}^{p} a_i x_i$$

classifies (X) into class 1 if $Z > c$ and into class 2 if $Z < c$ with a few misclassification as possible. The vector of coefficients $(\vec{a}_1, \vec{a}_2, \ldots \vec{a}_p)$ and threshold constant c are derived from the training set by maximizing the ratio of the between-class variation z to within-class variation and are equal to [49]:

$$\vec{a} = s^{-1}(\vec{m}_1 - \vec{m}_2)$$

and

$$c = \vec{a}(\vec{m}_1 + \vec{m}_2)/2,$$

where $\vec{m}_i$ are the sample mean vectors of characteristics for class 1 and class

2, respectively; *s* is the pooled covariance matrix of characteristics:

$$s = \frac{1}{n_1 + n_2 - 2}(s_1 + s_2)$$

s_i is the covariation matrix, and n_i is the sample size of class *i*. Thus, based on these equations we can calculate the coefficients of the LDF and the threshold constant c using the values of characteristics of site and non-site sequences from the training sets and then test the accuracy of the LDF on the test set data. The significance of a given characteristic or set of characteristics can be estimated by the generalized distance between two classes (called the Mahalonobis distance or D^2):

$$\vec{D}^2 = (\vec{m}_1 - \vec{m}_2)s^{-1}(\vec{m}_1 - \vec{m}_2)$$

which is computed based on values of the characteristics in the training sequences of classes 1 and 2. To find sequence features a lot of possible characteristics are generated, such as score of weight matrices, distances, oligonucleotide preferences within different sub-regions etc. Selection of the subset of significant characteristics *q* (among the tested *p*) is performed by a step-wise discriminant procedure including only characteristics, which significantly increase the Mahalonobis distance. The procedure to test this significance uses the fact that the quantity:

$$F = \frac{n_1 + n_2 - p - 1}{p - q} \frac{n_1 n_2 (D_p^2 - D_q^2)}{(n_1 + n_2)(n_1 + n_2 - 2) + n_1 n_2 D_q^2}$$

has an $F\ (p - q, n_1 + n_2 - p - 1)$ distribution [49] when testing hypothesis H0: $\Delta_p^2 = \Delta_q^2$, where Δ_m^2 is the population Mahalonobis distance based on *m* variables. If the observations come from multivariate normal populations, the posterior probability that the example belongs to class 1 may be computed as [49]

$$\Pr(\text{class}1/\vec{x}) = \frac{1}{1 + \dfrac{n_2}{n_1} \exp\{-z + c\}}.$$

The posterior probability $\Pr(\text{class}2/\vec{x}) = 1 - \Pr(\text{class}1/\vec{x})$.

3.3.7
Prediction of donor and acceptor splice junctions

Recognition of RNA splice sites by the spliceosome is very precise [50–52] indicating the presence of specific signals for their function. Splice site patterns are mainly defined by nucleotides at the ends of introns, because deletions of large intron parts often turn out not to effect their selection [7, 53]. A sequence of 8 nucleotides is highly conserved at the boundary between an exon and an intron (donor or 5′-splice site) and a sequence of 4 nucleotides, preceded by a pyrimidine-rich region, is also highly conserved between an exon and an intron (acceptor or 3′-splice site). The third less conserved sequence of about 5–8 nucleotides and containing an adenosine residue, lies within the intron, usually between 10 and 50 nucleotides upstream of the acceptor splice site (branch site). These sequences provide specific molecular signals by which the RNA splicing machinery can select the precise splice sites [18]. There are many efforts to analyze the sequences around these conserved regions [7, 18, 54]. It was shown that their consensus differ slightly between different classes of organisms [18, 55] and certain important information may be provided by the sequences outside the short conserved regions. Scoring schemes based on consensus sequences or weight matrices, which take into account information on open reading frames, free energy of base-pairing of RNA with snRNA and other peculiarities, yield an accuracy of about 80% for the prediction splice site positions [56, 57]. More accurate prediction is shown by neural network algorithms [58, 59]. The optimal network architecture (15 nucleotides window and 40 hidden units for donor sites and 41 nucleotides window and 20 hidden units) has an accuracy about of 94% (111/118) for predicting donor and 85% (100/118) for predicting acceptor splice sites in a test set of 118 examples [59]. Using joint coding/noncoding and splice site classification the prediction accuracy of about 95% with a low level of false positive sites was obtained. This work provides a good benchmark for testing new predictive algorithms, because the authors analysed large learning (331 examples) and testing (118 examples) sets. Because practically all donor sites contain the conserved dinucleotide GT and all acceptor sites contain AG, Mural et al. [60] used for splice sites selection the preferences of tabulated triplets in and about authentic splice junctions and also in and about pseudo-junctions which contain either a GT or an AG base pair. The authors obtained a good 91% accuracy for donor and 94% accuracy for acceptor splice sites prediction for primate genes. However, this data was based on small learning (about 150) and test (about 50) sets of splice junctions.

The overall view on the difference of triplet composition in splice and pseudosplice sequences is shown in Figure 3.6 [61]. This figure clearly demonstrates the various functional parts of splice sites. We see that only short regions around splice junctions have a great difference in triplet composition. Their sequences usually are used as the only determinants of donor or acceptor splice site positions. However, dissimilarity in many other regions can also be seen: for the donor site coding region, the G-rich intron region may be distinguished; for acceptor sites – intron G-rich region, branch point region, poly(T/C)-tract, and coding.

Splice site prediction methods using a linear function that combines triplet preferences around splice junction and preferences to be coding and intron of adjacent regions have been developed [45, 61]. 692 sequences with

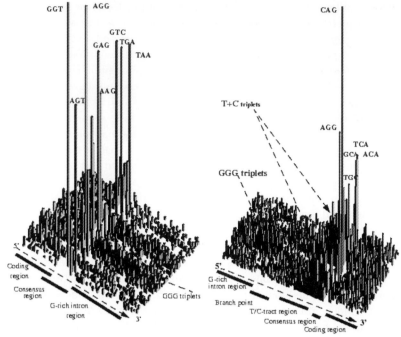

Fig. 3.6

Difference of the triplet composition within donor and GT-containing non-donor sequences (left); around acceptor and AG-containing non-acceptor sequences (right) in 692 human genes. Each column presents the difference of specific triplet numbers between sites and pseudosites in a specific position. For comparison the numbers were calculated for equal quantities of sites and pseudosites.

2037 donor splice sites and 2054 acceptor splice sites having the GT and AG conserved dinucleotide in flanking intron positions were extracted from GenBank [62]. Also, 89 417 pseudodonor and 134 150 pseudoacceptor sites that contain either a GT or an AG base pair (and are not annotated as splice sites) were extracted from these sequences. The characteristics of sequences around splice sites and pseudosites were used for developing and testing a human splice site recognition function to distinguish them. The training set includes 2/3 of all sequences, and the test set contains the remaining ones. The data set for computing octanucleotide preferences in coding and intron regions included 4 074 593 bp of coding regions and 1 797 572 bp of intron sequences.

3.3.7.1 **Donor splice site recognition**

The characteristics used for classifying donor site are: the triplet preferences (Eq. 13) in the potential coding region (-30 to -5); conserved consensus region (-4 to $+6$) and G-rich region ($+7$ to $+50$); the number of significant

Tab. 3.4
Significance of selected characteristics of donor splice sites.

	Characteristics	1	2	3	4	5	6	7
a	Individual D^2	9.3	2.6	2.5	0.01	1.5	0.01	0.4
b	Combined D^2	9.3	11.8	13.6	14.9	15.5	16.6	16.8

> 1, 2, 3 are the triplet preferences of consensus, intron G-rich and
> coding regions, respectively; 4 is the number of significant triplets in
> the consensus region, 5 and 6 are the octanucleotide preferences for
> being the coding 54 bp region on the left and for being the intron 54
> bp region on the right of donor splice site junction; 7 is the number
> of G bases, GG-doublets and GGG-triplets in intron G-rich region.

triplets in the conserved consensus region; octanucleotide preferences (Eq. 15) for being coding in the $(-60$ to $-1)$ region and being intron in the $(+1$ to $+54)$ region; the number of G-bases, GG-doublets and GGG-triplets in the ($|6$ to $|50)$ region. The values of these 6 characteristics of donor site were calculated for 1375 authentic donor site and for 60 532 pseudosite sequences from the learning set. The Mahalonobis distances showing the significance of each characteristic are given in Table 3.4a. We can see that the strongest characteristic for donor sites is the triplet composition in the consensus region $(D^2 = 9.3)$ and then the adjacent intron region $(D^2 = 2.6)$ and coding region $(D^2 = 2.5)$. Other significant characteristics are: the number of significant triplets in the conserved consensus region; the number of G-bases, GG-doublets and GGG-triplets; the quality of the coding and intron regions. The accuracy of the discriminant function based on these characteristics was tested on the recognition of 662 donor sites and 28 855 pseudosite sequences. The general accuracy of donor site prediction was 97%. This accuracy is better than in the neural network-based method, which has $CC = 0.61$ at 95% accuracy [59], comparing to $CC = 0.63$.

3.3.7.2 Acceptor splice site recognition

The characteristics used for acceptor splice sites are: the triplet preferences (Eq. 13) in the branch point region $(-48$ to $-34)$; poly(T/C)-tract region $(-33$ to $-7)$; conserved consensus region $(-6$ to $+5)$; coding region $(+6$ to $+30)$; and octanucleotide preferences (Eq. 15) of being coding in the $(+1$ to $+54)$ region and in the $(-1$ to $-54)$ region; and the number of T and C in poly(T/C)-tract region. The values of 7 characteristics of acceptor sites were calculated for 1386 authentic acceptor site and 89 791 pseudosite sequences from the learning set. The Mahalonobis distances showing individual significance for each characteristic are given in Table 3.5a. We can see that strongest characteristics for acceptor sites are: the triplet composition in poly(T/C)-tract region $(D^2 = 5.1)$; consensus region $(D^2 = 2.7)$; adjacent

Tab. 3.5
Significance of various characteristics of acceptor splice sites.

	Characteristics	1	2	3	4	5	6	7
a	Individual D^2	5.1	2.6	2.7	2.3	0.01	1.05	2.4
b	Combined D^2	5.1	8.1	10.0	11.3	12.5	12.8	13.6

1, 3, 4, 6 are the triplet preferences of poly(T/C)-tract, consensus,
coding and branch point regions, respectively; 7 is the number
of T and C in intron poly(T/C)-tract region, 2 and 5 are the
octanucleotide preferences for being coding 54 bp region on the left
and 54 bp region for being intron on the right of donor splice site
junction.

coding region ($D^2 = 2.3$); and branch point region ($D^2 = 1.0$). Some significance is found using the number of T and C in the adjacent intron region ($D^2 = 2.4$); and the quality of the coding region ($D^2 = 2.6$).

The general accuracy of acceptor site prediction was 96% ($C = 0.47$). Table 3.6 illustrates the performance for donor and acceptor sites and compares the results with the work of Mural et al. [60] who used triplet composition for splice site prediction, and the work of Brunak et al. [59] who used a complex neural network for site discrimination.

3.3.8
Recognition of promoter regions in human DNA

Eukaryotic polymerase II promoter sequences are the main regulatory elements of eukaryotic genes. Their recognition by computer algorithms will increase the quality of gene structure identification as well as provide the possibility to study gene regulation. The development of computer algorithms to recognize Pol II promoter sequences in genomic DNA is an extremely difficult problem in computational molecular biology. The 5'-flanking region of a promoter is very poorly described in general. It may contain dozens short motifs (5–10 bases) that serve as recognition sites for

Tab. 3.6
Splice site prediction accuracy.

Method	Solovyev et al. [61]	Mural et al. [60]	Brunak et al. [59]
Donor Sn	0.97	0.91	0.95
Donor CC	0.63	0.41	0.61
Acceptor Sn	0.96	0.93	0.96
Acceptor CC	0.47	0.36	0.40
Traning sites	2037	135	449
Test sites	662	50	118

proteins providing initiation of transcription as well as specific regulation of gene expression. These motifs differ among various groups of genes and even such a well known promoter element as the TATA-box is often absent in 5′-regions of many house-keeping genes. Each promoter has a unique selection and arrangement of these elements providing a unique program of gene expression. A comprehensive description of promoter recognition problems is considered in Chapter 4 of this book. Here we will consider some general features of PolII promoters that can be taken into account in gene prediction programs.

A review of the prediction accuracy of many general purpose promoter prediction program was presented by Fickett and Hatzigeorgiou [63]. This paper surveys oligonucleotide content based [64, 65], neural network [66, 67] and linear discriminant approaches [68] among others. Although the test set is relatively small (18 sequences) and has several problems that were noticed by [69], the results demonstrated that the programs can recognize no more than 50% of the promoters with a false positive rate of about 1 in 700–1000 bp. With the average size of a human gene being more than 5000 bases and many genes occupying hundreds of kilobases, we expect significantly more false positive predictions than real promoters. However, these programs can be used to find the promoter position (the start of transcription and the TATA-box) in a given 5′-region or to help select the correct 5′-exon in gene prediction approaches.

We will describe an improved version of the promoter recognition program TSSW (Transcription Start Site, W stands for using functional motifs from the Wingender [25] database) [68] to demonstrate sequence features that can be used to identify eukaryotic promoter regions. In this version it was suggested that TATA+ and TATA− promoters have very different sequence features and these groups were analyzed separately. Potential− TATA+ promoter sequences were selected by the value of score computed using the Bucher TATA box weight matrix [70] with the threshold closed to the minimal score value for the TATA+ promoters in the learning set. This choice of the threshold divides the learning set of known promoters into two approximately equal-size parts. Significant characteristics of both groups found by discriminant analysis are presented in Table 3.7. Values of Mahalanobis distances (D^2) of individual characteristics reflect the power of the feature to separate the signal from non-signal sequences. This analysis demonstrated that TATA− promoters have much weaker general features than TATA+ promoters. Probably the TATA− promoter possesses more gene specific structure and will be extremely difficult to predict by any general-purpose methods.

The recognition quality of the program was tested on 200 promoters, which were not included in the learning set. We provide the accuracy values for different levels of correctly predicted promoters in Table 3.8. The data demonstrate a poor quality of TATA− promoter recognition on long sequences and show that their recognition function can provide relatively

Tab. 3.7

Selected characteristics of promoter sequences used by TSSW programs for TATA+ and TATA− promoters.

Characteristiscs	D² for TATA + promoters	D² for TATA − promoters
• Hexaplets −200 to −45	2.6	1.4 (−100 to −1)
• TATA box score	3.4	0.9
• Triplets around TSS	4.1	0.7
• Hexaplets +1 to +40		0.9
• Sp1-motif content		0.9
• TATA fixed location	0.7	
• CpG content	1.4	0.7
• Similarity −200 to −100	0.3	0.7
• Motif Density(MD) −200 to +1	4.5	3.2
• Direct/Inverted MD −100 to +1	4.0	3.3 (−100 to −1)
Total Maxalonobis distance	11.2	4.3
Number promoters/non-promoters	203/4000	193/74 000

unambiguous predictions within regions less that 500 bp long. In contrast, 90% of the TATA+ promoters can be identified in the range about 2000 bp, which makes valid their incorporation into gene finding programs.

Recently several improvements of promoter prediction approaches have been published. Ohler et al. [69] used interpolated Markov chains in their approach and slightly improved the previous results. They identify 50% of the promoters in the Fickett and Hatzigeorgiou [63] promoter set, while having one false positive prediction every 849 bp. The old version of TSSW had an accuracy of 42% with a false positive rate of 1/789 bp. Another new program (Promoter 2.0) was designed by Knudsen [71] applying a combination of neural networks and genetic algorithms. Promoter 2.0 was tested on promoters in the complete Adenovirus genome, which is 35 937 bases long. The program predicted all 5 known promoter sites on the plus strand and 30 false positive promoters. The average distance of real and closest predicted promoter is about 115 bp. The TSSW program with the threshold

Tab. 3.8

Accuracy of promoter identification by TSSW program.

Type of promoter	Number of test sites	Correctly predicted	1 false positive per bp
TATA +	101	98%	1000
		90%	2200
		75%	3400
		52%	6100
TATA −	96	52%	500
		40%	1000

to predict all 5 promoters produces 35 false positives, but an average distance between the predicted TSS (transcription start site) and the real promoter is just 4 bp (2 predicted exactly, 1 with 1 bp shift, 1 with 5 bp shift and the weakest promoter was predicted with 15 bp shift).

3.3.9
Prediction of poly-A sites

The 3'-untranslated region (3'UTR) has several cytoplasmic functions affecting the localization, stability and translation of mRNAs [71]. Almost all eukaryotic mRNAs undergo 3'-end processing, which involves the endonucleotic cleavage followed by the polyadenylation of the upstream cleavage product [72, 73]. Recognition of essential sequences involves the formation of several large RNA-protein complexes [74]. RNA sequences directing binding of specific proteins are frequently poorly conserved and often recognized in a cooperative fashion [72]. Numerous experiments have revealed three types of RNA sequences defining a 3'-processing site [72, 75] (Figure 3.7). The most conserved is the hexamer signal AAUAAA (poly-A signal), situated 10–30 nucleotides upstream of the 3'-cleavage site. About 90% of the sequenced mRNAs have a perfect copy of this signal. Two other types, the upstream and the downstream elements are degenerate and have not been properly characterized. Downstream elements are frequently located within approximately 50 nucleotides 3' of the cleavage site [76]. These elements are often GU or U rich, although may have various base compositions and locations. On the basis of sequence comparisons McLachlan et al. [77] have suggested a consensus of the downstream element: YGUGUUYY.

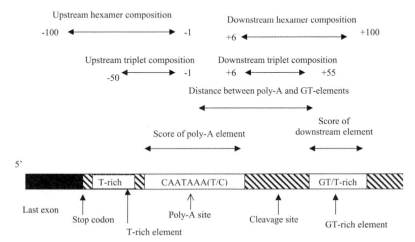

Fig. 3.7
Sequence characteristics selected to describe poly-A signal sequences.

The efficiency of polyadenylation in a number of genes can be also increased by sequences upstream of AAUAAA, which are generally U-rich [72]. All these RNA sequences serve as nucleation sites for the multicomponent protein complex catalyzing the polyadenylation reaction. Yada et al. [78] have conducted a statistical analysis of human DNA sequences in the vicinity of the poly-A signal in order to distinguish them from AATAAA sequences occurring in human DNA (pseudo polyA-signals) that are not involved in polyadenylation. They found that a base C frequently appears on the upstream side of the AATAAA signal and a base T or C often appears on the downstream side, implying that CAATAAA(T/C) can be regarded as a consensus of the poly-A signal. Kondrakhin et al. [79] constructed a generalized consensus matrix using 63 sequences of cleavage/polyadenylation sites in vertebrate pre-mRNA. The elements of the matrix were absolute frequencies of triplets at each site position. Using this matrix they have provided a multiplicative measure for recognition of polyadenylation regions. However this method has a very high false positive rate.

Salamov and Solovyev developed a LDF recognition function for the poly-A signal. The datasets for 3'-processing sites and "pseudo" poly-A signals were extracted from GenBank (Version 82). 3'-processing sites were taken from the human DNA entries, containing a description of the poly-A signal in the feature table. Pseudo sites were taken out of human genes as the sequences comprising (−100, +200) around the patterns revealed by the poly-A weight matrix (see below), but not assigned to poly-A sites in the feature table. Sequences submitted to GenBank before 1994 were included to the training set and those after 1994 to the test set. As a result there were 248 positive and 5702 pseudosites in the training set and respectively 131 and 1466 in the test set.

3.3.9.1 A model for recognition of 3'-processing sites

As the hexamer AATAAA is the most conservative element of 3'-processing sites it was considered as the main block in our complex recognition function. Although the hexamer is highly conserved, variants of this signal are observed. For example, in the training set 43 of 248 poly-A sites have hexamer variants of AATAAA with one mismatch. To consider such variants the position weight matrix for recognizing this signal has been used. The other characteristics such as content statistics of hexanucleotides, positional triplets in the upstream and downstream regions were defined relative to the position of the conservative hexamer sequence (Figure 3.7).

The Mahalonobis distances for each characteristic calculated on the training set are given in Table 3.9. The most significant characteristic is the score of the AATAAA pattern (estimated by the position weight matrix), which indicates the importance of occurrences of an almost perfect poly-A signal (AATAAA). The second most valuable characteristic is the hexanucleotide preferences of the downstream (+6 to +100) region. Although the discrim-

Tab. 3.9

Significance of selected characteristics of poly-A signal.

Characteristics	1	4	2	5	3	6	8	7
Individual D^2	7.61	3.46	0.01	2.27	0.44	1.61	0.16	0.17
Combined D^2	7.61	10.78	11.67	12.36	12.68	12.97	13.09	13.1

1 is the score of position weight matrix of poly-A signal, 2 is the score of the position weight matrix of the downstream GT/T-rich element, 3 is the distance between the poly-A signal and the predicted downstream GT/T-rich element, 4 is the hexanucleotide composition of the downstream (+6, +100) region, 5 is the hexanucleotide composition of the upstream (−100, −1) region, 6 is the positional triplet composition of the downstream (+6, +55) region, 7 is the positional triplet composition of the upstream (−50, −1) region, 8 is the positional triplet composition of the GT/T-rich downstream element.

inating ability of the GT-rich downstream element itself (characteristic 2) is very weak, combining it with the other characteristics significantly increases the total Mahalonobis distance.

According to the observation of Yada et al. [78], we take into account the hexamer AATAAA and its flanking nucleotides. Only sequences with a score that exceeds some threshold were considered as candidates of poly-A sites. Around the consensus we calculated the other characteristics (Figure 3.8) and the value of the linear discriminant function. The threshold was chosen as the minimal score observed for authentic poly-A signals in the training set. The poly-A signal weight matrix provides a higher score for typical poly-A signals (like CAATAAAT) and a smaller score for minor variants with other sequences. Such variants are often observed much more frequently among pseudosites and therefore the weight matrix is a good discriminant itself.

In general, characteristics from the downstream region of the AATAAA sequence are more informative than those from the upstream region. Such selected characteristics as positional triplet composition of the upstream and downstream regions as well as the distance between AATAAA pattern and GT-rich element might indicate the importance of definite location of sequence elements in the 3′-processing region. Altogether, these elements can create the bedding for cooperative interactions between proteins of multicomponent complex catalyzing the polyadenylation reaction. It is worth mentioning that we tested and selected only the most general characteristics of poly-A site regions.

The algorithm for 3′-processing site identification is realized in the POLYAH program (http://genomic.sanger.ac.uk). First, it searches for the AATAAA pattern using the position weight matrix and, after the pattern is found, it computes the value of the linear discriminant function defined by the characteristics around this signal. The poly-A site is predicted if the value of this function is greater than some empirically selected threshold. An accuracy of the method has been estimated on a set of 131 positive and 1466 negative sites, which were not used in training. When 86% of polyA-regions are predicted correctly, the fraction of correctly predicted sites among all predicted (specificity) is 51% and the correlation coefficient is 0.62.

Kondrakhin et al. [79] reported the error rates of their method at different thresholds for poly-A signal selection. If the threshold is set to predict 8 of 9 real sites their function also predicts 968 additional false sites. We have tested the POLYAH program with the same sequence of Ad2 genome and for 8 correctly predicted sites it gives only 4 false sites. A certain improvement of Poly-A recognition was reached in using a pair of quadratic discriminant function in Polyadq program [80]. This program outperform the POLYAH detection method and is the first that can detect significant numbers of ATTAAA-type signals.

3.4
Gene identification approaches

Most gene prediction systems combine information about functional signals and the regularities of coding and intron regions. Initially several algorithms predicting internal exons were constructed. The program SORFIND [81] was designed to predict internal exons based on codon usage and Berg & von Hippel [82] discrimination energy for intron–exon boundaries recognition. The accuracy of exact internal exons prediction (at both 5′- and 3′-splice junctions and in the correct reading frame) by the SORFIND program reaches 59% with a specificity of 20%. A dynamic programming approach (alternative to the rule-based approach) was applied by Snyder and Stormo (1993) to internal exon prediction in the GeneParser algorithm. It recognized 76% of the internal exons, but the structure of only 46% exons was exactly predicted when tested on entire GenBank entry sequences. HEXON (Human EXON) program [61] based on linear discriminant analysis was the most accurate in exact exon prediction that time.

Later a number of single gene prediction programs has been developed to assemble potential eukaryotic coding regions into translatable mRNA sequences selecting optimal combinations of compatible exons [47, 66, 83, 84]. Dynamic programming was suggested as a fast method to find an optimal combination of preselected exons [45, 85, 86], that is different from the approach suggested by Snyder and Stormo [46] in GeneParser algorithm to search for exon-intron boundary positions. The Genie algorithm uses a Generalized Hidden Markov Model. Genie is similar in design to GeneParser, but is based on a rigorous probabilistic framework [87]. The FGENEH (Find GENE in Human) algorithm includes 5′-, internal and 3′-exon identification linear discriminant functions and dynamic programming approach [61, 88]. A comprehensive test of FGENEH and the other gene finding algorithms has been carried out by Burset and Guigo [89]. The FGENEH program was one of the best in the tested group having the exact exon prediction accuracy 10% higher than for the others and the best level of accuracy on the protein level. The same best performance of FGENEH was shown by the developers of Genie [87].

3.5
Discriminative and probabilistic approaches for multiple gene prediction

International genome sequencing projects generate hundreds of megabases each year and require gene-finding approaches able to identify many genes encoded in the produced sequences. The value of sequence information for the biomedical community will strongly depend on availability of candidate genes computationally predicted in these sequences. Moreover, the initiative that created a 'rough draft' of the human genome can allow other scientists to proceed more rapidly with discovering disease genes [90]. The best multiple gene prediction programs include Genscan [32] and Fgenesh [21] (HMM based, probabilistic approach) and Fgenes (discriminative approach) [36]. Initially we will describe a general scheme of HMM based gene prediction (first realized by the Haussler group [87, 91]) as the most general description of a gene model. The pattern-based approach can be considered as a particular case where transition probabilities are not taken into account.

3.5.1
Multiple gene prediction using the HMM approach

A gene sequence can be considered as a succession of segments x_i representing exons, introns, $5'$ and $3'$-untranslated regions and the like.and These segments can be considered as different sequence states. There are 35 states describing the eukaryotic gene model considering direct and reverse chains as possible gene locations (Figure 3.8). In the current gene prediction approaches non-coding $5'$- and $3'$-exons (and introns) are not considered, because the absence of protein coding characteristics accounts for lower accuracy of their prediction. Also they do not code any protein sequences. The remaining 27 states include 6 exon states (first, last single and 3 types of internal exons in 3 possible reading frame) and 7 noncoding states (3 intron, noncoding $5'$- and $3'$-promoter and poly-A) in each chain plus noncoding intergenic region. The latter 27 states are connected with solid arrows in Figure 3.8, the eight unused states are connected with broken arrows.

The predicted gene structure can be considered as an ordered set of states/sequence pairs, $\phi = \{(q_1, x_1), (q_2, x_2), \ldots, (q_k, x_k)\}$, called the parse, such that probability $P(X, \phi)$ of generating X according to ϕ is maximal over

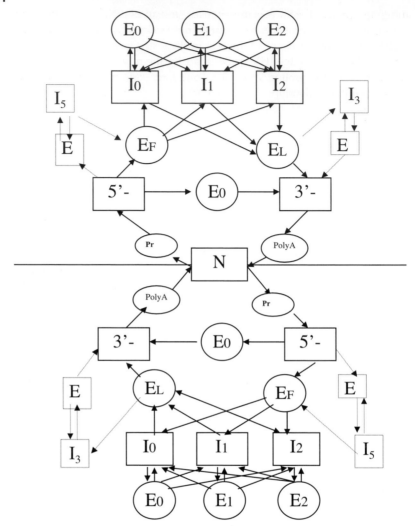

Fig. 3.8
Different sequence states and transitions in eukaryotic gene model. E_i and I_i are different exon and intron states, respectively ($i = 0, 1, 2$ reflect 3 possible different ORF). E marks non-coding exons and I5/I3 are 5'- and 3'-introns adjacent to non-coding exons.

all possible parses (or some score is optimal in some meaningful sense, i.e., best explains the observations [93]):

$$P(X, \varphi) = P(q_1) \left(\prod_{i=1}^{k-1} P(x_i \mid l(x_i), q_i) P(l(x_i) \mid q_i) (P(q_{i+1}, q_i) \right)$$
$$\times P(x_i \mid l(x_k), q_k) P(l(x_k) \mid q_k)$$

where $P(q_1)$ is denote the initial state probabilities; $P(x_i \mid l(x_1), q_i) P(l(x_i) \mid q_i)$

and $P(q_{i+1}, q_i)$ are the independent joint probabilities of generating the subsequence x_i of length l in state q_i and transiting to state $i + 1$, respectively.

Successive states of this HMM model are generating output according to the Markov process with inclusion of explicit state duration density. A simple technique based on the dynamic programming method for finding the optimal parse (or the single best state sequence) is called the Viterbi algorithm [92]. The algorithm requires on the order of $N^2 D^2 L$ calculations, where N is the number of states, D is the longest duration and L is the sequence length [93]. A helpful technique was introduced by Burge [28] to reduce the number of states and simplify computations by modeling non-coding state length with a geometric distribution. We consider shortly the algorithm of gene finding using this technique, which was initially implemented in the Genscan program [32] and used later in the Fgenesh program [21]. Since any valid parse will consist of only alternating series of Non-coding and Coding states: $NCNCNC, \dots, NCN$, we need only 11 variables, corresponding to the different types of N states. At each step corresponding to some sequence position we select the maximum joint probability to continue in the current state or to move to another non-coding state defined by a coding state (from a precomputed list of possible coding states), which is ending in analyzed sequence position.

Let us define the best score (highest joint probability) $\gamma_i(j)$ of the optimal parse of the subsequence $s_{1,j}$, which ends in state q_i at position j. Let we have a set A_j of coding states $\{-c_k-\}$ of lengths $\{d_k\}$, starting at positions $\{m_k\}$ and ending at position j, which have the previous states $\{b_k\}$. The length distribution of state c_k is denoted by $f_{ck}(d)$. The searching procedure can be stated as follows:

Initialization:

$$\gamma_i(1) = \pi_i P_i(S_1) p_i, \quad i = 1, \dots 11.$$

Recursion:

$$\gamma_i(j+1) = \max\Big\{ \gamma_i(j) p_i P_i(S_{j+1}),$$

$$\max_{c_k \in A_j}\{\gamma_i(m_k - 1)(1 - p_{bk}) t_{bk, ck} f_{ck}(d_k) P(S_{mk,j}) t_{ck, i} p_i P_i(S_{j+1})\}\Big\}$$

$$i = 1, \dots 11, \ j = 1, \dots, L - 1.$$

Termination:

$$\gamma_i(L+1) = \max\Big\{ \gamma_i(L), \max_{c_k \in A_j}\{\gamma_i(m_k - 1)(1 - p_{bk}) t_{bk, ck} f_{ck}(d_k) P(S_{mk,j}) t_{ck, i}\}\Big\}$$

$$i = 1, \dots 11.$$

For each step we record the location and type of transition maximizing the functional to restore the optimal set of states (gene structure) by a back-tracking procedure. Most parameters of these equations can be calculated on the learning set of known gene structures. Instead of scores of coding states $P(S_{mk,j})$ it is better to use log-likelihood ratios which do not produce scores below the limits of computer precision.

Genscan [32] was the first algorithm to predict multiple eukaryotic genes. Several similar HMM-based gene prediction programs were developed later, among them Veil [94], HMMgene [95], Fgenesh [21, 33], a variant of Genie [87] and GeneMark [96]. Fgenesh (Find GENES Hmm) is currently the most accurate program. It is different from Genscan in that in the model of gene structure signal terms (such as splice site or start site score) have some advantage over content terms (such as coding potentials). In log-likelihood terms this means that the splice sites and other exon functional signals have an additional score, depending on the environment of the sites. Also in computing the coding scores of potential exons, a priori probabilities of exons were taken into account according to Bayes' theorem. As a result, the coding scores of potential exons are generally lower than in Genscan. The donor recognition function is based on a second order position specific HMM (13) instead of 8 matrix, weight describing different groups of splice sites used in Genescan. Fgenesh works with separately trained parameters for each distinct model organism such as human, drosophila, nematode, yeast, arabidopsis monocot plants etc. Coding potentials were calculated separately for 4 isochores[2] (human) and for 2 isochores (other species). The run time of Fgenesh is practically linear and the current version has no practical limit on the length of analyzed sequence. The prediction about 800 genes in 34 MB of Chromosome 22 sequence takes about 1.5 minutes on a Dec-alpha processor EV6 for the latest Fgenesh version.

3.5.2
Pattern based multiple gene prediction approach

FGENES (Find GENES) is the multiple gene prediction program based on dynamic programming. It uses discriminant classifiers to generate a set of exon candidates. Similar discriminant functions were developed initially in Fexh (Find Exon), Fgeneh (Find GENE) (*h* stands for version to analyze human genes) and described in detail [35, 68, 88].

The following major steps describe analysis of genomic sequences by the Fgenes algorithm:

---

2) An isochore is a very long stretch ($>$ 300 kb) of DNA that is homogeneous in base composition and compositionally correlated with the coding sequence that it embeds

1) Create a list of potential exons, selecting all ORFs: ATG ... GT, AG ... GT, AG ... Stop with exons scores higher than the specific thresholds depending on GC content (4 groups);

2) Find the set of compatible exons with maximal total score. Guigo [41] described an effective algorithm of finding such a set. Fgenes uses a simpler variant of the algorithm: Order all exon candidates according to their 3'-end positions. Going from the first to the last exon select for each exon the maximally scoring path (compatible exons combination) terminated by this exon using the dynamic programming approach. Include to optimal gene structure either this exon or exon with the same 3'-splicing pattern ending at the same position or earlier (which has higher the maximal score path).

3) Take into account promoter or poly-A scores (if predicted) in terminal exons scores.

The run time of the algorithm grows approximately linearly with the sequence length. Fgenes is based on the linear discriminant functions developed for identifying splice sites, exons, promoter and poly-A sites [61, 68]. We consider these functions in the following sections to see which sequence features are important in exon prediction.

3.5.2.1 Internal exon recognition

For internal intron prediction we consider all open reading frames in a given sequence that flanked the AG (on the left) and GT (on the right) base pairs as potential internal exons. The structure of such exons is presented in Figure 3.9. The components of the recognition function for internal exon

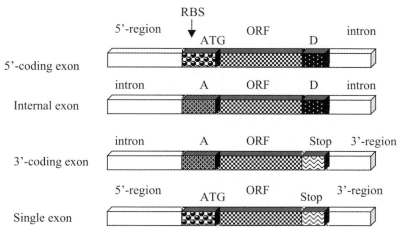

Fig. 3.9
Different functional regions of the first (a), internal (b), last (c) and single exons corresponding to components of recognition functions.

Tab. 3.10

Significance of selected internal exon characteristics. Characteristics 1 and 2 are the values of donor and acceptor site recognition functions; 3 is the octanucleotide preferences for being coding of potential exon region; 4 are the octanucleotide preferences for being intron 70 bp region on the left and 70 bp region on the right of potential exon region.

	Characteristics	1	2	3	4	5
a	Individual D^2	15.0	12.1	0.4	0.2	1.5
b	Combined D^2	15.0	25.3	25.8	25.8	25.9

prediction consist of the octanucleotide preferences for an intron 70 bp to the left of the potential intron region; the value of the acceptor splice site recognition function, the octanucleotide preferences for the coding ORF, the value of the donor splice site recognition function and the octanucleotide preferences for intron 70 bp to the right of potential intron region. The values of 5 characteristics were calculated for 952 authentic exons and for 690 714 pseudo-exon training sequences from the set. The Mahalonobis distances showing significance of each characteristic are given in Table 3.10. We can see that the strongest characteristics for exons are the values of recognition functions of flanking donor and acceptor splice sites ($D^2 = 15.04$ and $D^2 = 12.06$, respectively). The preference of ORF being a coding region has $D^2 = 1.47$ and adjacent left intron region has $D^2 = 0.41$ and right intron region has $D^2 = 0.18$.

The accuracy of the discriminant function based on these characteristics was calculated on the recognition of 451 exon and 246 693 pseudo-exon sequences from the test set. The general accuracy of exact internal exon prediction is 77% with a specificity of 79%. At the level of individual nucleotides, the sensitivity of exon prediction is 89% with a specificity of 89%; and the sensitivity of the intron prediction positions is 98% with a specificity of 98%. This accuracy is better than in the dynamic programming and neural network-based method [46], which has 75% accuracy of the exact internal exons prediction with a specificity of 67%. The method has 12% fewer false exon assignments with the better level of correct exon prediction.

3.5.2.2 Recognition of flanking exons

Figure 3.10 shows the three-dimensional histograms reflecting oligonucleotide composition of gene flanking regions created based on the graphical fractal representation of nucleotide sequences [97, 98, 99]. The clear differences in composition were exploited in the development of recognisers of these regions.

Fig. 3.10
Fractal graphical
representation of the
number of different
hexanucleotides in the
5'- (top) and 3'- (bottom)
gene regions. Each colon
represents the number of a
particular hexanucleotide
in the set of sequences.

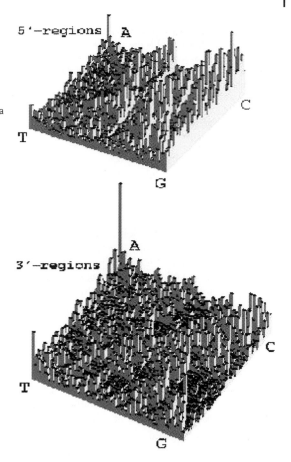

3.5.2.2.1 5'-terminal exon coding region recognition

For 5'-exon prediction, all open reading frames in a given sequence starting with an ATG codon and ending with a GT dinucleotide were considered as potential first exons. The structure of such exons is presented in Figure 3.9. The components of the 5'-exon recognition function included hexanucleotide preferences in the $(-150, -101)$, $(-100, -51)$ and $(-50, 1)$ regions to the left of the potential coding region; the average value of the positional triplet preferences in the $(-15, +10)$ region around the ATG codon; octanucleotide preferences of the coding ORF, the value of the donor splice site recognition function and the octanucleotide preferences to be intron 70 bp to the right of the potential coding region. The Mahalonobis distances showing the significance of each characteristic are given in Table 3.11. The accuracy of the discriminant function based on these characteristics was computed on the recognition of 312 the first exon and 246 693 pseudo-exon sequences. The gene sequences was scanned the 5'-exon with maximal weight were selected for each of them. The accuracy of the exact first coding

Tab. 3.11
Significance of selected 5'-exon characteristics. Characteristic 1 is the value of the donor site recognition function; 2 is the average value of positional triplet preferences in the $(-15, 10)$ region around the ATG codon; 4 is the octanucleotide preferences for being intron in the 70 bp region to the right of the potential exon; 3, 5 and 7 are the hexanucleotide preferences in the $(-150, -101)$, $(-100, -51)$ and $(-50, -1)$ regions to the left of the potential exon, respectively; 6 is the octanucleotide preferences for being coding in the exon region.

	Characteristics	1	2	3	4	5	6	7
a	Individual D^2	5.1	2.6	2.7	2.3	0.01	1.05	2.4
b	Combined D^2	5.1	8.1	10.0	11.3	12.5	12.8	13.6

exon prediction is 59%. It must be noted that the competition with the internal exons was not considered in this test.

3.5.2.2.2   3'-exon coding region recognition

All ORF regions that were flanked by GT (on the left) and finished with a stop codon were considered as potential last exons. The structure of such exons is presented in Figure 3.9. The characteristics of their discrimimant function and their Mahalonobis distances are presented in Table 3.12. The accuracy of the discriminant function was tested on the recognition of the last 322 exon and 247 644 pseudo-exon sequences. The gene sequences were scanned and the 3'-exon with the maximal weight was selected for each of them. The function can identify exactly 60% of the annotated last exons.

The recognition function of single exons combines the corresponding characteristics of 5'- and 3'-exons.

3.5.2.2.3   Combined prediction of the first, internal and last exons in human genes

The program FEX (FindEXon) predicts coding regions in a given sequence. The program initially predicts the internal exons based on the internal exon

Tab. 3.12
Significance of selected 3'-exon characteristics. Characteristic 1 is the value of the acceptor site recognition function; 2 is the octanucleotide preferences for being coding of ORF region; 3, 5 and 7 are the hexanucleotide preferences $(+100, 150)$, $(+50, +100)$ and $(+1, +50)$ regions to the left of the coding region, respectively; 4 is the average value of the positional triplet preferences in the $(-10, +30)$ region around the stop codon; 6 is the octanucleotide preferences for being intron in the 70 bp region to the left of the exon sequence.

	Characteristics	1	2	3	4	5	6	7
a	Individual D^2	10.0	3.2	0.8	2.2	1.2	0.2	1.6
b	Combined D^2	10.0	11.4	12.0	13.8	14.3	14.5	14.6

discriminant function. Then we search for the 5'-coding region starting from the beginning of the sequence until the end of the first predicted internal exon. In this region the 5'-coding exon with the maximal weight of the discriminant function for the first exon is selected. After that we search for the 3'-coding region starting from the beginning of the last predicted internal exon until the end of the sequence. In this region the 3'-coding exon with the maximal weight of the last exon discriminant function is selected. This program can be useful in producing a list of potential exons, which are not effected by their assembling, because exon assembling in some cases might significantly corrupt actual gene structure due to under-prediction of some small exons. However, exon assembling reduces significantly the number of false positive exons and this strategy is used in the most accurate gene prediction approaches.

3.5.3
Accuracy of gene identification programs

Most gene recognition programs were tested on a specially selected set of 570 one-gene sequences [89] of mammalian genes (Table 3.13). We can see that, on the average, the best programs predict accurately 93% of the exon nucleotides (Sn = 0.93) with just 7% false positive predictions. Because the most difficult task is to predict small exons and to exactly identify the 5'-

Tab. 3.13
Accuracy of gene prediction programs for single gene sequences from the [89] data set. Sn (sensitivity) is the number of exactly predicted exons/number of true exons (or nucleotides); Sp (specificity) is the number of exactly predicted exons/number of all predicted exons. Accuracy data for the programs developed before 1996 were estimated by Burset and Guigo [89]. The other data were received by the authors of the respective programs.

Algorithm	Sn (exons)	Sp (exons)	Sn nucleotides	Sp nucleotides	Reference
Fgenesh	0.84	0.86	0.94	0.95	[33]
Fgenes	0.83	0.82	0.93	0.93	[36]
GenScan	0.78	0.81	0.93	0.93	[32]
Fgeneh	0.61	0.64	0.77	0.88	[88]
Morgan	0.58	0.51	0.83	0.79	[108]
Veil	0.53	0.49	0.83	0.79	[94]
Genie	0.55	0.48	0.76	0.77	[87]
GenLang	0.51	0.52	0.72	0.79	[47]
Sorfind	0.42	0.47	0.71	0.85	[81]
GeneID	0.44	0.46	0.63	0.81	[66]
Grail2	0.36	0.43	0.72	0.87	[86]
GeneParser2	0.35	0.40	0.66	0.79	[46]
Xpound	0.15	0.18	0.61	0.87	[109]

and 3′-ends of an exon, the accuracy at the exon level is usually lower than at the nucleotide level.

The Table demonstrates that the modern multiple gene prediction programs as Fgenesh, Fgenes and Genescan significantly outperform the older approaches (some of them may have been improved since the initial publication). These data show definite progress in accuracy of gene recognition. The exon identification rate is actually even higher than the presented data since overlapping exons were not counted in exact exon predictions. However there is a lot of room for future improvements. Thus, the accuracy at the level of exact gene prediction is only 59% for Fgenesh, 56% for Fgenes and 45% for Genescan programs even on this relatively simple test.

The real challenge for *ab initio* gene identification is presented by long genomic sequences containing many genes in both DNA strands. There is not much information about real genes in such sequences. One example, which was experimentally studied in the Sanger Centre (UK), is the human BRACA2 region (1.4 MB) that contains 8 genes and 169 experimentally verified exons. This region is one of the most difficult cases in genome annotation, because it has genes with many exons and almost no genes encode amino acid sequences having similarity with known proteins. Moreover the region contains 4 pseudogenes and at least 2 genes have alternative splicing variants. The results of gene prediction initially provided by Hubbard and Bruskiewich (The Sanger Centre Genome Annotation Group) are shown in Table 3.14.

Fgenesh predicts 20% fewer false positive exons in this region than Gen-

Tab. 3.14

Accuracy of gene prediction programs for the BRACA2 1.4 MB human genomic sequence. When repeats have been defined by the RepeatMasker [100] program in the analyzed sequence these regions were masked for the prediction and excluded from the potential exon locations during prediction. The region consisted of 20 sequences with 8 verified genes, 4 pseudogenes and 169 exons. Later one sequence was constructed and 3 additional exons were identified. The results of the predictions on this sequence is marked in bold. CC is the correlation coefficient reflecting the accuracy of prediction at the nucleotide level. Snb, Spb – sensitivity and specificity at the base level (in %), Pe – number of predicted exons, Ce – number of correctly predicted exons, Sne, Spe – sensitivity and specificity at the exon level, Snep – exon sensitivity, including partially correct predicted exons (in %). Ov results by including as correctly predicted not only predicted exons which exactly coincide with the real exons, but count also predicted exons that overlapped with the real exons.

	CC	Snb	Spb	Pe	Ce/Ov	Sne	Sn_ov	Spe/Spe_ov
Genscan	0.68	90	53	271/**271**	109/**131**	65	80/**76**	40/**49**
Fgenesh	0.80	89	73	188/**195**	115/**131**	69	80/**76**	61/**67**
Fgenes	0.69	79	62	298/**281**	110/**136**	66	86/**78**	37/**48**
Genscan masked	0.76	90	66	217	109	65	80	50
Fgenesh masked	0.84	89	82	172/**168**	114/**131**	68	79/**73**	66/**76**
Fgenes masked	0.73	80	68	257/**228**	107/**133**	64	85/**75**	42/**58**

Tab. 3.15

Performance of multiple gene-finding for 38 genomic sequences. Me is missing exons and W is wrong exons. M_r lines provide predictions on sequences with masked repeats. Sn_o is exon prediction accuracy including overlapping exons.

Program	Sequences/ Genes	Accuracy per nucleotide			Accuracy per exon		Me	We	Genes/ Entries
		Sn	Sp	CC	Sn/Sn_o	Sp			
Fgenesh	38/77 M_r	0.94	0.87	0.90	0.85/0.93	0.80	0.08	0.14	0.36/0.11
		0.94	0.78	0.85	0.84/0.92	0.75	0.08	0.21	0.34/0.08
Genscan	38/77 M_r	0.93	0.82	0.87	0.80/0.90	0.74	0.10	0.18	0.29/0.03
		0.92	0.70	0.79	0.79/0.90	0.66	0.11	0.30	0.29/0.03
Fgenes	38/77 M_r	0.91	0.80	0.84	0.84/0.92	0.72	0.08	0.21	0.36/0.18
		0.92	0.76	0.83	0.84/0.93	0.68	0.07	0.30	0.39/0.21

scan with practically the same number of correctly predicted exons. Also, we can observe that even for such a difficult region, about 80% of the exons were identified by *ab initio* approaches. It is interesting to note that, when we take the subset of exons predicted by both programs (Fgenes+Fgenesh), then the predicted exons are correct in 93% cases. It is very important to know such reliable exons to start verification of a gene structure by experimental techniques.

For another test we selected a set of 19 long genomic sequences with between 26 000 and −240 000 bp and 19 multigene sequences with between 2 and 6 genes from GenBank. Table 3.15 demonstrates the results of gene prediction on these data. This results show that the accuracy is still pretty good on the nucleotide and exon level, but exact gene prediction is lower than for the test with short single gene sequences. Sensitivity for exact internal exon prediction is 85–90%, but 5′-, 3′- and single exons have a prediction sensitivity of ∼50–75%, which can partially explain the relatively low level of exact gene prediction. As a result we observe splitting up of some actual genes and/or joining some other multiple genes into a single one. A significant limitation of the current gene-finding programs is that they could not detect nested genes, i.e., genes located inside introns of another genes. While this is probably a rare event for the human genome, for organisms like *Drosophila* it presents a real problem. For example, annotators identified 17 examples of nested genes in the Adh region. [110]. Masking repeats seems important to increase (∼10%) the specificity of prediction.

3.5.4
Using protein or EST similarity information to improve gene prediction

Manual annotation experience shows that it is often advantageous to take into account all available information to improve gene identification. Automatic gene prediction approaches also take into account similarity information with known proteins or ESTs [101, 102]. Fgenesh+ [21] is a version of Fgenesh, which uses additional information from the available protein homolog. When exons initially predicted by Fgenesh show high similarity to some protein from the database, it is often advantageous to use this information to improve the prediction accuracy. Fgenesh+ requires an additional file with a protein homolog, and aligns all predicted potential exons with that protein using the SCAN 2 iterative local alignment algorithm [103]. To decrease the computation time, all overlapping exons in the same reading frame are combined into one sequence and aligned only once.

The main additions to the algorithm, relative to Fgenesh, include:

1) augmentation of the scores of exons with detected similarity by an additional term proportional to the alignment score.
2) additional penalty included for the adjacent exons in dynamic programming (Viterbi algorithm), if their corresponding aligned protein segments are not close in the corresponding similar protein.

Fgenesh+ was tested on the selected set of 61 GenBank human sequences, for which the Fgenesh predictions were not accurate (correlation coefficient $0.0 \leqslant CC < 0.90$) and which have protein homologs from other organisms. The percent sequence identity between encoded proteins and homologs was varied from 99% to 40%. The prediction accuracy on this set is presented in Table 3.16. The results show that if the alignment covers the whole length of both proteins, then Fgenesh+ usually increases the accuracy relative to Fgenesh. This phenomenon does not depend significantly on the level of sequence identity, as long as level exceed 40%. This property makes knowledge of proteins from even distant organisms valuable for improving the accuracy of gene identification. A similar approach of exploiting known EST/cDNA information was realized in Fgenesh_c program [21].

Tab. 3.16

Comparison of the accuracy of Fgenesh and Fgenesh+ on the set of human genes that are poorly predicted by *ab initio* methods, with known protein homologs from other organisms. The set contains 61 genes and 370 exons. CG – percent of correctly predicted genes; Sne, Spe are the sensitivity and specificity at the exon level (in %); Snb, Spb are the sensitivity and specificity at the base level (in %), respectively; CC – correlation coefficient.

	CG	Sne	Spe	Snb	Snb	CC
Fgenesh	0	63	68	86	83	0.74
Fgenesh+	46	82	85	96	98	0.95

3.6
Annotation of sequences from genome sequencing projects

The first task in analyzing these sequences is finding the genes. Knowledge of genes opens a new way of performing biological studies called 'functional genomics'. The other problem is to find out what all these new genes do, how they interact and are regulated [104]. Comparisons between genes of different genomes can provide additional insights into the details of the structure and function of genes.

We cannot predict exactly all gene components due to the limitation of our knowledge of the complex biological processes and signals regulating gene expression. In this respect, computer analysis of the genetically well studied Adh region of *D. melanogaster* by several gene-finding approaches [105] gives us a unique opportunity to define the reliability and limitations of our predictions and provide the strategy of right interpretation of prediction results in analysis of new genomic sequences. The predictions were evaluated by using two standards, one based on previously unreleased high quality full-length cDNA sequences and a second based on the set of annotations generated as part of an in-depth study of the region by a group of Drosophila experts. The performance of several accurate annotations is presented in Table 3.17. The CGG1 annotation comprised the non-

Tab. 3.17
Perfomance of several programs on the Adh region of *Drosophila*. The std3 contains 222 genes and 909 exons, the std1 set contains 43 genes and 123 exons. The annotated exons are taken from the set presented by organizers of the Genome Annotation Assessment Project (GASP1) at the time of the initial data analysis. Pe – number of predicted exons, Ce – number of correctly predicted exons. Pg – number of predicted genes, Pe – number of correctly predicted genes. Sn – sensitivity (in %), Sp – specificity (in %). At the exon level the second figure shows sensitivity taking into account exactly predicted and overlapped exons.

		CGG1	Fgenesh	Fgenesh pruned	Genie	Genie EST
Base Level	Sn std1	89	98	98	96	97
	Sn std3	87	92	88	79	79
	Sp std3	77	71	86	92	91
Exon Level	Pe	1115	1671	979	786	849
	Ce std1	80	100	100	86	95
	Sn std1	65/89	81/97	81/97	70	77
	Ce std3	544	601	565	447	470
	Sn std3	60/82	66/89	62/82	49	52
	Sp std3	49	36	58	57	52
Gene Level	Pg	288	530	262	241	246
	Cg std1	22	31	31	24	28
	Sn std1	51	72	72	56	65
	Cg std3	102	108	106	86	92
	Sn std3	46	49	48	39	41
	Sp std3	36	20	39	37	38

ambiguous gene set. The genes were included by the following rules (descending in priority):

a) all genes that were predicted by Fgenesh+;
b) genes predicted identically by both Fgenes (human parameters) and Fgenesh (drosophila parameters) programs;
c) in the regions of overlapped (but not exactly coincide) predictions, only one predicted gene was included with the priority given to the genes producing the longer protein.

Fgenesh+ was used to improve the accuracy of prediction for 49 genes. 37 of them were predicted using *D. melanogaster's* own proteins, already deposited in protein databases. Analysis of these predictions demonstrates that even for such cases prediction of accurate gene structure may not be trivial, although in most cases Fgenesh+ improved the prediction accuracy relative to *ab initio* methods.

The annotation CGG1 predicts about 87% of the real coding nucleotides and has just about 23% false positives (some of them might happen to be coding due to absence of experimental data in many regions). 89% of the exons are predicted exactly or with overlapping. These data show that *ab initio* predictions can provide information about practically all of protein coding genes (just 13% of coding region was not predicted) and can serve as a reasonable base for further experimental analysis.

It is interesting to note that the usage of two programs provided stable prediction accuracy on both (std1 and std3) sets. The Genie program, for example, demonstrated 20% decrease of sensitivity (Table 3.17). Because there is no version of Fgenes with all parameters computed on *Drosophila* genes, the optimal variant of automatic annotation was performed by using only the Fgenesh program.

It was found that the Fgenesh pruned predictions provided the best accuracy of annotation of the 2.9 MB Adh sequence. In this simple variant from the set of predicted genes all low-scoring genes (with average gene score less than 15) were discarded. This yields 98% of the coding nucleotide prediction on the set of 43 verified genes and 88% accurate coding nucleotide predictions with only 14% of false positives on a 222 gene set (where the significant parts of the genes were derived from Genscan predictions). The results demonstrate that most of the annotated genes in std3 are at least partially covered by predictions. For example, just 5 genes from the std3 set do not overlap with Fgenesh predictions (two of them are also included in the std1 set). From these 5 genes, 4 are located inside introns of other genes and 4 are single-exon genes. So one of the limitations of current gene-finding programs is that they cannot detect nested genes, i.e. genes located inside introns of other genes and this is one of the future directions for improvement of gene-finding software. Another drawback of the current gene-finding programs is that predictions of terminal exons are generally

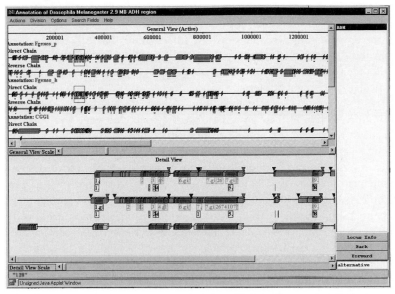

Fig. 3.11
InfoGene viewer [15] visualization of fgenes and fgenesh predictions in Adh region. In the lower panel fragments of annotations are presented that are marked in the upper panel. The last level presents the std3 manual annotation. Coding exons are marked by red color and introns by grey color. Green triangles show the start of transcription and red ones mark the poly-A signal.

much worse than the internal ones. This results in splitting up of some actual genes and/or joining some other multiple genes into a single one. Several examples of such a situation can be clear seen with the InfoGene Java viewer [15] (Figure 3.11) developed to present information about gene structures described in Genbank (collecting information about a gene from many entries) or annotated using gene prediction programs. On the std1 set Fgenesh predicts all internal exons correctly (100%), while only 72% of initial exons and 77% of terminal exons were predicted correctly. Thus better predicting the terminal exons and the related problem of better recognizing the beginnings (transcription start sites) and endings (polyA sites) of genes are the other areas of possible future improvements of gene-finding programs.

3.7
InfoGene: database of known and predicted genes

Genomic information is growing faster every day, but unfortunately the proportion of experimentally confirmed data is decreasing and the complexity of extraction of useful information is increasing. The InfoGene database [4, 5] provides one of the most complete gene-centered genomic databases practically with all coherent information that can be obtained from the GenBank feature tables. We can obtain the necessary information without looking into many GenBank entries, where the information about a particular gene might be stored.

The InfoGene database includes known and predicted gene structures with a description of their basic functional signals and gene components. All major organisms are presented in separate divisions. The information about a gene structure might be collected from dozens GenBank entries. This information can be applied to create different sets of functional gene component for extraction their significant characteristics used in gene prediction systems. InfoGene is realized under a JAVA interactive environment system [15] that provides visual analysis of known information about complex gene structure (Figure 3.12) and searches different gene components and signals. The database is available through WWW server of Computational Genomics Group at http://genomic.sanger.ac.uk/infodb.shtml.

The value of sequence information for the biomedical community will strongly depend on the availability of candidate genes that are computationally predicted in these sequences. Currently information about predicted genes is absent in sequence databases if the gene has no similarity on the protein level with a known protein. Using gene prediction the scientific community can start experimental work with most human genes, because

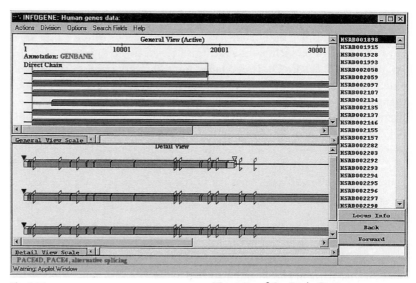

Fig. 3.12
InfoGene [15] presentation of human PACE4 (AB001898) gene. This gene has several alternative forms and described in 17 entries of GenBank. Continues sequences regions corresponding different GenBank entries are separated by the vertical bars.

gene-finding programs usually predict accurately at least the major part of the exons in a gene sequence. InfoGene includes all predicted genes for Human and *Drosophila* genome drafts and several chromosomes of the *Arabidopsis* genome. The database is available currently through WWW server at httpd://www.softberry.com/infodb.html. Recently, the similar project Ensembl was started as a collaboration between the Sanger Center and European Bioinformatics Institute (http://www.ensembl.org/).

3.7.1
Annotation of human genome draft

The nucleotide sequence of nearly 90% of the human genome (3 GB) has been determined by international sequencing effort. Assembly of the current draft of the human genome was performed by the Human Genome Project Team at UC Santa Cruz. On this sequence (with masked repeats) the FGENESH program was used to predict exons and assemble predicted genes. Annotation of similarity of each exon with the PfamA protein domain database [106] was produced by the Blast program [107]. Totally 49 171 genes and 282 378 coding exons were predicted. On average one gene was found per about 68 623 bp and one exon per 11 949 bp. A complete summary of this analysis including the gene and exon numbers in different human chromosomes is presented at http://www.softberry.com/inf/humd_an.html and can be viewed in the InfoGene database. Sequences of predicted exons and gene annotation data can be copied from this site also. 1154 types of PfamA different domains were found in the predicted proteins. The top part of the domain list is presented in Table 3.18.

3.8
Functional analysis and verification of predicted genes

Large scale functional analysis of predicted and known genes might be done using expression micro-array technology (see Chapter 5 of Volume 2 of this book). Often genes are presented on the chips by unique oligonucleotides close to the 3′-end of the mRNA. But there are many predicted new genes that have no known corresponding EST sequences. We can study the expression of such genes in a large number of human tissues using predicted exon sequences represented on one or several Affymetrix type DNA chips. As a result we will know not only expression properties of genes, but we can identify what exons are real. Observing coordinated expression of neighboring exons in different tissues it will often be possible to define gene boundaries, which is very difficult using *ab initio* gene prediction. Moreover such experiments might have additional value in defining disease

Tab. 3.18
PfamA domains identified in the predicted human genes. Domain of the same type localized in neighboring exons were counted only once.

Number	PfamA short name	Name
467	Pkinase	Eukaryotic protein kinase domain
372	7tm_1	7 transmembrane receptor (rhodopsin family)
308	Myc_N_term	Myc amino-terminal region
256	Topoisomerase_I	Eukaryotic DNA topoisomerase I
224	Ig	Immunoglobulin domain
183	Rrm	RNA recognition motif
182	PH	PH domain
180	Myosin_tail	Myosin tail
166	EGF	EGF-like domain
159	Filament	Intermediate filament proteins
154	Syndecan	Syndecan domain
143	Ras	Ras family
138	RNA_pol_A	RNA polymerase A/beta'/A" subunit
123	BTB	BTB/POZ domain
119	Granin	Granin (chromogranin or secretogranin)
119	Troponin	Troponin
113	Herpes_glycop_D	Herpesvirus glycoprotein D
111	Homeobox	Homeobox domain
110	SH3	SH3 domain
102	Trypsin	Trypsin
102	helicase_C	Helicases conserved C-terminal domain
100	KRAB	KRAB box
98	dehydrin	Dehydrins
96	ABC_tran	ABC transporter
95	ERM	Ezrin/radixin/moesin family
89	Collagen	Collagen triple helix repeat
87	Tryp_mucin	Mucin-like glycoprotein
84	Fn3	Fibronectin type III domain
81	pro_isomerase	Cyclophilin type peptidyl-prolyl cis-trans isomerase
81	HMG_box	HMG (high mobility group) box
79	SH2	Src homology domain 2

tissue specific genes, which can be used for the development of potential therapeutics.

The chip designed by EOS Biotechnology included all exons from Chromosome 22 predicted by Fgenesh and Genescan as well as exons from human genomic sequences of Phase 2 and 3 predicted by Fgenesh. It was found that the predicted exon sequences present a good alternative to EST sequences that open a possibility to work with predicted genes on a large scale.

In Figure 3.13 we have an example of expression behavior of five sequential exons along the Chromosome 22 sequence (expression data were received in EOS Biotechnology Inc.). Exons 2, 3 and 4 are Myoglobin gene exons. Tissue specific expression of them is clear seen with the major peaks located

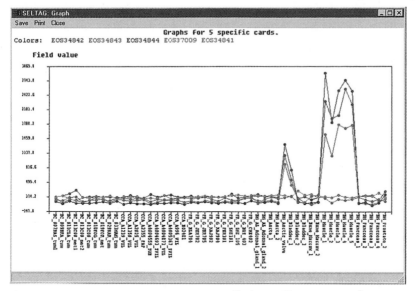

Fig. 3.13

Coordinative expression of three exons (EOS34842, EOS34842, EOS34842) of the Human Myoglobin gene from Chromosome 22 (exons were predicted by the Fgenesh program and used to design the EOS Biotechnology Human genome chip) in 50 different tissues. The high level of expression is observed only in several specific tissues. The two exons of Myoglobin predicted on the left side (EOS34841) and on the right side (EOS37009) respectively show completely different patterns of expression. The visualization is presented by the SELTAG program for analysis of expression data developed by Softberry Inc.

in skeletal muscle, heart and diaphragm tissues. The level of expression in these tissues is 10–100 times higher than the level of signals for other tissues as well as the average level of expression for randomly chosen exons. We found that three Myoglobin exons have expression level correlation coefficient 0.99, when for random exons it is about 0.06. These exons were predicted correctly by the Fgenesh program and were used for selection of oligonucleotide probes. From this result we can conclude that the predicted exons can be used as gene representatives. At the same time two flanking exons (1 and 5) from different genes show no correlation with the Myoglobin exons. This clearly demonstrates how expression data can be used to define gene boundaries. Another application of expression data is functional analysis and identification of alternatively spliced genes (exons), if in particular tissues some exons (or their parts) have very different expression intensities comparing with the other exons from the same gene. If 5′-alternative exons define different functional forms of genes (normal and disease-specific, for example), then the 3′-EST generated probes cannot be used for identification of disease specific gene variants.

3.9
Internet sites for gene finding and functional site prediction

Gene identification, and finding of ORFs, promoters and splice sites by the methods described above is available via World Wide Web. Table 3.19 presents a list of some useful programs but does not provide a comprehensive collection.

Tab. 3.19
Public web servers for eukaryotic gene and functional signal prediction.

Program/function	World Wide Web address
Fgenesh/HMM-based gene prediction (Human, Drosophila, Dicots, Monocots, C.elegans, S.pombe)	http://genomic.sanger.ac.uk/gf/gf.shtml http://searchlauncher.bcm.tmc.edu:9331/ seq-search/gene-search.html http://www.softberry.com/nucleo.html
Genscan/HMM-based gene prediction (Human, Arabidipsis, Maize)	http://genes.mit.edu/GENSCAN.html
HMM-gene/HMM-based gene prediction (Human, C.elegans)	http://www.cbs.dtu.dk/services/ HMMgene/
Fgenes/Disciminative gene prediction (Human)	http://genomic.sanger.ac.uk/gf/gf.shtml http://searchlauncher.bcm.tmc.edu:9331/ seq-search/gene-search.html
Fgenes-M/Prediction of alternative gene structures (Human)	http://genomic.sanger.ac.uk/gf/gf.shtml http://www.softberry.com/nucleo.html
Fgenesh + /Fgenesh_c/gene prediction with the help of similar protein/EST	http://genomic.sanger.ac.uk/gf/gf.shtml http://www.softberry.com/nucleo.html
Fgenesh-2/gene prediction using 2 sequences of close species	http://www.softberry.com/nucleo.html
BESTORF/Finding best CDS/ORF in EST (Human, Plants, Drosophila)	http://genomic.sanger.ac.uk/gf/gf.shtml http://www.softberry.com/nucleo.html
Mzef/internal exon prediction (Human, Mouse, Arabidopsis, Yeast)	http://argon.cshl.org/genefinder/
TSSW/TSSG/eukaryotic promoter prediction	http://searchlauncher.bcm.tmc.edu:9331/ seq-search/gene-search.html http://genomic.sanger.ac.uk/gf/gf.shtml
Promoter 2.0/promoter prediction	http://www.cbs.dtu.dk/services/ Promoter/
PromoterScan/eukaryotic promoter prediction	http://cbs.umn.edu/software/proscan/ promoterscan.htm
CorePromoter/promoter prediction	http://argon.cshl.org/genefinder/ CPROMOTER/index.htm
SPL/splice site prediction (Human, Drosophila, Plants, Yeast)	http://genomic.sanger.ac.uk/gf/gf.shtml http://www.softberry.com/nucleo.html
NetGene2/NetPGene/splice site prediction (Human, C.elegans, Plants)	http://www.cbs.dtu.dk/services/ NetPGene/
Dbscan/searching for similarity in genomic sequences and its visualization altogether with known gene structure	http://www.softberry.com/nucleo.html

Acknowledgements

I would like to acknowledge collaboration with Asaf Salamov and Igor Seledtsov in developing gene finding and sequence analysis algorithms and Moises Burset for collaboration in research presented in Section 3.2.

References

1 CAREY and SMALE **(1999)** *Transcriptional regulation in eukaryotes.* Cold Spring Harbor Laboratory Press, Cold Spring Harbor, New York.

2 LEVINE **(1998)** Genes, Oxford University Press, Oxford, UK.

3 BENSON D. A., BOGUSKI M. S., LIPMAN D. J., OSTELL J., OUELLETTE B. F., RAPP B. A., WHEELER D. L. **(1999)** GenBank. *Nucleic Acids Res.,* 27, 12–17.

4 SOLOVYEV V. V., SALAMOV A. A. **(1999)** INFOGENE: a database of known gene structures and predicted genes and proteins in sequences of genome sequencing projects. *Nucleic Acids Res.,* 27:1, 248–250.

5 SOLOVYEV V., SELEDTSOV I., SALAMOV A. **(1999)** *Infogene database of known and predicted genes.* httpd://genomic.sanger.ac.uk/infodb.shtml.

6 BREATHNACH R., BENOIST C., O'HARE K., GANNON F., CHAMBON P. **(1978)** Ovalbumin gene: evidence for a leader sequence in mRNA and DNA sequences at the exon-intron boundaries. *Proc. Natl. Acad. Sci.,* 75, 4853–4857.

7 BREATHNACH R., CHAMBON P. **(1981)** Organization and expression eukaryotic of split genes for coding proteins. *Annu. Rev. Biochem.,* 50, 349–393.

8 JACKSON I. J. **(1991)** A reappraisal of non-consensus mRNA splice sites. *Nucleic Acids Res.,* 19, 3795–3798.

9 HALL S. L., PADGETT R. A. **(1994)** Conserved sequences in a class of rare eukaryotic nuclear introns with non-consensus splice sites. *J. Mol. Biol.,* 239, 357–365.

10 TARN W. Y., STEITZ J. A. **(1996)** Highly diverged U4 and U6 small nuclear RNAs required for splicing rare AT-AC introns. *Science,* 273:5283, 1824–1832.

11 TARN W. Y., STEITZ J. A. **(1997)** Pre-mRNA splicing: the discovery of a new spliceosome doubles the challenge. *Trends Biochem. Sci.,* 22:4, 132–137.

12 SHARP P. A., BURGE C. B. **(1997)** Classification of introns: U2-type or U12-type. *Cell,* 91:7, 875–879.

13 SHAPIRO M. B., SENAPATHY P. **(1987)** RNA splice junctions of different classes of eukaryotes: sequence statistics and functional implications in gene expression. *Nucleic Acids Res.,* 15:17, 7155–7174.

14 BURSET M., SELEDTSOV I., SOLOVYEV V. **(2000)** Analysis of canonical and non-canonical splice sites in mammalian genomes. *Nucleic Acids Res.,* 28, 4364–4375.

15 SELEDTSOV I., SOLOVYEV V. **(1999)** *Genes_in_Pictures: Interactive Java viewer for Infogene database.* http://genomi.sanger.ac.uk/infodb.shtml.

16 STOESSER G., TULI M., LOPEZ R., STERK P. **(1999)** The EMBL Nucleotide Sequence Dabase. *Nucleic Acids Res.*, 27:1, 18–24.

17 PENOTTI F. E. **(1991)** Human Pre-mRNA splicing signals. *J. Theor. Biol.*, 150, 385–420.

18 SENAPATHY P., SAHPIRO M., HARRIS N. **(1990)** Splice junctions, brunch point sites, and exons: sequence statistics, identification, and application to genome project. *Methods in Enzymology*, 183, 252–278.

20 WU Q., KRAINER A. R. **(1997)** Splicing of a divergent subclass of AT-AC introns requires the major spliceosomal snRNAs. *RNA*, 3:6, 586–601.

21 SALAMOV A., SOLOVYEV V. **(2000)** Ab initio gene finding in Drosophila genomic DNA. *Genome Research* 10, 516–522.

22 SHAHMURIDOV K. A., KOLCHANOV N. A., SOLOVYEV V. V., RATNER V. A. **(1986)** Enhancer-like structures in middle repetitive sequences of the eukaryotic genomes. *Genetics (Russ.)*, 22, 357–368.

23 SOLOVYEV V., KOLCHANOV N. **(1999)** *Search for functional sites using consensus.* In: Computer analysis of Genetic Macromolecules. Structure, Function and Evolution. (eds. Kolchanov N. A., Lim H. A.), World Scientific, p. 16–21.

24 SHAHMURADOV and SOLOVYEV **(1999)** NSITE program for recognition of statistically significant regulatory motifs: http://softberry.com/nucleo.html.

25 WINGENDER E., DIETZE P., KARAS H., KNUPPEL R. **(1996)** TRANSFAC: a database of transcription factors and their binding sites. *Nucleic Acid. Res.*, 24, 238–241.

26 STADEN R. **(1984)** Computer methods to locate signals in nucleic acid sequences. *Nucleic Acids Res.*, 12, 505–519.

27 ZHANG M. Q. & MARR T. G. **(1993)** A weight array method for splicing signal analysis. *Comp. Appl. Biol. Sci.*, 9(5), 499–509.

28 BURGE C. **(1997)** *Identification of genes in human genomic DNA.* Ph.D. Thesis, Stanford, 152p.

29 STORMO G. D. & HAUSSLER D. **(1994)** *Optimally parsing a sequence into different classes based on multiple types of evidence.* In: Proceedings of the Second International Conference on Intelligent Systems for Molecular Biology, pp. 47–55, AAAI Press, Menlo Park, CA.

30 BALDI P., BRUNAK S. **(1998)** *Bioinformatics: the machine learning approach.* MIT Press, Cambrodge, Ma.

31 DURBIN R., EDDY S., KROGH A., MITCHISON G. **(1998)** Biological Sequence Analysis. Cambridge University Press, Cambridge, UK.

32 BURGE C., KARLIN S. **(1997)** Prediction of complete gene structures in human genomic DNA, *J. Mol. Biol.*, 268, 78–94.

33 SALAMOV A., SOLOVYEV V. **(1998)** Fgenesh multiple gene prediction program: httpd://genomic.sanger.ac.uk.

34 GREEN P., HILLIER L. **(1998)** GENEFINDER, unpublished software.

35 SOLOVYEV V. V., LAWRENCE C. B. **(1993a)** *Identification of Human gene functional regions based on oligonucleotide composition.* In: Proceedings of First International conference on Intelligent System for Molecular Biology (eds. Hunter L., Searls D., Shalvic J.), Bethesda, 371–379.

36 Solovyev V. **(1997)** Fgenes multiple gene prediction program: httpd://genomic.sanger.ac.uk/gf.html.

37 Shepherd J. C. W. **(1981)** Method to determine the reading frame of a protein from the purine/pyrimidine genome sequence and ist possible evolutionary justification. *Proc. Natl. Acad. Sci. USA*, 78, 1596–1600.

38 Borodovskii M., Sprizhitskii Yu., Golovanov E., Alexandrov N. **(1986)** Statistical patterns in the primary structures of functional regions of the genome in Escherichia coli. II. Nonuniform Markov Models. *Molekulyarnaya Biologia*, 20, 1114–1123.

39 Claverie J., Bougueleret L. **(1986)** Heuristic informational analysis of sequences. *Nucleic Acids Res.*, 10, 179–196.

40 Borodovsky M. & McIninch J. **(1993)** GENMARK: parallel gene recognition for both DNA strands. *Comp. Chem.*, 17, 123–133.

41 Guigo R. **(1999)** *DNA composition, codon usage and exon prediction.* In: Genetics Databases. Academic Press, 54–80.

42 Fickett J. W., Tung C. S. **(1992)** Assessment of Protein Coding Measures. *Nucleic Acids Res.*, 20, 6441–6450.

43 Uberbacher E., Mural J. **(1991)** Locating protein coding regions in human DNA sequences by a multiple sensorneural network approach. *Proc. Natl. Acad. Sci. USA*, 88, 11261–11265.

44 Farber R., Lapedes A., Sirotkin K. **(1992)** Determination of eukaryotic protein coding regions using neural networks and information theory. *J. Mol. Biol.*, 226, 471–479.

45 Solovyev V., Lawrence C. **(1993b)** *Prediction of human gene structure using dynamic programming and oligonucleotide composition.* In: Abstracts of the 4th Annual Keck Symposium. Pittsburgh, 47.

46 Snyder E. E., Stormo G. D. **(1993)** Identification of coding regions in genomic DNA sequences: an application of dynamic programming and neural networks. *Nucleic Acids Res.*, 21, 607–613.

47 Dong S., Searls D. **(1994)** Gene structure prediction by linguistic methods, *Genomics*, 23, 540–551.

48 Mathews B. W. **(1975)** Comparison of the predicted and observed secondary structure of T4 phage lysozyme, *Biochem. Biophys. Acta*, 405, 442–451.

49 Afifi A. A., Azen S. P. **(1979)** *Statistical analysis. A computer oriented approach.* Academic Press, New York.

50 Aebi M., Weissmann C. **(1987)** Precision and orderliness in splicing, *Trends in Genetics*, 3, 102–107.

51 Steitz et al. **(1988)** "Snurps". *Sci Am.*, 258, 56–60.

52 Green M. R. **(1991)** Biochemical mechanisms of constitutive and regulared pre-mRNA splicing. *Ann. Rev. Cell Biol.*, 7, 559–599.

53 Wieringa B., Meyer F., Reiser J., Weissmann C. **(1983)** Unusual splice sites revealed by mutagenic inactivation of an authentic splice site of the rabbit beta-globin gene. *Nature*, 301:5895, 38–43.

54 Mount S. **(1981)** A catalogue of splice junction sequences. *Nucleic Acids Res.*, 10, 459–472.

55 MOUNT S. M. **(1993)** *Messenger RNA splicing signal in Drosophila genes.* In: An atlas of Drosophila genes (ed. Maroni G.), Oxford.

56 NAKATA K., KANEHISA M., DeLISI C. **(1985)** Prediction of splice junctions in mRNA sequences. *Nucleic Acids Res.*, 13, 5327–5340.

57 GELFAND M. **(1989)** Statistical analysis of mammalian pre-mRNA splicing sites. *Nucleic Acids Res.*, 17, 6369–6382.

58 LAPEDES A., BARNES C., BURKS C., FARBER R., SIROTKIN K. **(1988)** Application of neural network and other machine learning algorithms to DNA sequence analysis. *Proc. Santa Fe Inst.*, 7, 157–182.

59 BRUNAK S., ENGELBREHT J., KNUDSEN S. **(1991)** Prediction of Human mRNA donor and acceptor sites from the DNA sequence. *J. Mol. Biol.*, 220, 49–65.

60 MURAL R. J., MANN R. C., UBERBACHER E. C. **(1990)** *Pattern recognition in DNA sequences: The intron-exon junction problem.* In: The first International Conference on Electrophoresis, Supercomputing and the Human Genome. (Cantor C. R., Lim H. A. eds), World Scientific, London, 164–172.

61 SOLOVYEV V. V., SALAMOV A. A., LAWRENCE C. B. **(1994)** Predicting internal exons by oligonucleotide composition and discriminant analysis of spliceable open reading frames. *Nucleic Acids Res.*, 22, 6156–5153.

62 CINKOSKY M., FICKETT J., GILNA P., BURKS C. **(1991)** Electronic data publishing and GenBank. *Science*, 252, 1273–1277.

63 FICKETT J., HATZIGEORGGIOU A. **(1997)** Eukaryotic promoter recognition. *Genome Research*, 7, 861–878.

64 HUTCHINSON G. **(1996)** The prediction of vertebrate promoter regions using differential hexamer frequency analysis. *Comput. Appl. Biosci.*, 12, 391–398.

65 AUDIC S, CLAVERIE J. **(1997)** Detection of eukaryotic promoters using Markov transition matrices. *Comput Chem*, 21, 223–227.

66 GUIGO R., KNUDSEN S., DRAKE N., SMITH T. **(1992)** Prediction of gene structure. *J. Mol. Biol.*, 226, 141–157.

67 REESE M. G., HARRIS N. L. and EECKMAN F. H. **(1996)** *Large Scale Sequencing Specific Neural Networks for Promoter and Splice Site Recognition.* Biocomputing: Proceedings of the 1996 Pacific Symposium (eds. L. Hunter and T. Klein), World Scientific Publishing Co, Singapore, 1996.

68 SOLOVYEV V. V., SALAMOV A. A. **(1997)** *The Gene-Finder computer tools for analysis of human and model organisms genome sequences.* In: Proceedings of the Fifth International Conference on Intelligent Systems for Molecular Biology (eds. Rawling C., Clark D., Altman R., Hunter L., Lengauer T., Wodak S.), Halkidiki, Greece, AAAI Press, 294–302.

69 OHLER U., HARBECK S., NIEMANN H., NOTH E., REESE M. **(1999)** Interpolated markov chains for eukaryotic promoter recognition. *Bioinformatics*, 15, 362–369.

70 BUCHER P. **(1990)** Weight matrix descriptions of four eukaryotic RNA polymerase II promoter elements derived from 502 unrelated promoter sequences. *J. Mol. Biol.*, 212, 563–578.

71 DECKER C. J. and PARKER R. **(1995)** Diversity of cytopasmatic functions for the 3′-untranslated region of eukaryotic transcripts. *Current Opinions in Cell Biology*, 1995, 386–392.

71 KNUDSEN S. **(1999)** Promoter 2.0: for the recognition of PolII promoter sequences. *Bioinformatics*, 15, 356–361.

72 WAHLE E. **(1995)** 3′-end cleavage and polyadelanytion of mRNA precursor. *Biochim. et Biophys. Acta*, 1261, 183–194.

73 MANLEY J. L. **(1995)** A complex protein assembly catalyzes polyadenylation of mRNA precursors. *Current Opin. Genet. Develop.*, 5, 222–228.

74 WILUSZ J., SHENK T., TAKAGAKI Y., MANLEY J. L. **(1990)** A multicomponent complex is required for the AAUAAA-dependent cross-linking of a 64-kilodalton protein to polyadenylation substrates. *Mol. Cell. Biol.*, 10, 1244–1248.

75 PROUDFOOT N. J., Poly(A) signals **(1991)** *Cell*, 64, 617–674.

76 WAHLE E., and KELLER W. **(1992)** The biochemistry of the 3′-end cleavage and polyadenylation of mRNA precursors. *Annu. Rev. Biochem.*, 1992, 61, 419–440.

77 MCLAUCHLAN J., GAFFNEY D., WHITTON J. L., CLEMENTS J. B. **(1985)** The consensus sequence YGTGTTYY located downstream from the AATAAA signal is required for efficient formation of mRNA 3′ termini. *Nucleic Acids Res.*, 13, 1347–1367.

78 YADA T., ISHIKAWA M., TOTOKI Y., OKUBO K. **(1994)** *Statistical analysis of human DNA sequences in the vicinity of poly(A) signal.* Institute for New Generation Computer Technology. Technical Report TR-876.

79 KONDRAKHIN Y. V., SHAMIN V. V., KOLCHANOV N. A. **(1994)** Construction of a generalized consensus matrix for recognition of vertebrate pre-mRNA 3′-terminal processing sites. *Comput. Applic. Biosci.*, 10, 597–603.

80 TABASKA J., ZHANG M. Q. **(1999)** Detection of polyadenilation signals in Human DNA sequences. *Gene*, 231, 77–86.

81 HUTCHINSON G. B., HAYDEN M. R. **(1992)** The prediction of exons through an analysis of splicible open reading frames. *Nucleic Acids Res.*, 20, 3453–3462.

82 BERG O. G., VON HIPPEL P. H. **(1987)** Selection of DNA binding sites by regulatory proteins. *J. Mol. Biol.*, 193, 723–750.

83 FIELDS C., SODERLUND C. **(1990)** GM: a practical tool for automating DNA sequence analysis. *Comp. Applic. Biosci.*, 6, 263–270.

84 GELFAND M. **(1990)** Global methods for the computer prediction of protein-coding regions in nucleotide sequences. *Biotechnology Software*, 7, 3–11.

85 GELFAND M., ROYTBERG M. **(1993)** Prediction of the exon-intron structure by a dynamic programming approach. *BioSystems*, 30, 173–182.

86 XU Y., J. R. EINSTEIN, R. J. MURAL, M. SHAH, and E. C. UBERBACHER **(1994)** *An improved system for exon recognition and gene modeling in human DNA sequences.* In: Proceedings of the 2nd International Conference on Intelligent Systems for Molecular Biology (eds. Altman R., Brutlag, Karp P., Lathrop R. and Searls D.) Stanford, AAAI Press, 376–383.

87 KULP D., HAUSSLER D., REES M., EECKMAN F. **(1996)** *A generalized Hidden Markov Model for the recognition of human genes in DNA*. In: Proceedings of the Fourth International Conference on Intelligent Systems for Molecular Biology (eds. States D., Agarwal P., Gaasterland T., Hunter L., Smith R.), St. Louis, AAAI Press, 134–142.

88 SOLOVYEV V. V., SALAMOV A. A., LAWRENCE C. B. **(1995)** *Prediction of human gene structure using linear discriminant functions and dynamic programming*. In: Proceedings of the Third International Conference on Intelligent Systems for Molecular Biology (eds. Rawling C., Clark D., Altman R., Hunter L., Lengauer T., Wodak S.), Cambridge, England, AAAI Press, 367–375.

89 BURSET M., GUIGO R. **(1996)** Evaluation of gene structure prediction programs, *Genomics*, 34, 353–367.

90 COLLINS F., PATRINOS A., JORDAN E., CHAKRAVARTI A., GESTELAND A., WALTERS L. **(1998)** New goals for the U.S. Human Genome Project: 1998–2003. *Science*, 282, 682–689.

91 KROGH A., MIAN I. S. & HAUSSLER D. **(1994)** A hidden Markov model that finds genes in E. coli DNA. *Nucleic Acids Res.*, 22, 4768–4778.

92 FORNEY, G. D. **(1973)** The Viterbi algorithm. *Proc. IEEE*, 61, 268–278.

93 RABINER L., JUANG B. **(1993)** *Fundamentals of speech recognition*. Prentice Hall, New Jersey.

94 HENDERSON J., SALZBERG S., FASMAN K. **(1997)** Finding genes in DNA with a Hidden Markov Model. *J. Comput. Biol.*, 4, 127–141.

95 KROG A. **(1997)** Two methods for improving performance of an HMM and their application for gene finding. *Ismb.*, 5, 179–186.

96 LUKASHIN A. V., BORODOVSKY M. **(1998)** GeneMark.hmm: new solutions for gene finding. *Nucleic Acids Res.*, 26, 1107–1115.

97 JEFFREY H. J. **(1990)** Chaos game representation of gene structure. *Nucleic Acids Res.*, 18, 2163–2170.

98 SOLOVYEV V. V. **(1993)** Fractal graphical representation and analysis of DNA and Protein sequences. *BioSystems*, 30, 137–160.

99 SOLOVYEV V. V., KOROLEV S. V., V. G. TUMANYAN, LIM H. A. **(1991)** A new approach to classification of DNA regions based on fractal representation of functionally similar sequences. *Proc. Natl. Acad. Sci. USSR (Russ.) (Biochemistry)*, 1991, 319(6), 1496–1500.

100 SMITH A., GREEN P. **(1997)** RepeatMasker for screening DNA sequences for interspersed repeats and low complexity DNA sequences: http://ftp.genome.washington.edu/RM/RepeatMasker.html.

101 GELFAND M., MIRONOV A., PEVZNER P. **(1996)** Gene recognition via spliced sequence alignment. *Proc Natl Acad Sci USA*, 93, 9061–9066.

102 KROGH A. **(2000)** Using database matches with HMMgene for automated gene detection in Drosophila. *Genome Res.* 10, 523–528.

103 SELEDTSOV I., SOLOVYEV V. **(2000)** SCAN2: Pairwise iterative alignment algorithm http://www.softberry.com/scan.html.

104 WADMAN M. **(1998)** Rough draft' of human genome wins researchers' backing. *Nature*, 393, 399–400.

105 REESE M. G., HARTZELL G., HARRIS N. L., OHLER U., & LEWIS S. E. **(2000a)** Genome annotation assessment in *Drosophila melanogaster. Genome Research (in press)*.

106 BATEMAN A., BIRNEY E., DURBIN R., EDDY S., HOWE K., SONNHAMMER E. **(2000)** The Pfam protein families database. *Nucleic Acids Res.*, 28, 263–266.

107 ALTSCHUL S. F., MADDEN T. L., SCHAFFER A. A., ZHANG J., et al. **(1997)** Gapped BLAST and PSI-BLAST: a new generation of protein database search programs. *Nucleic Acids Res.*, 25, 3389–3402.

108 SALSBERG S., DELCHER A., FASMAN K., HENDERSON J. **(1998)** A decision Tree System for finding genes in DNA. *J. Comp. Biol.*, 5, 667–680.

109 THOMAS A., SKOLNICK M. **(1994)** A probabilistic model for detecting coding regions in DNA sequences. IMA *J. Math. Appl. Med. Biol.*, 11, 149–160.

110 ASHBURNER M., MISRA S., ROOTE J., LEWIS S. E., BLAZEJ R., DAVIS T., DOYLE C., GALLE R., GEORGE R., HARRIS N., HARTZELL G., HARVEY D., HONG L., HOUSTON K., HOSKINS R., JOHNSON G., MARTIN C., MOSHREFI A., PALAZZOLO M., REESE M. G., SPRADLING A., TSANG G., WAN K., WHITELAW K., CELNIKER S., & RUBIN G. M. **(1999)** An exploration of the sequence of a 2.9-mb region of the genome of *Drosophila melanogaster*: the Adh region. *Genetics*, 153, 179–219.

4
Analyzing Regulatory Regions in Genomes

Thomas Werner

4.1
General features of regulatory regions in eukaryotic genomes

Regulatory regions share several common features despite their obvious divergence in sequence. Most of these common features are not evident directly from the nucleotide sequence but result from the restraints imposed by functional requirements. Therefore, understanding of the major components and events during the formation of regulatory DNA–protein complexes is crucial for the design and evaluation of algorithms for the analysis of regulatory regions. Transcription initiation from polymerase II (pol II) is the best understood example so far and will be a major focus of this chapter. However, the mechanisms and principles revealed from promoters are mostly valid for other regulatory regions as well.

Algorithms for the analysis and recognition of regulatory regions draw from the underlying biological principles, to some extent, in order to generate suitable computational models. Therefore, a brief overview over the biological requirements and mechanisms is necessary to understand what are the strengths and weaknesses of the individual algorithms. The choice of parameters and implementation of the algorithms largely control sensitivity and speed of a program. The specificity of software recognizing regulatory regions in DNA is determined, to a large extent, by how closely the algorithm follows what will be called the biological model from hereon. A detailed overview of this topic was recently published [1].

4.2
General functions of regulatory regions

The biological functionality of regulatory regions is generally not a property evenly spread over the regulatory region in total. Functional units usually are defined by a combination of defined stretches that can be delimited and possess an intrinsic functional property (e.g., binding of a protein or a curved DNA structure). Several functionally similar types of these stretches of DNA are already known and will be referred to as *elements*. Those elements are neither restricted to regulatory regions nor individually sufficient for the regulatory function of a promoter or enhancer. The function of the complete regulatory region is composed from the functions of the individual elements either in an additive manner (independent elements) or by synergistic effects (modules).

4.2.1
Transcription factor binding sites (TF-sites)

Binding sites for specific proteins are most important among these elements. They consist of about 10–30 nucleotides, not all of which are equally important for protein binding. As a consequence individual protein binding sites vary in sequence, even if they bind to the same protein. There are nucleotides contacted by the protein in a sequence-specific manner which are usually the best conserved parts of a binding site. Different nucleotides are involved in DNA backbone contacts, i.e., contacting the sugar-phosphate framework (not sequence specific as they do not involve the bases A, G, C, or T), and there are internal "spacers" not contacted by the protein at all. In general, protein-binding sites exhibit enough sequence conservation to allow for the detection of candidates by a variety of sequence similarity-based approaches. Potential binding sites can be found almost all over the genome and are not restricted to regulatory regions. Quite a number of binding sites outside regulatory regions are also known to bind their respective binding proteins [2]. Therefore the abundance of predicted binding sites is not just a shortcoming of the detection algorithms but reflects biological reality.

4.2.2
Sequence features

Regulatory DNA also contains several features not directly resulting in recognizable sequence conservation. For example, the sequences of two copies of a direct repeat (approximate or exact) are conserved in sequence with respect to each other but different direct repeats are not similar in sequence at all. Nevertheless, direct repeats are quite common within regulatory DNA regions. They consist either of short sequences which are repeated twice or more frequently within a short region or they can be complex repeats which repeat a pattern of two or more elements. Repeat structures are often associated with enhancers. Enhancers are DNA structures that enhance transcription over a distance without being promoters themselves. One example of a highly structured enhancer is the interleukin-2 enhancer [3]. Other sequence features that are hard to detect by computer methods include the relatively weak nucleosomal positioning signals [4], DNA stretches with intrinsic 3-D structures (like curved DNA [5]), methylation signals (if there are definite signals for methylation at all), and other structural elements.

4.2.3
Structural elements

Secondary structures are currently the most useful structural elements with respect to computer analysis. Secondary structrures are mostly known for RNAs and proteins but they also play important roles in DNA. Potential secondary structures can be easily determined and even scored via the negative enthalpy that should be associated with the actual formation of the hairpin (single strand) or cruciform (double strand) structure. Secondary structures are also not necessarily conserved in primary nucleotide sequence but are subject to strong positional correlation within the structure. Three-dimensional aspects of DNA sequences are without any doubt very important for the functionality of such regions. However, existing attempts to calculate such structures in reasonable time met with mixed success and cannot be used for a routine sequence analysis at present.

4.2.4
Organizational principles of regulatory regions

This paragraph will mainly concentrate on eukaryotic polymerase II promoters as they are currently the best-studied regulatory regions. The TF-sites within promoters (and likewise most other regulatory sequences) do not show any general patterns with respect to location and orientation within the promoter sequences. Even functionally important binding sites for a specific transcription factor may occur almost anywhere within a promoter. For example, functional AP-1 (Activating protein 1, a complex of two transcription factors, one from the fos and one from the Jun family) binding sites can be located far upstream, as in the rat bone sialoprotein gene where an AP-1 site located about 900 nucleotides upstream of the transcription start site (TSS) inhibits expression [6]. An AP-1 site located close to the TSS is important for the expression of Moloney Murine Leukemia Virus [7]. Moreover, functional AP-1 sites have also been found inside exon 1 (downstream of the TSS) of the proopiomelanocortin gene [8] as well as within the first intron of the fra-1 gene [9], both locations outside the promoter. Similar examples can be found for several other TF-sites, illustrating why no general correlation of TF-sites within specific promoter regions can be defined. TF binding sites can be found virtually everywhere in promoters but in individual promoters possible locations are much more restricted. A closer look reveals that the function of an AP-1 binding site often depends on the relative location and especially on the sequence context of the binding site. The AP-1 site in the above mentioned rat bone sialoprotein gene overlaps with a set of glucocorticoid responsive element (GRE, the DNA sequence that is bound by the glucocorticoid receptor which is a transcription factor) half sites which are crucial for the suppressive function.

The context of a TF-site is one of the major determinants of its role in transcription control. As a consequence of context requirements, TF sites are often grouped together and such functional groups have been described in many cases. A systematic attempt of collecting synergistic or antagonistic pairs of TF binding sites has been made with the COMPEL database [10, 11]. In many cases, a specific promoter function (e.g., a tissue-specific silencer) will require more than two sites. Promoter subunits consisting of groups of TF binding sites that carry a specific function independent of the promoter will be referred to as *promoter modules*. A more detailed definition of promoter modules has been given recently by Arnone and Davidson [12]. In summary, promoter modules contain several transcription factor binding sites which act together to convey a common function like tissue-specific expression. The organization of binding sites (and probably also of other elements) of a promoter module appears to be much more restricted than the apparent variety of TF sites and their distribution in the whole promoter suggests. Within a promoter module both sequential order and distance can be crucial for function indicating that these modules may be the critical determinants of a promoter rather than individual binding sites. Promoter modules are always constituted by more than one binding site. Since promoters can contain several modules that may use overlapping sets of binding sites, the conserved context of a particular binding site cannot be determined from the primary sequence. The corresponding modules must be detectable separately before the functional modular structure of a promoter or any other regulatory DNA region can be revealed by computer analysis. One well known general promoter module is the core promoter, which will be discussed in more detail below. However, the basic principles of modular organization are also true for most if not all other regulatory regions and are neither peculiar nor restricted to promoters.

4.2.4.1 Module properties of the core promoter

The core promoter module can be defined functionally by its capability to assemble the transcription initiation complex (see Figure 4.3) and orient it specifically towards the TSS of the promoter [13], defining the exact location of the TSS. Various combinations of about four distinguishable core promoter elements that constitute a general core promoter can achieve this. This module includes the TATA box, the initiator region (INR), an upstream activating element, and a downstream element. (The TATA box is a basic

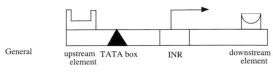

Core promoter module

Fig. 4.1
General structure of a polymerase II core promoter. Simultaneous presence of all four elements is not always essential.

INR = initiator region. The shapes above the bar symbolize additional protein binding sites and the arrow indicates the transcription start site.

transcription element, which is located about 20–30 nucleotides upstream of the actual transcription start site and is known to bind to the TATA box binding protein TBP.) However, this is also where the straightforward definition of a core promoter module ends because not all four elements are always required, or some elements can be too variable to be recognizable by current computer tools.

4.2.4.2 First group: TATA box containing promoters without known initiator

Successful positioning of the initiation complex can start at the TATA box containing promoters by the TFIID complex, which contains the TATA box binding protein as well as several other factors. Together with another complex of general transcription factors, termed TFIIB, this leads to the assembly of an initiation complex [14]. If an appropriate upstream TF binding site cooperates with the TATA box, no special initiator or downstream sequences might be required, which allows for the assembly of a functional core promoter module from just two of the four elements (Figure 4.1). This represents one type of a distinct core promoter that contains a TATA box, common among cellular genes in general.

4.2.4.3 Second group: TATA-less promoters with functional initiator

However, as it is known from a host of TATA-less promoters, the TATA box is by no means an essential element of a functional core promoter. An INR combined with a single upstream element has also been shown to be capable of specifically initiating transcription [15], although initiators cannot be clearly defined on the sequence level so far. Generally, a region of 10–20 nucleotides around the transcription start site is thought to represent the initiator. A remarkable array of four different upstream TF sites (SP1, AP-1, ATF, or TEF1) was shown to confer inducibility by T-antigen to this very simple promoter. T-antigen is a potent activating protein from a virus called SV40. This is an example of a TATA-less distinct promoter that can be found in several genes from the hematopoietic lineage (generating blood cells).

The third combination is called a composite promoter and consists of both a TATA box and an initiator. This combination can be found in several viral promoters and it has been shown that an additional upstream TF binding site can influence whether the TATA box or the initiator element will be determining the promoter properties [16]. The authors showed that upstream elements can significantly increase the efficiency of the INR in this combination while especially SP-1 sites made the TATA box almost obsolete in their example. The combination of TATA box with an INR had the general effect to induce resistance against the detrimental effects of a TFIIB mutant, which interfered with expression from TATA-only promoters. This is also an example for the more indirect effects of specific arrangements in promoters that may not be apparent unless special conditions occur.

The last group consists of so-called null-promoters which have neither a TATA box nor an initiator and rely exclusively on upstream and downstream elements [17].

Basically, at least the four different core promoter types detailed above have been identified so far, which all represent valid combinations of core promoter sites (reviewed in [17]). If the combinations involving upstream and downstream elements are also considered, a total of 7 possible core promoter modules are possible (most of which can be actually found in genes and consist of the four variants in Figure 4.2 adding upstream or downstream elements or both).

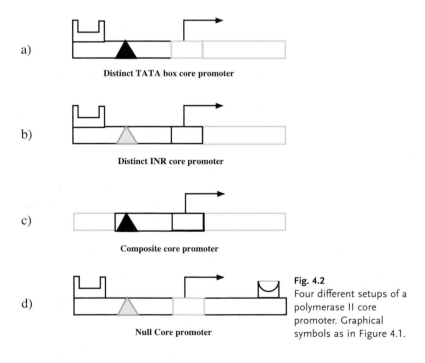

a)

Distinct TATA box core promoter

b)

Distinct INR core promoter

c)

Composite core promoter

d)

Null Core promoter

Fig. 4.2
Four different setups of a polymerase II core promoter. Graphical symbols as in Figure 4.1.

The only apparent common denominator of transcription initiation within a promoter would be that there must be at least one (or more) core promoter element within a certain region. This assumption is wrong. Both spacing and/or sequential order of elements within the core promoter module are of utmost importance regardless of the presence or absence of individual elements (as a rule, there appear to be some exceptions). Moreover, many distinct promoters have requirements for specific upstream or downstream elements and will function with their specific TF. Moving around the initiator, the TATA box and, to some extent, also upstream elements can have profound effects on promoter functions. For example, insertion of just a few nucleotides between the TATA box and an upstream TF binding site (MyoD) in the desmin gene promoter cuts the expression levels by more than half [18]. Moreover, the promoter structure can affect later stages of gene expression like splicing [19]. It was also shown for the rat beta-actin promoter that a few mutations around the transcription start site (i.e. within the initiator) can render that gene subject to translational control [20].

As a final note, the mere concept of one general TATA box and one general INR is an oversimplification. There are several clearly distinguishable TATA boxes in different promoter classes [21, 22, 23, 24] and the same is true for the INR region which also has several functionally distinct implementations as the glucocorticoid-responsive INR in the murine thymidine kinase gene [25], the C/EBP binding INR in the hepatic growth factor gene promoter [26], or the YY-1 binding INR [27].

Most of the principles of variability and restrictions detailed above for the core promoter modules are also true for other promoter modules that modify transcriptional efficiency rather that determining the start point of transcription as the core promoter does. The bottom line is that the vast majority of alternative arrangements of the elements that can be seen in a particular promoter might not contribute to the function of the promoter. Module-induced restrictions are not necessarily obvious from the primary sequences. Figure 4.3 shows a schematic pol II promoter with the initiation complex assembled which illustrates that it matters where a specific protein is bound to the DNA in order to allow for proper assembly of the molecular jigsaw puzzle of the initiation complex. This is not immediately obvious from inspection of promoter sequences because there exist several (but a strictly limited set of) alternative solutions to the assembly problem. As complicated as Figure 4.3 may appear, it still ignores all aspects of chromatin rearrangements and nucleosomal positions which also play an important role in transcription regulation. Stein et al. [28] has detailed an example of the profound influence of these effects on promoter-protein complex assembly and function for the ostecalcin promoter in a study. However, chromatin-related effects are not yet considered in any of the promoter prediction methods. Therefore, we do not go into any more details here.

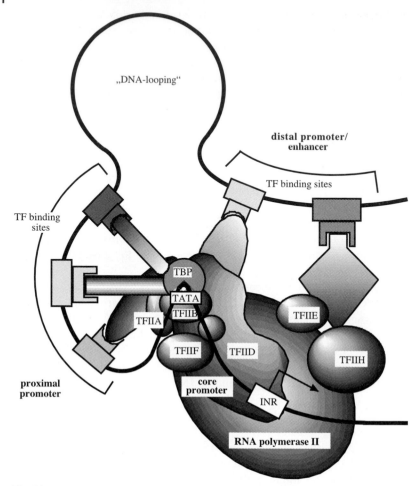

"DNA-looping"

distal promoter/
enhancer

TF binding sites

TF binding
sites

TBP

TATA

TFIIB

TFIIA

TFIIE

TFIIF

TFIID

TFIIH

proximal
promoter

core
promoter

INR

RNA polymerase II

Fig. 4.3
Transcription initation complex bound to a schematic
promoter

4.2.5
Bioinformatics models for the analysis and detection of regulatory regions

Algorithms used to analyze and detect regulatory regions are necessarily
based on some kind of usually simplified model of what a regulatory region
should look like. All of these models are inevitably compromising between
accuracy with respect to the biological model (the standard of truth) and
computational feasibility of the model. For example, a computational model
based on *a priori* three-dimensional structure prediction derived from

molecular dynamics using sophisticated force fields may be the most accurate model for a region but cannot be used for the analysis of real data due to excessive demand on computational resources. On the other hand a model based on simple sequence similarities detected by IUPAC consensus (see Chapter 3 for detailed definition) sequences can be easily used on a PC but results will usually not match the biological truth in an acceptable manner.

4.2.5.1 Statistical models
It has been noted several years ago that promoters and most likely also other regulatory regions like enhancers contain more transcription factor binding sites that non-regulatory sequences. An analysis of the relative frequencies of such sites within a sliding window therefore can yield some information about the potential regulatory character of a stretch of DNA, which is the prototype of simple statistical models. Several programs exist that rely to some extent on this type of statistics.

4.2.5.2 Mixed models
Of course, it is quite clear from section 4.1 that a pure statistical model is an oversimplification that will adversely affect the accuracy of prediction despite its attractive ease of implementation. Therefore, mixed models are also used that take at least some regional information into consideration and can be seen as statistical models split into compartments. Within the compartments solely statistical features are considered, but promoter organization is somewhat reflected by the arrangement of the compartments which represent different promoter regions.

4.2.5.3 Organizational models
The last category consists of models that try to closely follow the organizational principles of real regulatory regions. In order to accomplish that, individual promoter elements like transcription factor binding sites as well as their relative order and distances are encoded in a formal model which reflects the setup of a single promoter or a small group of functionally similar promoters. Although they are matching the biological situation best they are not yet suitable for widespread application. High quality sets of training sequences are required to generate such models because many parameters concerning nature, order and distance of elements have to be determined. This is a process far from being an easy task and limits application of such models to relatively few examples that require elaborate analysis.

4.3
Methods for element detection

4.3.1
Detection of transcription factor binding sites

Transcription factor binding sites are the most important elements within regulatory DNA regions like promoters or enhancers. The majority of the known transcription factors recognize short DNA stretches of about 10–15 nucleotides in length that show different degrees of internal variation. Successful detection of protein binding sites in DNA sequences always relies on precompiled descriptions of individual binding sites. Such descriptions are usually derived from a training set of four or more authentic binding sites. However, the criteria applied for the decision whether a site is authentic or not vary considerably among authors of different publications. One of the first approaches to define protein binding sites used IUPAC consensus sequences, which indicate the predominant nucleotide or nucleotide combination at each position in a set of example sequences. The IUPAC string TGASTCA indicates that the first three positions are most frequently T, G, and A while the fourth position may be C or G followed by T, C, and A in most cases. IUPAC consensus sequences became very popular as they are extremely easy to define from even a small set of sequences and their definition does not require more than a pencil and a sheet of paper.

However, IUPAC consensus sequences strongly depend on the sequence set used for definition. The final IUPAC consensus sequence remains arbitrary depending on the rules used to determine the consensus. Cavener defined some rules that we have used at GSF for several years now and, in our experience, IUPAC consensus sequences defined that way can be useful [29]. However, IUPAC consensus sequences may reject biologically functional binding sites due to a single mismatch (or an ill-defined IUPAC sequence).

The concept of nucleotide weight matrix (NWM) descriptions has been developed in the 1980s as an alternative to IUPAC strings [30, 31]. However, although weight matrices proved to be generally superior to IUPAC strings their biggest disadvantage is the absolute requirement for predefined matrices, which are more complicated to construct than IUPAC strings and require specific software. This delayed widespread use of weight matrices for almost a decade although the methods were principally available. They remained mostly unused because only a few special matrices had been defined [32]. The situation changed when in 1995 two (overlapping) matrix libraries for TF sites were compiled and became widely available almost simultaneously [33, 34]. Matrix Search [34] transformed the TRANSFAC database as complete as possible (starting at two binding sites for one factor) into matrices using a log-odds scoring approach. The MatInspector library is largely based on a stringent selection from the matrix table of the TRANSFAC database, including the matrices derived from the ConsInspector library [35, 36, 37]. and several genuine matrices. The Information Matrix Database was compiled from the TRANSFAC matrix table and the TFD. The MatInspector library is updated with each new TRANSFAC release while IMD has not been updated recently. Available matrix detection programs have been reviewed recently [36] and a comparison of these methods by application to a test set of sequences was published [37]. For convenience Table 4.1 summarizes which methods for the detection of transcription factor binding sites are available in the internet with emphasis on programs featuring a WWW-interface.

4.3.2
Detection of structural elements

Regulatory sequences are associated with a couple of other individual elements or sequence properties in addition to the factor binding sites. Among these are secondary structure elements like the HIV-1 TAR region (Trans-ActivatingRegion, which constitutes an RNA enhancer [38], cruciform DNA structures [39], or simple direct repeats [40]. Three-dimensional structures like curved DNA [41] also influence promoter function. Most of these elements can be detected by computer-assisted sequence analysis [42, 43, 44, 45] but none of them is really promoter specific and all such elements can be found frequently outside of promoters. The promoter or enhancer function arises from the combination of several elements that need to cooperate to exert transcription control which none of them can achieve alone. This also illustrates the main problem of promoter recognition. It is necessary to compile several individually weak signals into a composite signal, which then indicates a potential promoter without being overwhelmed by the combinatorial complexity of potential element combinations.

4.3.3
Assessment of other elements

Several methods employ statistical measures of sequence composition to include features of regulatory sequences, which cannot be described by the three types discussed above. This includes frequencies of oligonulceotides (dinucleotides, trinucleotides, and hexamers are used most frequently), CpG islands (CG dinucleotides are usually underrepresented in mammalian genomes except in part of coding and regulatory sequences. CpG islands are regions where the dinucleotide is NOT underrepresented [46], and periodicity of weak sequence patterns (AA, TT etc). Definitions of such elements are usually too weak to allow any significant contribution to current prediction programs. However, this situation might well change due to

Tab. 4.1
Internet accessible methods to detect promoter elements (transcription factor binding sites)

Program	Availability	Comments
MatInspector	http://www.gsf.de/cgi-bin/matsearch.pl	MatInspector matrix
	http://genomatix.gsf.de/cgi-bin/matinspector/matinspector.pl	library (includes TRANSFAC matrices)
SIGNAL SCAN	http://bimas.dcrt.nih.gov/molbio/signal/	IUPAC consensus library
MATRIX SEARCH	http://bimas.dcrt.nih.gov/molbio/matrixs/	IMD matrix library (TRANSFAC + TFD)
TFSearch	http://pdap1.trc.rwcp.or.jp/research/db/TFSEARCH.html	TRANSFAC matrices
TESS	http://agave.humgen.upenn.edu/utess/tess/	TRANSFAC matrices

the unprecedented amounts of continuous genomic sequences that become available in the course of the current genome sequencing projects.

4.4
Analysis of regulatory regions

Basically two different tasks can be distinguished in the analysis of regulatory regions. The first task is analysis aiming at the definition of common features based on sets of known regulatory sequences. This is a prerequisite for the definition of descriptions suitable for large-scale application for prediction of potential regulatory regions within new anonymous sequences which can be regarded as the second task.

4.4.1
Training set selection

One of the most important steps in comparative sequence analysis is the selection of suitable training sets of sequences. If a training set of promoters consists only of constitutively expressed sequences (constant level of expression, no or little regulation) little can be learned about any kind of tissue-specific expression regardless of the methods applied. Also inclusion of too many wrong sequences (e.g., that are no promoters at all) may prevent any meaningful analysis. Although this appears a bit trivial at first, it is a real issue when data are scarce and less well-characterized sequences have to used.

Control sets known not to be functionally similar to the training sets are about as important as the training sets themselves. However, true negative regions are even scarcer than known regulatory regions. Negative often means just "no positive functions found" which can also be due to failures or simply means that the sequences have not been tested at all. Therefore, statistical negative control sequences are often required. Random sequences can be easily generated but often are of limited use, as they do not represent several important features of natural DNA correctly. This includes underrepresented features (e.g., CpG islands), asymmetric features (e.g., strand specificity), or repetitive DNA elements. Selection of appropriate control sequences can be a major effort, but is also crucial for the validity of the evaluation of any method. Common problems with controls are either known or unknown biases in the control set or circularity problems, i.e. the training and the test sets of sequences are related or overlap. The availability of large continuous stretches of genomic DNA from the genome sequencing projects constantly improves this situation. Genomic sequences should always be the first choice for controls as they reflect the natural situation.

4.4.2
Statistical and biological significance

The quality of sequence pattern recognition is often optimized to improve the correlation of the methods with the data (positive and negative training sets). However, in most cases it is not possible to collect sufficient data to perform a rigorous correlation analysis. Therefore, bioinformatics methods often rely on statistical analysis of their training sequences and optimize for statistically most significant features. Unfortunately, this kind of optimization does not always reflect the evolutionary optimization of regulatory sequences that is always optimizing several features at once.

The dynamics of biological function often necessitates suboptimal solutions. For example, real sequences usually do not contain binding sites with the highest affinity for their cognate protein because binding AND dissociation of the protein are required for proper function. The perfect binding site would interfere with the dissociation and is therefore strongly selected against.

4.4.3
Context dependency

The biological significance of any sequence element is defined by the regulatory function it can elicit. This is usually dependent on a functional context rather than being a property of individual elements. Therefore, statistical significance of the features or scores of individual elements is neither necessary nor sufficient to indicate biological significance. Recognition of the functional context in an essentially linear molecule like DNA can be achieved by correlation analysis of individual elements, which became an important part of all semi-statistical or specific modeling approaches discussed below.

4.5
Methods for detection of regulatory regions

There are several methods available for the prediction of regulatory DNA regions in new sequence data. Table 4.2 lists methods available with a special focus on programs that provide a WWW-interface. Unfortunately, there is no "one-does-it-all" method and all methods have their individual strong and weak points. A program doing an excellent job in one case might be a complete failure in another case where other methods are successful. Therefore, we will describe a whole lot of methods without intending any rank by order of discussion. We will rather follow the functional hierarchy that appears to apply to the different regulatory regions.

Tab. 4.2
Available promoter/promoter regions prediction tools

Program	Availability	Comments
Promoter prediction		**WWW-accessible**
FunSiteP	http://transfac.gbf.de/dbsearch/funsitep/fsp.html	includes proximal promoter
NNPP	http://www.hgc.lbl.gov/projects/promoter.htm	core promoter
PromFD	http://beagle.colorado.edu/~chenq/Hypertexts/PromFD.html chenq@beagle.colorado.edu	includes proximal promoter for further infomation
Promoter Scan	http://biosci.umn.edu/software/proscan/promoterscan.htm	includes proximal promoter
TSSG/TSSW	http://dot.imgen.bcm.tmc.edu:9331/gene-finder/gf.html	includes proximal promoter triplet & hexamer frequencies
Core Promoter Finder	http://sciclio.cshl.org/genefinder/CPROMOTER/human.htm	discrimination analysis
	Download/inquire	
Audic/Claverie	audic@newton.cnrs-mrs.fr	Markov models
PromFind	ftp://iubio.bio.indiana.edu/molbio/ibmpc http://www.rabbithutch.com	hexamer frequencies for further information
XGRAIL	ftp://arthur.epm.ornl.gov/pub/xgrail	core promoter in gene context
Promoter module/region recognition		**WWW-accessible**
FastM library	http://www.gsf.de/cgi-bin/fastm.pl http://genomatix.gsf.de/cgi-bin/fastm2/fastm.pl	module of 2 TF sites uses MatInspector
TargetFinder	http://gcg.tigem.it/TargetFinder.html	module of 1 TF site combined with 1 annotated feature
	Download/inquire	
GenomeInspector	http://www.gsf.de/biodv/genomeinspector.html ftp://ariane.gsf.de/pub/unix/genomeinspector/	correlation analysis, e.g 2 TF binding sites
Muscle-specific regions	ficketjw@molbio.sbphrd.com	contact for download
Xlandscape	ftp://beagle.colorado.edu/pub/Landscape/xland.v1. tar.Z	word frequencies, not promoter specific

4.5.1
Types of regulatory regions

4.5.1.1 **Matrix attachment regions**

A chromatin loop is the region of chromosomal DNA located between two contact points of the DNA with the nuclear matrix marked by so-called Scaffold/Matrix Attachment Regions (S/MARs). The nuclear matrix is a mesh of proteins lining the inner surface of the nuclear envelope. Transcriptional regulation requires the association of DNA with this nuclear matrix, which retains a variety of regulatory proteins. S/MARs are composed of several elements including transcription factor binding sites, AT-rich stretches, potential cruciform DNA, and DNA-unwinding regions to name a few of the most important S/MAR elements. There is an excellent review on chromatin domains and prediction of MAR sequences by Boulikas [47], explaining S/MARs and their elements in detail. Kramer et al. [48] published a method to detect potential S/MAR elements in sequences and made the method available via WWW (http://www.ncgr.org/MAR-search/) [49]. Their method is based on a statistical compilation of the occurrence of a variety of S/MAR features (called rules). Accumulation of sufficient matches to these rules will be predicted as potential S/MAR regions. The specificity of the method depends critically on the sequence context of the potential S/MAR sequences. Therefore, results are difficult to evaluate by comparisons. However, so far this is the only method available to predict S/MAR regions. We developed another approach to define especially AT-rich MARs called SMARTest which is available on the web at http://genomatix.gsf.de-. SMARTest is based on a library of MAR-associated nucleotide weight matrices and determines S/MARs independent of any larger sequence context. Therefore, the method is suitable to test isolated S/MAR fragments. MARFinder and SMARTest are complementary and should be seen in combination rather than as alternatives.

4.5.1.2 **Enhancers/silencers**

Enhancers are regulatory regions that can significantly boost the level of transcription from a responsive promoter regardless of their orientation and distance with respect to the promoter as long as they are located within the same chromatin loop. Silencers are basically identical to enhancers and follow the same requirements but exert a negative effect on promoter activities. At present there are no specific programs to detect enhancers and silencers. However, programs designed to detect the internal organization of promoters are probably also suitable to detect enhancers and silencers since these elements often also show a similar internal organization as promoters.

4.5.1.3 **Promoters**

Promoters are DNA regions capable of specific initiation of transcription (start of RNA synthesis) and consist of three basic regions (See Figure 4.2). The part determining the exact nucleotide for transcriptional initiation is called the core promoter and is the stretch of DNA sequence where the RNA polymerase and its cofactors assemble on the promoter.

The region immediately upstream of the core promoter is called the proximal promoter and usually contains a number of transcription factor binding sites responsible for the assembly of an activation complex. This complex in turn recruits the polymerase complex. It is generally accepted that most proximal promoter elements are located within a stretch of about 250–500 nucleotides upstream of the actual transcription start site (TSS).

The third part of the promoter is located even further upstream and is called the distal promoter. This region usually regulates the activity of the core and the proximal promoter and also contains transcription factor binding sites. However, distal promoter regions and enhancers exhibit no principal differences. If a distal promoter region acts position and orientation independent it is called an enhancer.

4.5.2
Programs for recognition of regulatory sequences

There are several ways promoter recognition tools can be categorized. We will focus on the main principles and intended usage of the programs rather than technical details, as this will also be the main interest of experimentally working scientists. Two generally distinct approaches have been used so far in order to achieve *in silico* promoter recognition [50]. The majority of programs focused on *general promoter recognition*, which represents the first category. One group of programs in this category (see below) concentrates on recognition of core promoter properties and infers promoter location solely on that basis while the other group consists of programs that take into account also the proximal promoter region of about 250–300 nucleotides upstream of the TSS. General recognition models are usually based on training sets derived from the Eukaryotic Promoter Database (EPD [51]) and various sets of sequences without known promoter activities. The EPD is an excellent collection of DNA sequences that fulfil two conditions: They have been shown experimentally to function as promoters and the transcription start site is known. The beauty of these approaches is their generality which does not require any specific knowledge about a particular promoter in order to make a prediction. This appears ideal for the analysis of anonymous sequences for which no *a priori* knowledge is available. The bad news is that the specificity of all such general approaches implemented so far is limited. The inevitable huge burden of false-positive predictions in longer sequences precludes large scale application of such approaches to genomic sequences.

The second category of tools aims at *specific promoter recognition* relying on more detailed features of promoter subsets like combinations of individual elements. The beauty of this approach is excellent specificity, which is extremely helpful if only promoters of a certain class are of interest or megabases of sequences have to be analyzed. The bad news here is limited applicability, i.e., each promoter group or class requires a specifically pre-defined model before sequences can be analyzed for these promoters. This may result in a huge number of false negatives in large-scale analysis.

We will briefly discuss individual methods in these two categories with emphasis on the implementation of the biological principles of promoter features. Recently, a practical comparison of the majority of available tools based on general promoter models has been carried out, which has shown that none of these methods is clearly superior to its peers [50]. Therefore, I will not go into details about performance of the methods here (see Chapter 3 for details).

4.5.2.1 Programs based on statistical models (general promoter prediction)

These programs aim at the detection of pol II promoters by a precompiled general promoter model that is part of the method. Despite the complicated modular structure of promoters outlined above there is a solid rational basis for this general model. All promoters must have a functional core promoter module often containing a TATA box which is the prime target of the majority of the general promoter prediction tools. This is also one of the reasons why some programs confine their analysis to the core promoter region which avoids problems with the much more diverse proximal regions. General models that include the proximal region consequently treat this part of the promoter as a purely statistical problem of TF binding site accumulations, sometimes fine tuned by some sort of weighting based on occurrence frequencies of TF binding sites in promoters as compared to a negative sequence set.

However, the cost of generality without exception is a huge number of false positive predictions (typically about one match in 10 000–30 000 nucleotides). Sacrifice of a considerable percentage of true promoters (30% or more) is also a necessity to maintain at least some specificity. There also is inevitably no clue what kind of promoter was detected in case of a match.

4.5.2.1.1 PromoterScan

Several of the general promoter prediction programs followed the basic design of Prestridge who used the Eukaryotic Promoter Database (EPD) by Bucher [51] to train his software for promoter recognition. His program *PromoterScan* was the first published method to tackle this problem [52].

He utilized primate non-promoter sequences from GenBank as a negative training set and included the proximal promoter region in the prediction. The program uses individual profiles for the TF binding sites indicative of their relative frequency in promoters to accumulate scores for DNA sequences analyzed. PromoterScan employs the SignalScan IUPAC library of TF binding sites [53], introducing a good deal of biological knowledge into the method, although modular organization of the proximal region is necessarily ignored. Results of the first version were combined with the Bucher NWM for the TATA box, which served as a representation of the core promoter module [32].

There is now a new version of *PromoterScan* available, PromoterScan 2.0. This new version is supposed to provide more information from inside the "black box" which a promoter used to be for version 1.0 and is also able to compare a predicted promoter to EPD promoter sequences on the basis of the pattern of TF sites. Although this moves promoter comparisons with PromoterScan 2.0 effectively closer to the specific recognition of individual elements this approach is not used for the initial promoter prediction. PromoterScan is available via WWW, which is a definitive advantage for occasional promoter testing.

4.5.2.1.2 **PromFD**

This program [54], extends the model behind *PromoterScan* in two ways. One major difference is use of the IMD matrix library [34] instead of the SIGNAL SCAN IUPAC strings. The other feature is the inclusion of patterns of strings of 5–10 nucleotides in length that were found to be overrepresented in the training set of promoters. Basically, the intrinsic model is the same as in *PromoterScan* but the overrepresented strings may account for some so far unknown binding sites that are missing in the libraries. PromFD requires local installation and is not available via a WWW-interface. There have been no updates so far.

4.5.2.1.3 **TSSG/TSSW**

Two other methods (*TSSG and TSSW*) also including proximal promoter regions are available via WWW and share the basic algorithm. They center on detection of a TATA box as most prominent part of the core promoter. Promoter prediction is then based on the score of the TATA box, and nucleotide triplet distributions around the putative TSS. In analogy to Prom-Find (see below) hexamer frequencies in three 100 nucleotide wide windows upstream of the putative TSS are also considered. These data are combined with potential TF binding sites, which are predicted either based on a TFD derived compilation by Prestridge (in TSSG), or on the TRANSFAC database [11] in TSSW. Fickett and Hatzigeorgiou did not report significant improvements in the predictive capabilities by inclusion of triplets and hexamers [50]. However, the general aim of these programs is not independent promoter prediction but to assist in finding exon-intron structures of complete genes.

4.5.2.1.4 XLandscape

The group of Stormo developed another method, which is called *Xlandscape* and essentially determines nucleotide strings of various lengths called words which are specifically associated with promoters, exons or introns [55]. Then a score for a sequence is determined indicating by the different promoter, exon, and intron scores whether a particular region is likely to belong to one of these groups. Although this method was not developed for promoter prediction it fares about as well as all other general promoter prediction programs although it completely ignores even the core promoter module. The program requires local installation.

4.5.2.1.5 PromFind

This program [56] is similar to *Xlandscape*. It relies on the difference in hexamer frequencies between promoters and regions outside of promoters which is a more restricted view as compared to the landscaping approach. The advantage is less computational complexity. The program was also trained on sequences from the EPD and corresponding coding and non-coding regions outside the promoters taken from GenBank. Any region in which the ratio of promoter to non-promoter hexamer-frequencies reaches a threshold is considered and only the region where this measure is maximal is defined as the promoter. This again relies on pure statistics disregarding any organizational features of promoters. I am not aware of any updates since the initial publication.

4.5.2.1.6 NNPP

This program [57] utilizes time delay neural networks to locate a TATA box combined with an initiator region and thus is a representative of the second subgroup which focuses on the minimum promoter region. Although the program does allow for variable spacing between the elements, especially distinct and null promoters will pose principal problems for this method as it includes some modeling of a TATA box. An improved version of the program was published recently and the program features a lower number of false negatives (about 1 match/kb). This is good news if a short region of DNA is to be analyzed (length should be less than 2 kb). However, long DNA regions pose a problem, as the number of false positives becomes overwhelming. The program is available via a WWW-interface (http://www-hgc.lbl.gov/projects/promoter.html).

4.5.2.1.7 CorePromoter Search

Michael Zhang [58] published a new method to detect TATA-box containing core promoters by discrimination analysis. This method is available via a WWW-interface, which already requires restriction of the sequence length to 1 kb. Core Promoter Search and NNPP are alternative implementations of a similar general promoter model and can be applied in parallel.

4.5.2.2 Programs utilizing mixed models

These programs also rely on statistical promoter models but include directly or indirectly some organizational features of promoters placing them in between the pure statistical models and attempts to approximate the biologically important structured organization of promoters.

4.5.2.2.1 FunSiteP

This program [59] also takes into account proximal promoter regions and utilizes a collection of TF binding sites [60] with which a promoter set taken from EMBL (472 promoters) was analyzed. From this analysis a weight matrix of TF binding site localization was derived representing regions in promoters with lower or higher concentrations of TF binding sites. FunSiteP not only reports potential promoter matches but also assigns them to one of seven promoter classes. These were taken from Bucher's definition (from EPD) and consist of small nuclear RNAs, structural proteins, storage and transport proteins, enzymes, hormones, growth factors, and regulatory proteins, stress or defense related proteins, and unclassified proteins. Although these classes are very broad as defined by biological function, they represent an attempt towards more specific promoter recognition. FunSiteP is also available via a WWW-interface.

4.5.2.2.2 Audic/Claverie

A program designed by Audic and Claverie [61] uses Markov models of vertebrate promoters generated again by training on the EPD and non-promoter sequences outside of the promoters. Markov models principally allow for the inclusion of organizational features and the nature of the training set determines whether this becomes part of the model or not. We describe the program in this section because the many different promoters in EPD most likely cause these Markov models to be more general than specific.

4.5.2.3 Programs relying on organizational models

This category of methods introduces the functional context in form of heuristic rules or tries to learn the context from comparative sequence analysis.

These methods emphasize specific modeling of promoters or promoter substructures rather than general recognition. Therefore, it is not possible to directly assess the promoter prediction capabilities of these methods. However, in many cases recognizing a common substructure between promoters can be very helpful especially for experimental design.

4.5.2.3.1 **FastM**

This method was derived from the program ModelGenerator [22] and takes advantage of the existence of NWM libraries. It can be accessed via a WWW-interface (http://www.genomatix.de "free services") and allows for a straightforward definition of any modules of two TF binding sites by simple selection from the MatInspector Library [33]. This now enables definition and detection of wide variety of synergistic TF binding site pairs. These pairs often are functional promoter modules conferring a specific transcriptional function to a promoter [10, 62]. *FastM* models of two binding sites can successfully identify promoters sharing such composite elements but are not promoter specific. Composite elements can also be located in enhancers or similar structures. A commercial version of FastM is available that enables definition of complete, highly specific promoter class models including up to 10 individual elements.

4.5.2.3.2 **TargetFinder**

Another approach aiming at modeling of promoter substructures consisting of two distinct elements is TargetFinder [63]. This method combines TF binding sites with features extracted from the annotation of a database sequence to allow selective identification of sequences containing both features within a defined length. The advantage is that TargetFinder basically also follows the module-based philosophy but allows inclusion of features that have been annotated by experimental work for which no search algorithm exists. Naturally, this excludes analysis of new anonymous sequences. The program is accessible via a WWW-interface (http://gcg.tigem.it/TargetFinder.html).

It should be mentioned here that Fickett also employed the idea of a two TF binding site module to successfully detect a subclass of muscle-specific regulatory sequences governed by a combination of MEF2 and MyoD [64]. However, this was also a very specific approach and no general tool resulted out of that work. The MEF2/MyoD model can be used to define a corresponding module with FastM. Wasserman and Fickett [65] recently published a modeling approach based on clustering of a preselected set of NMW (defined in this study) correlated with muscle-specific gene expression. They were able to detect about 25% of the muscle-specific regulatory regions in sequences outside their training set and more than 60% in their training set. They classify their method as regulatory module detection. However, their results suggest that they probably detect a collection of different more specific modules with respect to the definition given above. Although the method is not promoter-specific and the specificity is moderate, it is a very interesting approach, which has potential for further development. The authors will make the non-commercial software used in their approach available on request (contact J. Fickett).

Generally, this group of methods achieves much higher specificity than programs following general models. However, the price for this increase in specificity is usually restriction of the promoter models to a small subset (class) of promoters.

4.5.2.4 The organizational model of histone H1 promoters

The specific modeling of a promoter class can be demonstrated on a well-known example from the cell-cycle regulation. Histone genes are required during the DNA replication and they show up as a group of coregulated genes in array analyses [66]. This suggests that they might have a common promoter structure. Histone H1 gene promoters for example can be found in the Eukaryotic Promoter Database (EPD, [51]).

The model for Histone H1 genes (Figure 4.4) was based on only 9 training sequences taken from homology group 17 of the EPD. They share a 100 bp fragment around the transcription start site but not much similarity elsewhere in the 600 nucleotides (standard length of EPD promoter sequences). A model containing 5 different transcription factor binding sites was derived from the set of sequences shown in Figure 4.4 using GEMS Launcher (Genomatix Software, Munich).

Notably, the model generated contains one factor, E2F that is known to be involved in cell cycle regulation of genes [67]. The model appears to be very selective and only one match per 12 million base pairs of the mammalian sections of the EMBL database was found. Almost all matches (total of 59) are known Histone H1 genes, except in the human section where the majority of matches (25) are within anonymous sequences. However, given the extraordinary specificity throughout the mammalian sections, it seems safe to assume that most of the unknown matches within the human database section actually identified new so far unannotated genes, that are subject to

Fig. 4.4
Organizational model of histone H1 promoters containing six transcription factor binding sites.

a similar regulation as histones (There is no direct evidence that these are histone genes). However, this is a working hypothesis which needs to be verified.

4.6
Annotation of large genomic sequences

Almost all of the methods discussed above were developed before the databases started to be filled with sequence contigs exceeding 100 000 nucleotides in length. The complete human genome draft now contains more than 3 billion nucleotides. This changes the paradigm for sequence annotation. While complete annotation remains an important goal, specific annotation becomes mandatory when even individual sequences exceed the capabilities of researchers for manual inspection. Annotation of genomic sequences has to be fully automatic in order to keep pace with the rate of generation of new sequences. Simultaneously, annotations are embedded into a large natural context rather than residing within relatively short isolated stretches of DNA. This has several quite important consequences.

4.6.1
The balance between sensitivity and specificity

I will confine the discussion here to regulatory regions but the problems are general. A very sensitive approach will minimize the amount of false negative predictions and thus is oriented towards a complete annotation. However, this inevitably requires accepting large numbers of false positive hits, which easily outnumber the true positive predictions by one order of magnitude.

In order to avoid this methods can be designed to yield the utmost specificity (e.g., specific promoter modeling as discussed above). Here, the catch is that inevitably a high number of false-negative results, which also may obscure 70% to 90% of the true positive regions.

A little thought experiment demonstrates the dilemma of current methods for annotation of sequences. Assume we are analyzing a region of 3 billion bp of contiguous human genomic DNA (the total genome). The human genome is estimated to contain about $60 000 \pm 30 000$ promoters. On average general promoter prediction programs detect one promoter per 1000 to 10 000 nucleotides according to the respective authors. Assuming an optimistic value (1/10 000) this would result in about 300 000 predictions. Given the true match rate determined by Fickett and Hatzigeorgiou [50] of less than 20% on average the result would be 20 000 true predictions versus 280 000 or more than 90% false positives.

On the other hand specific methods were shown to produce more that 50% true positive matches in their total output [21] but recognize just a small fraction of all promoters. A single specific model like the actin class model [21] matches about once every 2.5 million bps and thus would yield a total of 1200 matches, 600 of which can be expected to be true. This is great in terms of specificity, but loosing more than 90% of the true promoters present is certainly far from what sequence annotation is aiming at. The existence of 100 or, most likely, even more promoter models of the specificity of the actin model would be required to achieve specific recognition of most of the promoters present within the genome. Definition of the required number of specific models based on current technologies is not a feasible task. Therefore, new developments have already been initiated to overcome the current obstacles.

The numbers mentioned above are necessarily very rough estimates. However, two- or three-fold variations would not change the general results. It is quite evident that functional promoter analysis in laboratories is capable of dealing with several hundred or even thousand predicted regions while predicting several hundred thousand or even millions of regions remains out of reach. It is safe to assume that further improvement of laboratory high throughput technologies and enhancement of the specificity of promoter recognition *in silico* will meet somewhere in the future to close the gap in our knowledge about the functional regulation of the genome. It is also quite clear from the past and present developments that bioinfomatics will probably cover significantly more than half of that path.

4.6.2
The larger context

There will be help on the way towards more specific modeling of functional regions in genomic sequences. The almost unlimited natural context of regions in genomic sequences will allow for completely new approaches to comparative sequences analysis, which has already proven to be the most powerful approach in bioinformatics of genomic sequences (e.g., in detection of new sequence elements). Comparative analysis will be instrumental in determining the anatomy of regulatory networks including MARs, LCRS, enhancers, and silencers in addition to the promoter sequences. It can be safely assumed that context-sensitive sequence analysis will prevail in the long run over any methods dealing with short isolated sequences.

4.6.3
Aspects of comparative genomics

The context information is by no means restricted to other regions within the same genome. Approaches based on comparative genomics employing sequence information from two or more genomes proved to be very powerful as they introduce phylogenetic aspects into the analysis. Phylogenetic footprints have already shown this [67]. This way, discrimination between noise and functionally conserved regions can be improved resulting in easier detection of regions useful for predictive efforts.

4.6.4
Analysis of data sets from high throughput methods

Another field to which the bioinformatics of regulatory DNA regions can be expected to contribute significantly is the analysis of results from high throughput experiments in expression analysis (e.g., all forms of expression arrays). Due to the discontinuous nature of regulatory regions there is no way of deducing common regulatory features from the expression data directly that are usually based on coding regions. However, the general availability of the corresponding genomic regulatory regions for many (and very soon all) of the genes analyzed in an expression array experiment enables attempts to elucidate the genomic structures underlying common expression patterns of genes. Expression arrays (described in detail in Chapter 6, Volume 2) directly deliver information, *which* genes are *where* expressed under the conditions tested. However, they cannot provide any clue to *why* this happens or how the same genes would behave under yet untested conditions. Identification of functional features by comparative sequence analysis (e.g., promoter modules) can reveal different functional subgroups of promoters despite common regulation under specific conditions. Consequently, the detection of known functional modules can suggest expression patterns under yet untested conditions. Moreover, the organizational structures of promoters can also be used to identify additional potential target genes either within the same organism in other genomes or via comparative genomics. Given the exponential number of possibilities for combinations of conditions, bioinformatics of regulatory sequences will also become instrumental for the rational design of expression arrays as well as for selection of experimental conditions.

4.7
Conclusions

The experimental dissection of functional mechanisms of transcription control has gained an enormous momentum during recent years. The ever rising number of publications on this topic bears witness to this development which found one hallmark manifestation in the introduction of a new section in the Journal of Molecular and Cellular Biology entirely devoted to analysis of transcription control. The complex interleaved networks of transcription control certainly represent one of the cornerstones on which to build our understanding of how life functions, in terms of embryonic development, tissue differentiation, and maintenance of the shape and fitness of adult organisms throughout life. This is also the reason why both the experimental analysis as well as the bioinformatics of transcription control will move more and more into the focus of medical/pharmaceutical research. A considerable number of diseases are directly or indirectly connected to alterations in cellular transcription programs (e.g., most forms of cancer). Furthermore, many drugs influence transcription control via signaling pathways (triggering transcription factors) which could also be connected to certain side effects of drugs. The human genome sequencing project will provide us with a complete catalog of the components of a human probably within a year. This will constitute a blueprint of the material basis of a human. However, only the analysis of the regulatory part of the genome and the corresponding expression patterns and the complex metabolic networks will provide deeper insight into how the complex machinery called life is working. Definition and detection of regulatory regions by bioinformatics will contribute to this part of the task.

References

1 WERNER, T. **(1999)** Identification and characterization of promoters in eukaryotic DNA sequences Mammalian Genome, 10, 168–175.

2 KODADEK, T. **(1998)** Mechanistic parallels between DNA replication, recombination and transcription. *Trends Biochem. Sci.* 23, 79–83.

3 ROTHENBERG, E. V., WARD, S. B. **(1996)** A dynamic assembly of diverse transcription factors integrates activation and cell-type information for interleukin 2 gene regulation. *Proc. Natl. Acad. Sci. USA* 93, 9358–9365.

4 IOSHIKHES, I., BOLSHOY, A., DERENSHTEYN, K., BORODOVSKY, M., TRIFONOV, E. N. **(1996)** Nucleosome DNA sequence pattern revealed by multiple alignment of experimentally mapped sequences. *J. Mol. Biol.* 262, 129–139.

5 SLOAN, L. S., SCHEPARTZ, A. **(1998)** Sequence determinants of the intrinsic bend in the cyclic AMP response element. *Biochemistry* 37, 7113–7118.

6 YAMAUCHI, M., OGATA, Y., KIM, R. H., LI, J. J., FREEDMAN, L. P., SODEK, J. **(1996)** AP-1 regulation of the rat bone sialoprotein gene transcription is mediated through a TPA response element within a glucocorticoid response unit in the gene promoter. *Matrix Biol.* 15, 119–130.

7 SAP, J., MUÑOZ, A., SCHMITT, J., STUNNENBERG, H., VENNSTRÖM, B. **(1989)** Repression of transcription mediated at a thyroid homone response element by the v-erb-A oncogene product. *Nature* 340, 242–244.

8 BOUTILLIER, A. L., MONNIER, D., LORANG, D., LUNDBLAD, J. R., ROBERTS, J. L., LOEFFLER, J. P. **(1995)** Corticotropin-releasing hormone stimulates proopiomelanocortin transcription by cFos-dependent and -independent pathways: Characterization of an AP1 site in exon 1. *Molecular Endocrinol* 9, 745–755.

9 BERGERS, G., GRANINGER, P., BRASELMANN, S., WRIGHTON, C., BUSSLINGER, M. **(1995)** Transcriptional activation of the fra-1 gene by AP-1 is mediated by regulatory sequences in the first intron. *Mol Cell Biol* 15, 3748–3758.

10 KEL, O. V., ROMASCHENKO, A. G., KEL, A. E., WINGENDER, E., KOLCHANOV, N. A. **(1995)** A compilation of composite regulatory elements affecting gene transcription in vertebrates. *Nucleic Acids Res* 23, 4097–4103.

11 HEINEMEYER, T., WINGENDER, E., REUTER, I., HERMJAKOB, H., KEL, A. E., KEL, O. V., IGNATIEVA, E. V., ANANKO, E. A., PODKOLODNAYA, O. A., KOLPAKOV, F. A., PODKOLODNY, N. L., KOLCHANOV, N. A. **(1998)** Databases on transcriptional regulation: TRANSFAC, TRRD and COMPEL. *Nucleic Acids Res* 26, 362–367.

12 ARNONE, M. I., DAVIDSON, E. H. **(1997)** The hardwiring of development: organization and function of genomic regulatory systems. *Development* 124, 1851–1864.

13 ZAWEL, L., REINBERG, D. **(1995)** Common themes in assembly and function of eukaryotic transcription complexes. *Annu. Rev. Biochem.* 64, 533–561.

14 CONAWAY, J. W., CONAWAY, R. C. **(1991)** Initiation of Eukaryotic Messenger RNA Synthesis. *J Biol Chem* 266, 17721–17724.

15 GILINGER, G., ALWINE, J. C. **(1993)** Transcriptional activation by simian virus-40 large T-Antigen – requirements for simple promoter structures containing either TATA or initiator elements with variable upstream factor binding sites. *J Virol* 67, 6682–6688.

16 COLGAN, J., MANLEY, J. L. **(1995)** Cooperation between core promoter elements influences transcriptional activity in vivo. *Proc Natl Acad Sci USA* 92, 1955–1959.

17 NOVINA, C. D., ROY, A. L. **(1996)** Core promoters and transcriptional control. *Trends in Genetics* 12, 351–355.

18 LI, H., CAPATANAKI, Y. **(1994)** An E box in the desmin promoter cooperates with the E-box and MEF-2 sites of a distal enhancer to direct muscle-specific transcription. *EMBO J* 13, 3580–3589.

19 CRAMER, P., PESCE, C. G., BARALLE, F. E., KORNBLIHTT, A. R. **(1997)** Functional association between promoter structure and transcript alternative splicing. *Proc Natl Acad Sci USA* 94, 11456–11460.

20 BIBERMAN, Y., MEYUHAS, O. **(1997)** Substitution of just five nucleotides at and around the transcription start site of rat beta-actin promoter is sufficient to render the resulting transcript a subject for translational control. *FEBS Lett* 405, 333–336.

21 FRECH, K., QUANDT, K., WERNER, T. **(1998)** Muscle actin

genes: A first step towards computational classification of tissue specific promoters. *In Silico Biol.* 1, 129–138.

22 FRECH, K., DANESCU-MAYER, J., WERNER, T. **(1997a)** A novel method to develop highly specific models for regulatory units detects a new LTR in GenBank which contains a functional promoter. *J Mol Biol* 270, 674–687.

23 FRECH, K., WERNER, T. **(1996a)** Specific modelling of regulatory units in DNA sequences. In Pacific Symposium on Biocomputing '97, R. B. Altman, A. K. Dunker, L. Hunter and T. E. Klein eds. (World Scientific), 151–162.

24 FRECH, K., BRACK-WERNER, R., WERNER, T. **(1996b)** Common modular structure of lentivirus LTRs. *Virology* 224, 256–267.

25 RHEE, K., THOMPSON, E. A. **(1996)** Glucocorticoid regulation of a transcription factor that binds an initiator-like element in the murine thymidine kinase (Tk-1) promoter. *Molecular Endocrinol* 10, 1536–1548.

26 JIANG, J. G., ZARNEGAR, R. **(1997)** A novel transcriptional regulatory region within the core promoter of the hepatocyte growth factor gene is responsible for its inducibility by cytokines via the C/EBP family of transcription factors. *Mol Cell Biol* 17, 5758–5770.

27 USHEVA, A., SHENK, T. **(1996)** YY1 transcriptional initiator: Protein interactions and association with a DNA site containing unpaired strands. *Proc Natl Acad Sci USA* 93, 13571–13576.

28 STEIN, G. S., VANWIJNEN, A. J., STEIN, J., LIAN, J. B., MONTECINO, M. **(1995)** Contributions of nuclear architecture to transcriptional control. *Int. Rev. Cytol.* 162A, 251–278.

29 CAVENER, D. R. **(1987)** Comparison of the consensus sequence flanking translational start sites in Drosophila and vetrebrates. *Nucleic Acids Res* 15, 1353–1361.

30 STADEN, R. **(1984)** Computer methods to locate signals in nucleic acid sequences. *Nucleic Acids Res.* 12, 505–519.

31 STORMO, G. D., HARTZELL III, G. W. **(1989)** Identifying protein-binding sites from unaligned DNA fragments. *Proc Natl Acad Sci USA* 86, 1183–1187.

32 BUCHER, P. **(1990)** Weight matrix description of four eukaryotic RNA polymerase II promoter elements derived from 502 unrelated promoter sequences. *J Mol Biol* 212, 563–578.

33 QUANDT, K., FRECH, K., KARAS, H., WINGENDER, E., WERNER, T. **(1995)** MatInd and MatInspector: New fast and versatile tools for detection of consensus matches in nucleotide sequence data. *Nucleic Acids Res* 23, 4878–4884.

34 CHEN, Q. K., HERTZ, G. Z., STORMO, G. D. **(1995)** MATRIX SEARCH 1.0: A computer program that scans DNA sequences for transcriptional elements using a database of weight matrices. *Comp Appl Biosci* 11, 563–566.

35 FRECH, K., HERRMANN, G., WERNER, T. **(1993)** Computer-assisted prediction, classification, and delimitation of protein binding sites in nucleic acids. *Nucleic Acids Res.* 21, 1655–1664.

36 FRECH, K., QUANDT, K., WERNER, T. **(1997b)** Finding protein-binding sites in DNA sequences: The next generation. *Trends Biochem Sci* 22, 103–104.

37 FRECH, K., QUANDT, K., WERNER, T. (1997c) Software for the analysis of DNA sequence elements of transcription. *Comp Appl Biosci* 13, 89–97.

38 BOHJANEN, P. R., LIU, Y., GARCIABLANCO, M. A. (1997) TAR RNA decoys inhibit Tat-activated HIV-1 transcription after preinitiation complex formation. *Nucleic Acids Res* 25, 4481–4486.

39 WANG, W. D., CHI, T. H., XUE, Y. T., ZHOU, S., KUO, A., CRABTREE, G. R. (1998) Architectural DNA binding by a high-mobility-group/kinesin-like subunit in mammalian SWI/SNF-related complexes. *Proc Natl Acad Sci USA* 95, 492–498.

40 BELL, P. J. L., HIGGINS, V. J., DAWES, I. W., BISSINGER, P. H. (1997) Tandemly repeated 147 bp elements cause structural and functional variation in divergent MAL promoters of Saccharomyces cerevisiae. *Yeast* 13, 1135–1144.

41 KIM, J., KLOOSTER, S., SHAPIRO, D. J. (1995) Intrinsically bent DNA in a eukaryotic transcription factor recognition sequence potentiates transcription activation. *J Biol Chem* 270, 1282–1288.

42 CHETOUANI, F., MONESTIE, P., THEBAULT, P., GASPIN, C., MICHOT, B. (1997) ESSA: an integrated and interactive computer tool for analysing RNA secondary structure. *Nucleic Acids Res* 25, 3514–3522.

43 SCHUSTER, P., STADLER, P. F., RENNER, A. (1997) RNA structures and folding: From conventional to new issues in structure predictions. *Curr Op Struct Biol* 7, 229–235 STADEN, R. (1984) Computer methods to locate signals in nucleic acid sequences. *Nucleic Acids Res* 12, 505–519.

44 NAKAYA, A., YAMAMOTO, K., YONEZAWA, A. (1995) RNA secondary structure prediction using highly parallel computers. *Comp Appl Biosci* 11, 685–692.

45 NIELSEN, D. A., NOVORADOVSKY, A., GOLDMAN, D. (1995) SSCP primer design based on single-strand DNA structure predicted by a DNA folding program. *Nucleic Acids Res* 23, 2287–2291.

46 SHAGO, M., GIGUERE, V. (1996) Isolation of a novel retinoic acid-responsive gene by selection of genomic fragments derived from CpG-island-enriched DNA. *Mol Cell Biol* 16, 4337–4348.

47 BOULIKAS, T. (1996) Common structural features of replication origins in all life forms. *J Cell Biochem* 60, 297–316.

48 KRAMER, J. A., SINGH, G. B., KRAWETZ, S. A. (1996) Computer-assisted search for sites of nuclear matrix attachment. *Genomics* 33, 305–308.

49 SINGH, G. B., KRAMER, J. A., KRAWETZ, S. A. (1997) Mathematical model to predict regions of chromatin attachment to the nuclear matrix. *Nucleic Acids Res.* 25, 1419–1425.

50 FICKETT, J. W., HATZIGEORGIOU, A. C. (1997) Eukaryotic promoter recognition. *Genome Res* 7, 861–878.

51 PERIER, R. C., JUNIER, T., BUCHER, P. (1998) The Eukaryotic Promoter Database EPD. *Nucleic Acids Res* 26, 353–357.

52 PRESTRIDGE, D. S. (1995) Predicting Pol II promotor sequences using transcription factor binding sites. *J Mol Biol* 249, 923–932.

53 Prestridge, D. S. (1991) SIGNAL SCAN: a computer program that scans DNA sequences for eukaryotic transcriptional elements. *Comp Appl Biosci* 7, 203–206.

54 Chen, Q. K., Hertz, G. Z., Stormo, G. D. (1997) PromFD 1.0: A computer program that predicts eukaryotic pol II promoters using strings and IMD matrices. *Comp Appl Biosci* 13, 29–35.

55 Levy, S., Compagnoni, L., Myers, E. W., Stormo, G. D. (1998) Xlandscape: the graphical display of word frequencies in sequences. *Bioinformatics* 14: p. 74–80.

56 Hutchinson, G. B. (1996) The prediction of vertebrate promoter regions using differential hexamer frequency analysis. *Comp Appl Biosci* 12, 391–398.

57 Reese, M. G., Harris, N. L., Eeckman, F. H. (1996) Large Scale Sequencing Specific Neural Networks for Promoter and Splice Site Recognition. In: Proceedings of the Pacific Symposium on Biocomputing, Eds Hunter l., Klein T.

58 Zhang, M. Q. (1998) Identification of human gene core promoters in silico. *Genome Res.* 8, 319–326.

59 Kondrakhin, Y. V., Kel, A. E., Kolchanov, N. A., Romashchenko, A. G., Milanesi, L. (1995) Eukaryotic promoter recognition by binding sites for transcription factors. *Comp Appl Biosci* 11, 477–488.

60 Faisst, S., Meyer, S. (1992) Compilation of Vertebrate-Encoded Transcription factors. *Nucleic Acids Res* 20, 3–26.

61 Audic, S., Claverie, J. M. (1997) Detection of Eukaryotic promoters using Markov transition matrices. *Comp Chem* 21, 223–227.

62 Klingenhoff, A., Frech, K., Quandt, K., Werner, T. (1999) Functional promoter modules can be defected by formal models independent of overall nucleoside sequence similarity. *Bioinformatics* 15, 180–186.

63 Lavorgna, G., Guffanti, A., Borsani, G., Ballabio, A., Boncinelli, E. (1999) TargetFinder: searching annotated sequence databases for target genes of transcription factors. *Bioinformatics* 15, 172–173.

64 Fickett, J. W. (1996) Coordinate positioning of MEF2 and myogenin binding sites (Reprinted from Gene-Combis, vol 172, pg GC19–GC32, 1996). *Gene* 172, GC19–GC32.

65 Wassermann, W. W., Fickett, J. W. (1998) Identification of regulatory regions which confer muscle-specific gene expression. *J Mol Biol* 278, 167–181.

66 Eisen, M. B., Spellman, P. T., Brown, P. O., Botstein, D. (1998) Cluster analysis and display of genome-wide expression patterns. *Proc Natl Acad Sci USA* 95, 14863–14868.

67 Duret, L., Bucher, P. (1997) Searching for regulatory elements in human noncoding sequences. *Curr. Op. Struct. Biol.* 7, 399–406.

68 Yang, R., Muller, C., Huynh, V., Fung, Y. K., Yee, A. S., Koeffler, H. P. (1999) Functions of cyclin A1 in the cell cycle and its interactions with transcription factor E2F-1 and the Rb family of proteins. *Mol. Cell. Biol.* 19, 2400–2407.

5
Homology Modeling in Biology and Medicine

Roland L. Dunbrack, Jr.

5.1
Introduction

5.1.1
The concept of homology modeling

To understand basic biological processes such as cell division, cellular communication, metabolism, and organismal development and function, knowledge of the three-dimensional structure of the active components is crucial. Proteins form the key players in all of these processes, and study of their diverse and elegant designs is a mainstay of modern biology. The Protein Databank (PDB) of experimentally determined protein structures [1, 2] now contains some 18 000 entries, which can be grouped into between 300 and 700 related families based either on structure or on sequence similarity [3–5]. The fact that proteins that share very little or no sequence similarity can have quite similar structures has led to the hypothesis that there are in fact on the order of 1000–7000 different families [6, 7] which have been adapted by a process of duplication, mutation, and natural selection to perform all the biological functions that proteins perform.

Since it was first recognized that proteins can share similar structures [8], computational methods have been developed to build models of proteins of unknown structure based on related proteins of known structure [9]. Most such modeling efforts, referred to as homology modeling or comparative modeling, follow a basic protocol laid out by Greer [10, 11]: 1) identify a *template* or *parent* structure related to the *target* sequence of unknown structure, and align the target sequence to the template sequence and structure; 2) for core secondary structures and all well-conserved parts of the alignment, borrow the backbone coordinates of the template according to the sequence alignment of the target and template; 3) for segments of the target sequence for which coordinates cannot be borrowed from the template because of insertions and deletions in the alignment (usually in loop regions of the protein), build these segments using some construction method based on our knowledge of the determinants of protein structure; 4) build side chains determined by the target sequence on to the backbone model built from the template structure and loop construction. In practice, identifying a structural homologue requires aligning the sequences, and so steps 1 and 2 are performed together. Step 2 is often a manual adjustment of the alignment, but automated sequence alignment methods different from those used to make the identification may be used. Also, Steps 3 and 4, backbone and side-chain modeling, may be coupled, since certain backbone conformations may be unable to accommodate the required side chains in any low-energy conformation. An alternative strategy has been developed by Blundell and colleagues, based on averaging a number of template structures, if these exist, rather than using a single structure [12–14]. More complex procedures based on reconstructing structures (rather than perturbing a starting structure) by satisfying spatial restraints using distance geometry [15] or molecular dynamics and energy minimization [16–19] have also been developed.

Many methods have been proposed to perform each of the steps in the homology modeling process. There are also a number of research groups that have developed complete packages that take as input a sequence alignment or even just a sequence and develop a complete model. In this Chapter, we describe some of the basic ideas that drive loop and side-chain modeling individually and the programs publicly available that implement them. We also discuss some of the programs that perform all of the steps in modeling. The identification and alignment steps will be covered in the next Chapter of this volume.

5.1.2
How do homologous proteins arise?

By definition, homologous proteins arise by evolution from a common ancestor. But there are several different mechanisms in play, and these are illustrated in Figure 5.1. The first is random mutation of individual nucleotides that change protein sequence, including missense mutations (changing the identity of a single amino acid) as well as insertions and deletions of a number of nucleotides that result in insertion and deletion of amino acids. As a single species diverges into two species, a gene in the parent species will continue to exist in the divergent species and over time will gather mutations that change protein sequence. In this case, the genes in the different organisms will usually maintain the same function. These genes are referred to as orthologues of one another. A second mechanism is duplication of a gene or of a gene segment within a single organism or germ line cell. As time goes by, the two copies of the gene may begin to gather mutations. If the parent gene performed more than one function, for example similar catalytic activity on two different substrates, one of the

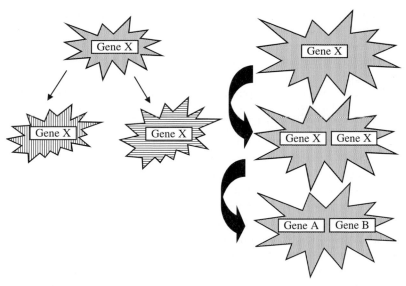

Fig. 5.1
Orthologues vs. paralogues. Schematic of
evolutionary process that gives rise to
homologous proteins. Left: A single gene X
in one species is retained as the species
diverges into two separate species. The
genes in these two species are *orthologous*.
Right: A single gene X in one species is
duplicated. As each gene gathers
mutations, it may begin to perform new
functions, or the two genes may specialize
in carrying out two or more functions of
the ancestral gene, thus improving the
fitness of the organism. These genes in
one species are *paralogous*. If the species
diverges, each daughter species may
maintain the duplicated genes, and
therefore each species contains an
orthologue and a paralogue to each gene
in the other species.

duplicated genes may gain specificity for one of the reactions, while the
other gene gains specific activity for the other. If this divergence of specificity in the two proteins is advantageous, the duplication will become fixed
in the population. These two genes are paralogues of one another. If the
species with the pair of paralogues diverges into two species, each species
will contain the two paralogues. Each gene in each species will now have an
orthologue and a paralogue in the other species.

5.1.3
The purposes of homology modeling

Homology modeling of proteins has been of great value in interpreting the
relationships of sequence, structure, and function. In particular, orthologous proteins usually show a pattern of conserved residues that can be interpreted in terms of three-dimensional models of the proteins. Conserved

residues often form a contiguous active site or interaction surface of the protein, even if they are distant from each other in the sequence. With a structural model, a multiple alignment of orthologous proteins can be interpreted in terms of the constraints of natural selection and the requirements for protein folding, stability, dynamics, and function.

For paralogous proteins, three-dimensional models can be used to interpret the similarities and differences in the sequences in terms of the related structure but different functions of the proteins concerned [20]. In many cases, there are significant insertions and deletions and amino acid changes in the active or binding site between paralogues. But by grouping a set of related proteins into individual families, orthologous within each group, the evolutionary process that changed the function of the ancestral sequences can be observed. Indeed, homology models can serve to help us identify which protein belongs to which functional group by the conservation of important residues in the active or binding site [21]. A number of recent papers have been published that use comparative modeling to predict or establish protein function [22–26].

Another important use of homology modeling is to interpret point mutations in protein sequences that arise either by natural processes or by experimental manipulation. Now that the human genome project has produced a rough draft of the complete human genome sequence, there are starting to be collected voluminous data concerning polymorphisms and other mutations related to differences in susceptibility, prognosis, and treatment of human disease. There are now many such examples, including the Factor V/Leiden R506Q mutation [27] that causes increased occurrence of thrombosis, mutations in cystathionine beta synthase that cause increased levels of homocysteine in the blood, a risk factor for heart disease [28], and BRCA1 for which many sequence differences are known, some of which may lead to breast cancer [29]. At the same time, there are many polymorphisms in important genes that have no discernible effect on those who carry them. At least for some of these, there may be some effect that has yet to be measured in a large enough population of patients, and the risk of cancer, heart disease, or other illness to these patients is unknown. This is yet another important application of homology modeling, since a good model may indicate readily which mutations pose a likely risk and which do not.

Homology models may also be used in computer-aided drug design, especially when a good template structure is available for the target sequence. For enzymes which maintain the same catalytic activity, the active site may be sufficiently conserved that a model of the protein provides a reasonable target for computer programs which can suggest the most likely compounds that will bind to the active site (see also Chapter 7 of this volume). This has been used successfully in the early development of HIV protease inhibitors [30, 31] and in the development of anti-malarial compounds that target the cysteine protease of *P. falciparum* [32].

5.1.4
The effect of the genome projects

The many genome projects now completed or underway have greatly affected the practice of homology modeling of protein structures. First, the many new sequences have provided a large number of targets for modeling. Second, the large amount of sequence data makes it easier to establish remote sequence relationships between proteins of unknown structure and those of known structure on which a model can be built. The most commonly used methods for establishing sequence relationships such as PSI-BLAST [33] are dependent on aligning many related sequences to compile a pattern or profile of sequence variation and conservation for a sequence family. This profile can be used to search among the sequences in the PDB for a relative of the target sequence. The more numerous and more varied sequences are in the family the more remote are the homologous relationships that can be determined, and the more likely it is that a homologue of known structure for a target sequence can be found. Third, it is likely that the accuracy of sequence alignments between the sequence of unknown structure that we are interested in (referred to as the target sequence) and the protein sequence of known structure used for model building (referred to as the template structure) are also greatly improved with profiles established from many family members of the target sequence [34]. Fourth, the completion of a number of microbial genomes has prodded a similar effort among structural biologists to determine the structures of representatives of all common protein sequence families, or all proteins in a prototypical genome, such as *E. coli* or yeast [35]. Protein structures determined by X-ray crystallography or NMR spectroscopy are being solved at a much faster pace than was possible even 10 years ago. The great increase in the number of solved protein structures has a great impact on the field of homology modeling, since it becomes ever more likely that there will be a template structure in the PDB for any target sequence of interest.

Given the current sequence and structure databases, it is of interest to determine what fraction of sequences might be modeled and the range of sequence identities between target sequences and sequences of known structure. In Figure 5.2, we show histograms of sequence identities of the sequences in several genomes and their nearest relatives of known structure in the Protein Data Bank. These relationships were determined with PSI-BLAST as described in the legend. PSI-BLAST is fairly sensitive in determining distant homology relationships [34, 36, 37]. The results indicate that on average 30–40% of genomic protein sequences are related to proteins of known structure, which presents a large number of potential targets for homology modeling. But it should also be pointed out that the average sequence identity between target sequences and template structures in the PDB is less than 25%. It is likely that as sequence comparison methods improve, our ability to identify increasingly remote homologies will in-

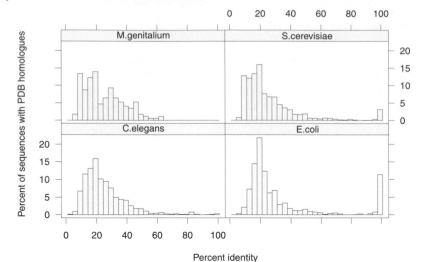

Fig. 5.2

Distribution of sequence identities between proteins in four genomes and their closest homologues in the Protein Databank for those sequences in genomes with homologues in the PDB. PSI-BLAST was used to search the non-redundant protein sequence database with a representative set of PDB sequences as queries. The program was run for four iterations, with a maximum expectation value of 0.0001 (see Chapter 2 for an explanation of this value) used to determine sequences which are included in the position-specific similarity matrix. After four iterations, each matrix was used to search each of the four genomes. Coiled-coil and low-sequence complexity sequences were removed from each genome and the non-redundant sequence database. All hits in the genomes with E-values less than 0.001 were saved, and the histograms were built from the PSI-BLAST derived sequence identities.

crease, and we will be able to determine relationships at even lower sequence identity with confidence [38].

The low sequence identity between target and template sequences in Figure 5.2 presents a major challenge for homology modeling practitioners, since a major determinant in the accuracy of homology modeling is the sequence identity between the target sequence and the sequence of the template structure. At levels below 30% sequence identity, related protein structures diverge significantly and there may be many insertions and deletions in the sequence [39]. At 20% sequence identity, the average RMSD of core backbone atoms is 2.4 Å [39]. But as demonstrated in Figure 5.2, it is likely that we will most often face a situation where the target and template sequences are remotely related. Most widely used homology modeling methods have been predicated on much higher sequence identities between template and target, usually well above 30% [40–42]. What methods should be used at sequence identities in the 10–30% range is of crucial importance in this so-called post-genomic era.

5.2
Input data

To produce a protein model that will be useful and informative requires more than placing a new sequence onto an existing structure. A large amount of sequence data and other kinds of experimental data can often be gathered on the target sequence and on its homologue of known structure to be used for model building. This information can be used to build a better model *and* as the data to be interpreted in light of the model. The goal is to forge an integrated model of the protein sequence, structure, and function, not merely to build a structure. In Table 5.1, we list the kinds of information which might be available for a target protein and how these data might be processed. With the large amount of sequence information available, it is almost always possible to produce a multiple alignment of

Tab. 5.1
Input information for homology model building

Target sequence
- Target orthologous relatives (from PSI-BLAST)
- Target paralogous relatives (from PSI-BLAST)
- Multiple sequence alignment of orthologues and paralogues (either BLAST multiple alignment or (preferably) other multiple alignment program)
- Sequence profile of ortho/paralogues

Template sequence
- Homologue(s) of known structure (template(s)) determined by database search methods (BLAST, PSI-BLAST, intermediate-sequence-search methods, HMM's, fold recognition methods)
- Template orthologous sequences
- Template paralogous sequences
- Multiple sequence alignment of parent orthologues and paralogues

Alignment of target sequence to template sequence and structure
- Pairwise alignment
- Profile alignment
- Multiple sequence alignment of target and template sequence relatives
- Profile-profile alignment
- Fold recognition alignment
- Visual examination of proposed alignments and manual adjustment
- Assessment of confidence in alignment by residue (some regions will be more conserved than others)

Structure alignment of multiple templates, if available
- Align by structure (fssp, VAST, CE, etc.)
- Compare sequence alignments from structure to sequence alignments from multiple sequence alignments (see above)

Experimental information
- Mutation data (site-directed, random, naturally occurring)
- Functional data – e.g. DNA binding, ligands, metals, catalysis, etc.

sequences related to the target protein. The first step in modeling therefore is to use a database search program such as PSI-BLAST [33] against a non-redundant protein sequence database such as NCBI's *nr* database [43] or the curated SwissProt database [44]. With some care, a list of relatives to the target sequence can be gathered and aligned. PSI-BLAST provides reasonable multiple alignments, but it may be desirable to take the sequences identified by the database search and realign them with a multiple sequence alignment program such as CLUSTAL W [45], prrp [46], and multalin [47]. Another source of protein sequences related to the target is the database of expressed sequence tags (ESTs) available at NCBI. ESTs are derived from sequencing DNA transcriptions of mRNAs derived from cellular samples from various species or tissue types. While these sequences contain some non-coding nucleotides, the bulk of the sequence represents transcribed codons. This DNA sequence database can be searched with protein sequence queries using the TBLASTN program (http://www.ncbi.nlm.nih.gov/blast). NCBI also makes available the sequences of unfinished microbial genomes for TBLASTN searches (http://www.ncbi.nlm.nih.gov/Microb_blast/unfinishedgenome.html), and a search of these may provide additional sequences related to the target.

It may be that a database search consisting of several rounds of PSI-BLAST will provide one or more sequences of known three-dimensional structure. If this is not the case then more sensitive methods based on fold recognition [48–79] (see Chapter 6 of this volume) or hidden Markov models (see Appendix to this volume) [80–87] of protein superfamilies may identify a suitable template structure. Once a template structure is identified, a sequence database search will provide a list of relatives of the template, analogous to searches for relatives of the target. At this stage it is useful to divide the sequences related to the target into orthologues and paralogues of either the target or the template (or both). The sequence variation within the set of proteins that are orthologous to the target provides information as to what parts of the sequence are most conserved and therefore likely to be most important in the model. Similarly variation in the set of proteins that are orthologous to the template provide a view of the template protein family that can be used to identify features in common or distinct in the template and target families. These features can be used to evaluate and adjust a joint multiple alignment of both families.

If there are multiple structures in the Protein Data Bank that are homologous to the target sequence, then it is necessary to evaluate them to determine which PDB entry will provide the best template structure and whether it will be useful to use more than one structure in the modeling process. In the case of a single sequence that occurs in multiple PDB entries, it is usually a matter of selecting the entry with the highest resolution or the most appropriate ligands (DNA, enzyme inhibitors, metal ions). In other cases, there may be more than one homologue related to the target sequence, and the task is to select the one more closely related to the target or to combine information from more than one template structure to build the model. To do this, a structure alignment of the potential templates can be performed by one of a number of computer programs available on the Internet (Dali [88], CE [89], etc., see Table 5.6). From alignments of the target to the available templates, the location of insertions and deletions can be observed and often it will be clear that one template is better than others. This may not be uniform, however, such that some regions of the target may have no insertions or deletions with respect to one template, but other regions are more easily aligned with the other template. In this case, a hybrid structure may be constructed [13].

Finally, any other experimental information available on the target or template proteins may be very helpful in producing and interpreting a structural model. This can include inhibitor studies, DNA binding and sequence motifs, proteolysis sites, metal binding, mutagenesis data, and so forth. A number of databases are available on the web that summarize information on particular genes, or that collect information on mutations and polymorphisms linked to disease including: the Cancer Genome Anatomy Project [90]; the Online Mendelian Inheritance in Man (OMIM) [91]; YDB (Yeast database), WormDB (*C. elegans* database), and PombeDB (*Schizosaccharomyces pombe* database), all available from http://www.proteome.com/databases/); and the Human Gene Mutation Database [92].

Finally, we note that many methods in homology modeling rely on statistical data on protein conformations. We present a primer on some aspects of protein structure in the Appendix to this Chapter.

5.3
Methods

5.3.1
Modeling at different levels of complexity

Once an alignment is obtained between the target and a protein of known structure (as described in "Input Data" or in Chapter 6 of this volume), it is possible to build a series of models of increasing sophistication.

5.3.1.1 Simple model
Keep backbone and conserved side chains by renaming and renumbering coordinates in the template structure with the new sequence using the alignment of target and template; rebuild other side chains using a side-chain modeling program (e.g., SCWRL [93, 94]); do not model insertions or deletions; that is, do not build new loops and do not close up gaps. For this purpose, we have made available a Perl program called *blast2model* that

uses BLAST alignments as input and the SCWRL program to build model structures (http://www.fccc.edu/research/labs/dunbrack/software.html).

5.3.1.2 Stepwise model

Borrow core backbone from template structure, minus coil regions with insertions or deletions in the sequence alignment; rebuild core side chains; rebuild coil regions with loop prediction method in conjunction with side-chain prediction method. Core backbone and side chains may or may not be held fixed during loop prediction (e.g., CONGEN [95]).

5.3.1.3 Global model

Build entire protein from spatial restraints drawn from known structure(s) and sequence alignment (e.g., MODELLER [16, 17]).

5.3.1.4 Choosing a model

It is not always the case that more sophisticated models are better than simpler, less complete ones. If elements of secondary structures are allowed to move away from their positions in the template and large changes are made to accommodate insertions and deletions, it may be the case that the model is further away from the target structure (if it were known) than the template structure was to begin with. This is the "added value" problem discussed by John Moult at the CASP meetings [96–98]. We would like methods that change the template structure closer to the target structure, such that they "add value" to a simple model based on an unaltered template structure, perhaps with side chains replaced. Extensive energy minimization or molecular dynamics simulations often bring a model further away from the correct structure than toward it [99].

The simple model is sometimes justified when there are no insertions and deletions between the template and target or when these sequence length changes are far from the active site or binding site of the protein to be modeled. This often occurs in orthologous enzymes that are under strong selective pressure to maintain the geometry of the active site. Even in non-orthologous enzymes, sometimes we are most interested in an accurate prediction of the active site geometry and not in regions of the protein distant from the active site. Unless loop modeling is accurate, it is possible that modeled loops in more sophisticated models may get in the way of side-chain modeling. If uncertain parts of the chain are not modeled, then side chains may find their correct positions and the model may be superior to a more complete version.

A stepwise model is probably the most common method used in homology modeling, since it is conceptually simpler than the more complex models and since each piece can be constructed and examined in turn. Some programs therefore proceed by taking the sequence–structure alignment, removing all regions where there are insertions and deletions, and reconstructing loops and side chains against the fixed template of the remaining atoms. Some methods may also allow all parts of the template structure to adjust to the changes in sequence and insertions and deletions. This usually takes the form of a Monte Carlo or molecular dynamics simulation [18]. A global model, as described above, rebuilds a structure according to constraints derived from the known template structure or structures. This is in contrast to stepwise models that proceed essentially by replacing parts of the template structure and perhaps perturbing the structure.

Many computer programs for homology modeling are developed to solve a single problem – such as loop or side-chain building, and may not be set up to allow all atoms of the protein to adjust or to model many components simultaneously. In many cases these methods have been tested by using simplified modeling situations with the newly developed software. Such examples include experiments with removing and rebuilding loops onto single protein structures, and stripping and rebuilding all side chains. Therefore in the next sections we review some of the work in these two areas.

5.3.2
Loop modeling

5.3.2.1 Input information

In stepwise construction methods, backbone segments which differ in length between the template and target (according to the sequence alignment) need to be rebuilt. In some situations, even when the sequence length of a coil segment is maintained, it may be necessary to consider alternative conformations to accommodate larger side chains or residues with differing backbone conformational requirements, Gly $\leftrightarrow$ non-Gly, or Pro $\leftrightarrow$ non-Pro mutations (see Appendix). Most such loop construction methods have been tested only on native structures from which the loop to be built has been removed. But the reality in homology modeling is more complicated, requiring several choices to be made in building the complete structure. These include how much of the template structure to remove before loop building; whether to model all side-chains of the core before rebuilding the loops; and whether to rebuild multiple loops simultaneously or serially.

Deciding how much of the template structure to remove before loop building depends on examination of the sequence alignment and the template structure. Sequence alignments with insertions and deletions are usually not unambiguous. Most sequence alignment methods ignorant of structure will not juxtapose a gap in one sequence immediately adjacent to a gap in another sequence. That is, they will produce an alignment that looks like this alignment:

```
AGVEPMENYKLS

SG---LDDFKLT
```

rather than like this one:

```
AGVEPMEN---YKLS

SGL-----LDDFKLT
```

However, the latter alignment is probably more realistic [100], indicating that a 5 amino acid loop in the first sequence and structure is to be replaced with a 3 amino acid loop in the second sequence. The customary practice is to remove the whole segment between two conserved secondary structures units. Even with this practice, ambiguity remains, since the ends of secondary structures, especially α-helices, are not well determined. If loop building methods were accurate, then removing more of the segment would be a good idea. But long loops (longer than 6 amino acids) are difficult to rebuild accurately, and hence there is cause to preserve as much of the starting structure as possible. Once the backbone has been borrowed from the template in stepwise modeling, one has to decide the order of building the core side chains, the backbone of loops to be built, and their side chains. They may be built sequentially, or allowed to vary simultaneously. Side chains from the core may guide the building of the loop, but at the same time may hinder correct placement. It is certainly the case that in the final structure there must be a reasonably low-energy conformation that can accommodate all loops and side chains simultaneously. Different authors have made different choices, and there has been little attempt to try vary the procedure while keeping the search algorithm and potential energy function used [74, 101–105] fixed.

5.3.2.2 **Loop conformational analysis**

Loop structure prediction is always based in one way or another on an understanding of loop conformations in experimentally determined structures. Loop conformational analysis has been performed on a number of levels, ranging from classification of loops into a number of distinct types to statistical analysis of backbone dihedral angles. Loop classification schemes have usually been restricted to loops of a particular size range: short loops of 1–4 residues, medium loops of 5–8 residues, and long loops of 9 residues or longer.

Thornton and coworkers have classified β-turns, which are short loops of 2–5 amino acids that connect two anti-parallel β-sheet strands [106–109]. These loops occur in a limited number of conformations that depend on the sequence of the loop, especially on the presence of glycine and proline residues at specific positions. The backbone conformation can be characterized by the conformations of each amino acid in terms of regions of the Ramachandran map occupied (usually defined as α_R, β_P, β_E, γ_R, α_L, and γ_L; see Appendix) [108]. Usually one or more positions in the loops requires an α_L conformation and therefore a glycine, asparagine, or aspartic acid residue. One useful aspect of this analysis is that if a residue varies at certain positions or there are short insertions at certain positions, the effect on the loop can be predicted [109] since the number of possibilities for each length class is small. The program BTPRED is available (see Table 5.6) to predict the locations of specific types of β-turns from protein sequences and secondary structure predictions derived from other programs [110]. Single amino acid changes tend to maintain the loop conformations, except when Pro residues substitute for residues with $\phi > 0°$ (see Appendix to this Chapter), while insertions change the class of the loop.

In recent years with a larger number of structures available, medium length loops have also been classified [111–117] by their patterns of backbone conformation residue by residue (α_R, β_P, etc.). A number of regularly occurring classes have been found, depending on length, type of secondary structure being connected, and sequence. These classes cover many but by no means all of the loops seen in non-β turn contexts. The work of Oliva et al. [114] is probably the most thorough classification to date.

Longer loops (> 8 amino acids) have been investigated by Martin et al. [118] and Ring et al. [119]. Martin et al. found that long loops fall into 2 classes, those that connect spatially adjacent secondary structures and those that connect secondary structures separated by some distance. Ring et al. provided a useful classification of longer loops as either strap (long extended loops), Ω loops (similar to those described by Leszczynski [113]) which resemble the Greek letter, and ζ loops, which are non-planar and have a zigzag appearance. The different loop types were found to have different distributions of virtual $C\alpha$-$C\alpha$-$C\alpha$-$C\alpha$ dihedrals to accommodate their shapes.

Swindells et al. [120] have calculated the intrinsic ϕ, ψ propensities of the 20 amino acids from the coil regions of 85 protein structures. The distribution for coil regions is quite different than for the regular secondary structure regions, with a large increase in β_P and α_L conformations and much more diverse conformations in the β_E and α_R regions. Their results also indicate that the 18 non-Gly, Pro amino acid type are in fact quite different from each other in terms of their Ramachandran distributions, despite the fact that they are usually treated as identically distributed in prediction methods [95, 121]. Their analysis was divided into the main broad regions of the Ramachandran map, ignoring the α_L region. The results are intriguing, in that the probability distributions are distinct enough even when calculated from a relatively small protein dataset. See the Appendix to this Chapter for examples.

5.3.2.3 Loop prediction methods

Loop prediction methods can be analyzed for a number of important factors in determining their usefulness: 1) method of backbone construction; 2) what range of lengths are possible; 3) how widely is the conformational space searched; 4) how are side chains added; 5) how are the conformations scored (i.e., the potential energy function); and 6) how much has the method been tested (length, number, self/non-self). We summarize a large number of published methods in Table 5.2. Only a subset of these methods are available as programs that can be downloaded or as web-servers that can perform calculations on input structures. This is indicated in Table 5.2, and the addresses for available programs are listed in Table 5.6.

5.3.2.3.1 Database methods

The most common approach to loop modeling involves using "spare parts" from other (unrelated) protein structures [11, 13, 119, 122–133]. These database methods begin by measuring the orientation and separation of the backbone segments flanking the region to be modeled, and then search the PDB for segments of the same length which span a region of similar size and orientation. This work was pioneered by Jones and Thirup [122]. They defined a procedure in which Cα-Cα distances were measured among six residues, three on either side of a backbone segment to be constructed. These 15 Cα-Cα distances were used to search structures in the Protein Databank for segments with similar Cα-Cα distances and the appropriate number of intervening residues. Other authors have used the same method for locating potential database candidates for the loop to be constructed [124, 127, 128, 130].

In recent years, as the size of the PDB has increased, database methods have continued to attract attention. With a larger database, recurring structural motifs have been classified for loop structures [111, 114, 116, 129, 134], including their sequence dependence. Database methods have been applied only for loops of up to 8 residues. Fidelis et al. [126] found that for loops of length greater than 7 there is not likely to be a segment in the PDB that corresponds to the correct loop conformations for the 58 protein they looked at. Some authors report that when the database contains a loop of similar structure to the target to be modeled, their methods perform well, with RMSD values around 1 Å or better. Otherwise they tend to fail [124, 127, 130].

Although many methods have been published, they have usually only been tested on a small number of loops, and then usually in the context of rebuilding loops onto their own backbones, rather than in the process of homology modeling. These numbers are listed in Table 5.2. One exception to the small numbers of tests is the work of Fechteler et al. [128], who developed a database method (implemented in the program BRAGI) and tested it on 71 insertion and 74 deletion regions of 1–3 amino acids in length. Another is the work of Rufino et al., who used their loop classification scheme predictively on 1785 loops [129]. Unfortunately, they found their database not particularly successful in making predictions. Their database could only be used to make a prediction in 63% of cases, and only 54% of these predictions were considered correct.

5.3.2.3.2 **Construction methods**

The main alternative to database methods is construction of loops by random or exhaustive search mechanisms. Moult and James [135] early on used a systematic search to predict loop conformations up to 6 residues long. They pioneered several useful concepts in loop modeling by construction: the use of a limited number of ϕ, ψ pairs for construction; construction from each end of the loop simultaneously; discarding conformations of partial loops that cannot span the remaining distance with those residues left to be modeled; using side-chain clashes to reject partial loop conformations; and the use of electrostatic and hydrophobic free energy terms in evaluating predicted loops. Their method successfully predicted the structures of two loops in trypsin. The CONGEN program of Bruccoleri and Karplus [18, 95, 136, 137] is also based on an exhaustive search algorithm, this time based on an evenly spaced grid of backbone conformations (either 15° or 30°) in accessible regions of the Ramachandran map. Gly and Pro are treated as separate classes, while the remaining 18 amino acids are treated identically in terms of their backbone energies and allowed range. Amino acids are built in turn, with partial conformations discarded if it will be impossible to close the loop with the remaining amino acids to be built. Energies are evaluated with the CHARMM polar-hydrogen force field [138]. Side chains are built by exhaustive search on 30° increments of side-chain dihedrals, as each residue is built. Since the method has been tested only on a small number of immunoglobulin CDR loops, it is difficult to tell how generally successful the algorithm is likely to be.

A particularly interesting loop construction algorithm is the scaling-relaxation method of Zheng et al. [139–141]. In this method, a full segment is sampled and its end-to-end distance is measured. If this distance is longer than the segment needs to be, then the segment is scaled in size so that it fits the end-to-end distance of the protein anchors. This results in very short bond distances, and unphysical connections to the anchors. From

Tab. 5.2
Loop methods

Authors	Availability	Side chains	Potential energy	Search method	Test set + criteria	Claims
Abagyan [147, 149, 292]	ICM ($3250 per year)	Searched with Monte Carlo probability function	Molecular mechanics + solvation terms	Construction: Internal coordinate Monte Carlo from biased probability distribution function for dihedrals	5 CASP2 deletion loops of 4–5 aa and 2 CASP2 insetion loops (10,13 aa)	Deletion loops 1.19–2.55 Å RMSD; Insertion loops 8.1, 11.4 Å RMSD
Bruccoleri, Karplus [95, 136, 201]	CONGEN	Generated on grid of χ angles	CHARMM	Construction: Grid-based search + closure with Go-Scheraga + minimization with CHARMM; unproductive branches are truncated	5 self-proteins, 12 loop, 5–9 aa [95]; 12 Ig loops [136]	Ig loops: 0.6–2.6 Å for backbone atoms
Carlacci, Chou, Englander [293, 294]		Part of simulation	ECEPP/2; vdW + electrostatic, hydrogen bonds, torsional potential; ε = 2; united-atom + polar hydrogens; harmonic force on N, Cα, C, O atoms of terminal residues on x-ray positions	Construction: MC/simulated annealing; fixed template; fixed bond lengths and angles; random translations and rotations of modeled segment + random changes in dihedrals in simulated annealing	helix, sheet, loops 5–9 aa of BPTI only	5 res loop = 1.0 Å backbone rms; 7 aa loop = 1.6 Å backbone rms; 9 aa loop = 1.87 Å backbone rms

Collura, Higo, Robson, Garnier [295–297]	Part of simulation	Robson-Platt potential [298]; vdW + electrostatic; torsions on side chains; r-dependent dielectric; united-atom + polar hydrogens	Construction: MC/simulated annealing; increasing weight of different energy terms non-uniformly; anchors fixed; initial conformation is extended peptide with harmonic potential centered on 4 backbone atoms at each end; rigid geometry except dihedrals	8 self-protein loops in Ig, BPTI, trypsin, 6–8 aa	RMS 1 Å for backbone; more complicated solvent models not necessarily helpful
Dudek, Palmer, Scheraga [299–301]	Can be implemented in ECEPP ($195)	ECEPP + hydration function	Added after backbone deformation + minimization	Construction: Random deformation (in acceptable $\phi\psi$ regions spec. for each aa) + minimization + add side chains + minimization	5 BPTI self-loops 1 BT self-loop, all length 7	Depends on knowing initial structure. Low RMS 0.16, 0.27, 0.91, 1.15, 0.32, 0.89. Deformations are random $\phi\psi$ pairs for 3 residues and analytic calculation of remaining $\phi\psi$ to close loop

Tab. 5.2 (continued)

Authors	Availability	Side chains	Potential energy	Search method	Test set + criteria	Claims
Janardhan, Vajda [102]	Can be implementeed in CONGEN	Built with CONGEN	CHARMM + surface-solvation term + side-chain entropy term	Construction: CONGEN: systematic search	6 CASP1 loops	Evaluating free energies for CASP1 predictor loops + CONGEN side chains; importance of internal energy terms; side-chain entropy in loop predictions
Moult, James, Fidelis [126, 135]		Rotamer library, complete search	Electrostatic term, image-charge solvation term, van der Waals, hydrophobic term	Construction: by 11 $\phi\psi$ pairs; truncate unproductive branches; build many copies from each end and link; allow closer than normal contacts	2SGT loops based on bovine trypsin, 4, 5 residues; 11 homology loops in 4 proteins; 4–6 aa in length	Hydrophobic term very useful; database methods are not adequate even for short loops
Rao, Teeter [148]	Can be implemented in XPLOR	Part of MD simulation	AMBER $\varepsilon = 4r$; counterions present	Construction: AMBER 50 ps simulation	One 7-residue loop in purothionin built from the corresponding 7-residue loop in crambin	Successfully reproduced purothionin loop with MD and simulated annealing

Rapp, Friesner [104]	Can be implemented in MacroModel, ($995)	Added with RSA of Shenkin, rotamer based, after initial generated loops	AMBER, AMBER94 + Born solvation implemented in MacroModel	Construction: Initial loops generated randomly with $\phi\psi$ library. Loops that fit are saved. 10 000 step substructure minimization; then MD and simulated annealing; minimization	2 selfloops, 8 and 12 aa from ribonuclease A	r-dependent dielectric should fail badly; 12 aa loop at 0.81 Å; AMBER94 needed, not AMBER united atom model
Ring, Cohen [119, 302]	Drawbridge	Isoteric placement + Ponder & Richards rotamers	Simple steric clash check	Construction: genetic algorithm used to search among statistical distribution of tetrapeptide conformers; TWEAK algorithm used to refine model [143, 144]	4 CDR loops in antibody HY-HEL5	Better results than other methods on this antibody
Shenkin, Fine, Wang, Yarmush, Levinthal [143, 144]	Implemented in InsightII as "TWEAK"	None in early steps; added for low energy loops and searched on 10° grid, then minimized with PAKGGRAF	DISCOVER, CHARMM	Construction: Random-tweak: starting random structure is constrained by Lagrange multipliers to fit the Jones-Thirup base distances; max change in dihedrals = 10° at each step, until convergence; followed by MD/min or min.	CDR's of MCPC603	Success on short loops; multiple minima for longer CDR; need to add side chains to restrict some of the mimima

Tab. 5.2 (continued)

Authors	Availability	Side chains	Potential energy	Search method	Test set + criteria	Claims
Sowdhamini, Ramakrishnan, Balaram [145]		Added after search and disulfide bond formation, minimized; Janin rotamers	MODIP force field	Construction: Uniform random $\phi\psi$ (G, P, non-GP) in allowed regions; if can form disulfide, it is made, followed by energy minimization	Disulfide bonded structures, 15 aa	Disulfide loop predictions: conotoxin, endothelin, 3–5 Å rms; generate 350 000 conformations to find loops for 15 aa
Sudarsanam, Srinivasan [303, 304]		None	Simple steric checks.	Construction: Random draws from statistical distribution of ψ_I-ϕ_{I+1} for all 400 possible pairs of aa types, further subdivided by neighboring aa types (resulting in tetramer distributions). 10 000 conformers generated and clustered and those that fit anchors used as prediction	7 peptides, 6–20 aa in length	Various conformations can be generated, including helix and sheet.

Reference	Software	Side chains	Energy	Method	Test set	Results
Zheng, Delisi, Kyle [139–141]	Can be implemented in CHARMM and CONGEN	Included in simulation	CHARMM, no solvation; $\varepsilon = 4r$	Construction: Scaling-relaxation method in CONGEN: reduce random loops to fit anchors and then scale to size gradually with ABNR minimization; random loops from $\phi\psi$ distribution	8 self-loops from 6 proteins, 7 aa each	Lowest E is lowest RMS 7/8 times; Zheng93 0.83–1.44 Å RMS bb Zheng96 has 0.38 Å
Bates, Oliva, Sternberg [114, 131, 305]		Summers procedure to construct similar side chains; otherwise McGregor ss-dep	CHARMM	Database: [114] like Jones-Thirup based on $C\alpha$ positions; followed by annealing in Quanta	9 CASP2 loops, 2–9 aa in length	CASP2: best loop may be most probable, not one with lowest energy because of secondary structure shifts
Fechteler, Lessel, Schomburg [128, 132, 134]	BRAGI	Not predicted	Contact distances only	Database: clustered fragments to reduce library size [134]; fit by comparison of internal distance $C\alpha$ matrices	71 insertions, 74 deletions, 1–8 aa	Good results on short indel regions
Greer [11]		Maximal overlap with old side chain	Hagler potential	Database: Alignment of multiple structures related to target to define variable regions (VR's); database fragments aligned to $C\alpha$ steric check, build side chains with overlap, restrained minimization	No validation	No comparison to known structures

Tab. 5.2 (continued)

Authors	Availability	Side chains	Potential energy	Search method	Test set + criteria	Claims
Koehl, Delarue [127]	Homology program	handled with Tuffery library and SCMF search	van der Waals only	Database: Database of 81 proteins scanned for Jones-Thirup distances; minimized first; mean field theory	8 gaps in BPTI, each 4 aa; same 8 based on homology; 6 other non-self loops in homo.	Loop closure: 0.28–1.8 Å RMS; 67% χ_1 (as in sc SCMF); 0.19–0.90 for homology loops
Pellequer, Chen [306]	Can be implemented in CHARMM	Tuffery rotamer library [165]	CHARMM + finite difference Poisson Boltzmann electrostatics	Database: Placement from database followed by energy minimization; selection based on free energy cycle calculation with CHARMM + Poisson-Boltzmann electrostatics	6 CDR loops from 1 antibody	Gas phase energy most discriminative; solvation free energy helped in some cases.
Rooman, Wodak, Thornton [115]		None	None	Database: Sequence pattern to predict loops	1464 loops	Sequence alone in loop prediction is not very powerful.

Rufino, Donate, Canard, Blundell [111, 129]	Sloop	Not included	None. Loops chosen by sequence and anchor positions in comparison to database loops.	Database: by class of loop, 1–8 aa long, based on length, secondary structure, vectors, sequence compared to class [111]	1785 loops	Predictions for loops that can be matched to a class in database: under lax criteria, predictions can be made for 63% of loops, of which 54% are correct.
Samudrala, Moult [198, 199]	RAMP	Side-chain rotamer sampled and evaluated with database potential	Interatomic potential from database	Database: Graph theory and clique finding among backbone segments derived from PDB	22 CASP2 loops in 3 proteins; 4 CDR loops in antibody	Database method remains limited; improvement over CASP1
Shepherd, Gorse, Thornton [110]	BTPRED	None	None	Database: Prediction of β-turn loop type by sequence in comparison to conformational analysis of β turns in structures with a neural network	3359 β turns	Correct prediction of turn type in 47/59 cases (80%).

Tab. 5.2 (continued)

Authors	Availability	Side chains	Potential energy	Search method	Test set + criteria	Claims
Summers, Karplus [124]		Isosteric replacements	CHARMM, r-dependent dielectric	Database: uses Jones-Thirup 15 Cα-Cα distances of 3 residues on either side of loop; 31 protein database; constrained energy minimization of fragments	18 non-self loops, 0–4 aa in length	2/3 of loops were well-modeled
van Vlijmen, Karplus [130]		Reduced side-chain model of Levitt; CB-only in minimization of initial loops; later bbdep rotamer library	CHARMM	Database: Jones-Thirup distances + Cβ atoms	18 self-loops	8/18 up to 9 aa < 1.07 Å; 17/18 had correct loop in top 3 conformations
Wilmot and Thornton [107, 108]		None	None	Database: Prediction of β-turn loop type by sequence in comparison to conformational analysis of β turns in structures	59 loops	Correct prediction of turn type in 47/59 cases (80%).

| Wojcik, Mornon, Chomilier [116, 133] | No side chains | CHARMM (X-PLOR) | Database: Classification database of 3–8 aa loops [116]; choice determined by sequence + distance of ends; let anchors rotate ψ minimize to close loops; minimization with X-PLOR | 3–8 aa in length; 13 CASP3 loops | 0.5–2.8 Å backbone RMS for CASP3 loops |

Methods for loop prediction are listed. When publicly available programs can be identified, they have been listed by name. Some methods have been implemented in publicly available programs, but no specific software or scripts are available.

Abbreviations used in table: MC – Monte Carlo; aa – amino acid; vdW – van der Waals potential; Ig – immunoglobulin or antibody; CDR – complementarity-determining regions in antibodies; RMS – root-mean-square deviation; r-dependent dielectric – distance-dependent dielectric constant; ε – dielectric constant; MD – molecular dynamics simulation; selfloops – prediction of loops performed by removing loops from template structure and predicting their conformation with template sequence; bbdep – backbone-dependent rotamer library; SCMF – self-consistent mean field; PDB – Protein Data Bank; Jones-Thirup distances – interatomic distances of 3 Cα atoms on either side of loop to be modeled.

there, energy minimization is performed on the loop, slowly relaxing the scaling constant, until the loop is scaled back to full size. The method is efficient, since many loops can be generated which span the anchors, closing the loop with reasonable dihedral angles. Again, this method has been tested on only a few proteins, mostly immunoglobulins. Using the same methodology, Rosenbach and Rosenfeld have addressed the important issue of simultaneously modeling two or more loops in proteins [142]. In one test case, they found it advantageous to model two loops simultaneously, so that each felt the presence of the other in energy calculations

Other methods have built chains by sampling Ramachandran conformations randomly, keeping partial segments as long as they can complete the loop with the remaining residues to be built [143–145] For longer loops, these methods seem to be much more promising, since they spend less time in unlikely conformations searched in the grid method. Many of these methods are based on Monte Carlo or molecular dynamics simulations with simulated annealing to generate many conformations which can then be energy minimized and tested with some energy function to choose the lowest energy conformation for prediction [143, 144, 146–148]. Several authors have developed Monte Carlo methods that draw ϕ, ψ values from probability distributions derived from the PDB [149, 150] for other purposes, such as NMR structure refinement and *ab initio* peptide folding. These methods are promising because they are faster than exhaustive calculations, but have not been designed or tested on loop generation.

One important aspect in the development of a prediction method based on random (or exhaustive) construction of backbone conformations is the free energy function used to discriminate among those conformations which successfully bridge the anchors (see Table 5.2). Janardhan and Vajda [102] have found that a free energy function including a molecular mechanics term for the conformational energy, and a hydrophobic surface free energy term for the solvation is able to discriminate between decoy loops and real loops and to locate the correct conformation. Rapp and Friesner [104] reported recently that the AMBER94 force field and a generalized Born solvation model coupled with molecular dynamics and Monte Carlo simulations was able to regenerate conformations close to the crystal structure for one 8 and one 12 amino acid loop segment of ribonuclease. Their energy function was able to discriminate between good and poor conformations.

5.3.2.4 **Available programs**

Apart from complete modeling packages (discussed below), only a small number of programs are freely and publicly available from the many methods that have been published (see Table 5.6). Among database methods, this includes only Swiss-PDBViewer [151], and BRAGI which implements the method of Fechteler et al. [128]. Several loop databases are available. Among construction methods, CONGEN is currently available [95]. CODA runs both a database and an *ab initio* construction method [152].

5.3.3
Side-chain modeling

5.3.3.1 Input information

Side-chain modeling is a crucial step in predicting protein structure by homology, since side-chain identities and conformations determine the specificity differences in enzyme active sites and protein binding sites. The problem has been described as "solved" [153], although new methods [154] or improvements on older ones [94] continue to be published. Some side-chain prediction methods stand on their own, and are meant to be used with a fixed backbone conformation and sequence to be modeled given as input. Other methods have been developed in the context of general homology modeling methods, including the prediction of insertion-deletion regions. Even when using general modeling procedures, such as MODELLER, it may be worthwhile subsequently to apply a side-chain modeling step with other programs optimized for this purpose. This is especially the case when side-chain conformations may be of great importance to interpretation of the model. It is also often the case that insertion-deletion regions are far away from the site of interest, and loop modeling may be dispensed with. Indeed, significant alterations of the backbone of the template, if they are not closer to the target to be modeled (if it were known) than the template, may in fact result in poorer side-chain modeling than if no loop modeling were performed. As described above, the choice of the template may depend not only on sequence identity but also on the absence of insertions and deletions near the site of interest. If this is successful, side-chain modeling rises in importance in relation to loop prediction.

Side-chain prediction methods described in detail in the literature have a long history, although only a small number of programs are currently publicly available (see Tables 5.3 and 5.6). Nearly all assume a fixed backbone, which may be from a homologous protein of the structure to be modeled, or may be the actual X-ray backbone coordinates of the protein to be modeled. Many methods have in fact only been tested by replacing side chains onto backbones taken from the actual three-dimensional coordinates of the proteins being modeled ("self-backbone predictions"). Nevertheless, these methods can be used for homology modeling by substituting the target sequence onto the template backbone. When a protein is modeled from a known structure, information on the conformation of some side chains may be taken from the template. This is most frequently the case when the template and target residue are identical, in which case the template residue's Cartesian coordinates may be used. These may be kept fixed as the other side chains are placed and optimized, or they may be used only as a starting conformation and optimized with all other side chains. Only a small number of methods use information about non-identical side chains borrowed from the template. For instance, Phe ↔ Tyr substitutions only require the building or removal of a hydroxyl group while Asn ↔ Asp substitutions require changing one of δ atoms from NH2 to O or vice versa. Summers and Karplus [155, 156] used a more detailed substitution scheme, where for instance the χ_1 angle of very different side-chain types (e.g., Lys ↔ Phe) might be used in building side chains (see Appendix for definition of side-chain dihedral angles). In the long run, this is probably not advantageous, since the conformational preferences of non-similar side-chain types may be quite different from each other.

5.3.3.2 Rotamers and rotamer libraries

Nearly all side-chain prediction methods depend on the concept of side-chain *rotamers*. From conformational analysis of organic molecules, it was predicted long ago [157, 158] that protein side chains should attain a limited number of conformations because of steric and dihedral strain within each side chain and between the side chain and the backbone (dihedral strain occurs because of Pauli exclusion between bonding molecular orbitals in eclipsed positions) [159]. For sp^3–sp^3 hybridized bonds, the energy minima for the dihedral are at the staggered positions that minimize dihedral strain at approximately 60°, 180°, and −60°. For sp^3–sp^2 bonds, the minima are usually narrowly distributed around +90° or −90° for aromatics and widely distributed around 0° or 180° for carboxylates and amides (e.g., Asn/Asp χ_2 and Glu/Gln χ_3).

As crystal structures of proteins have been solved in increasing numbers, a variety of rotamer libraries have been compiled with increasing amounts of detail and greater statistical soundness; that is, with more structures at higher resolution [160–169]. The earliest rotamer libraries were based on a small number of structures [160–163]. Even the widely used Ponder & Richards library was based on only 19 structures, including only 16 methionines [163]. The most recent libraries are based on over 600 structures with resolution of 1.8 Å or better and mutual sequence identity less than 50% between any two chains used.

Most rotamer libraries are backbone-conformation-independent. In these libraries, the dihedral angles for side chains are averaged over all side chains of a given type and rotamer class, regardless of the local backbone conformation or secondary structure. These libraries include two in common use in side-chain conformation prediction methods, that of Ponder & Richards and that of Tuffery et al. [165]. It should be noted that the Ponder-Richards library is based on a very small sample of proteins and should *not* be used for conformation prediction (which was not its intended use anyway). The Tuffery library is based on 53 structures, which is also a very small sample compared to the PDB now available. Kono and Doi also published a rotamer library based on a cluster analysis of 103 structures [170]. Richardson et al. have compiled rotamers recently from proteins with resolution better than 1.7 Å (compared to the more common 2.0 Å resolution cutoff), while excluding residues whose conformation is likely to be poorly determined, for example, partially disordered residues (high X-ray temperature factors) and those with unfavorable steric overlaps. This winnowing process leaves a smaller sample but one with more tightly clustered dihedral values [171].

Several libraries have been proposed that are dependent on the conformation of the local backbone [164, 166–169]. McGregor et al. [164] and Schrauber et al. [167] compiled rotamer probabilities and dihedral angle averages in different secondary structures. To my knowledge these libraries are not used in any available side-chain conformation prediction programs. We have used Bayesian statistical methods to compile a backbone-dependent rotamer library with rotamer probabilities and average angles and standard deviations at all values of the backbone dihedral angles ϕ and ψ in $10°$ increments [166, 168, 169]. The current version of this library is based on 699 chains with resolution better than 1.8 Å and less than 50% mutual sequence identity. The library is described in greater detail in the appendix.

5.3.3.3 Side-chain prediction methods

Side-chain prediction methods can be classified in terms of how they treat side-chain dihedral angles (rotamer library, grid, or continuous dihedral angle distribution), potential energy function used to evaluate proposed conformations, and search strategy. These factors are summarized in Table 5.3 for nearly all side-chain prediction methods published to date. It is also useful to know how well each method has been tested, and this information is also given in the table.

As demonstrated in Table 5.3, the potential energy functions in side-chain prediction methods have varied tremendously from simple steric exclusion terms to full molecular mechanics potentials. In most cases, the potential energy function is a standard Lennard-Jones potential:

$$E(r) = 4\varepsilon \left[\left(\frac{\sigma}{r}\right)^{12} - B\left(\frac{\sigma}{r}\right)^{6} \right] \tag{5.1}$$

In this equation, r is the distance between two non-bonded atoms and ε and σ are parameters that determine the shape of the potential. This potential has a minimum at the distance $r = 2^{1/6}\sigma$ and a well depth of ε. Different values of σ and ε may be chosen for different pairs of atom types. Some potential energy functions for side chains may also include a hydrogen bond term. Depending on the potential parameters, these potentials may not accurately model the relative energies of rotamers for each side-chain type that are determined from local interactions within each side chain and between the side chain and the local backbone. For instance, in molecular

Tab. 5.3
Side-chain methods

Authors	Availability	Rotamers & Flexibility	Potential energy	Search method	Test set + criteria	Claims
Bower, Cohen, Dunbrack [93, 94]	SCWRL	Backbone-dependent rotamer library [169]	Log(rotamer prob) + linear-truncated steric term	Combinatorial search of clashing clusters	300 high-resolution self-bb proteins; 1490 non-self proteins; whole PDB as self-bb predictions	80% χ_1 correct
Chinea, Vriend [194, 284]	WHAT IF	Position specific rotamers extracted from PDB, 5–7 aa	Packing + HB + vdw	Ordered placement by entropy	15 non-self proteins	No statistical results given
Desmet, Lasters, DeMaeyer, Hazes [181, 182, 184, 307]		Backbone-independent rotamer library: 437 rotamers, DeMaeyer = Ponder & Richards [163] + Schrauber [167] + $2 \times 10°$ steps for Aro $\chi 1 \times 2 \times 20°$ Aro $\chi 2$	Force field based on CHARMM implemented in BRUGEL program	Dead-end elimination	2 self-bb proteins [181]; 19 self-bb proteins [307]	72% χ_1 correct ~70% in [307]
Dunbrack & Karplus [166]	Can be implemented in CHARMM	Backbone-dependent rotamer library	CHARMM force field with no electrostatics	Placement with most probable backbone-dependent rotamer, followed by energy minimization of clashing pairs of residues	6 self-bb proteins	82% χ_1 correct

Reference	Program (price)	Rotamer representation	Energy function	Search method	Test set	Results
Esenmenger, Abagyan, Argos [308]	ICM ($3250 per year)	Canonical dihedral angles, −60°, 180°, +60°	vdW + HB + torsions from ECEPP/2	Local energy search, followed by global energy minimization; ICM program	6 self-bb proteins, 2 non-self proteins	80% χ_1, 67% χ_{12}; 60% χ_1 for homology models
Holm, Sander [176]	TORSO	Tuffery library [165]	vdW	MC: side-chain/backbone energy > 10 kcal/mol removes rotamer	33 self-bb proteins, 5 non-self proteins	72% aver χ_1; 81% buried χ_1; non-self 54–73% correct χ_1 with fixed conserved χ_1
Hwang, Liao [179]		Neural network on 20° χ grid	Levitt vdW truncated at 7 kcal/mol	Simulated annealing, Markov chain Monte Carlo	12 self-bb proteins	74% χ_1, 60% χ_{12}
Keller et al. [309]		Tuffery library	6–12 potential of Lee and Subbiah [190]	Dead-end elimination	15 self-bb proteins	73.4% χ_1, 60% χ_{12}
Koehl. Delarue [173]	Confmat	Tuffery library + modifications	vdW, disulfides, truncated to 10 kcal/mol (Levitt, 1983)	Mean field theory	30 self-bb proteins 6 non-self proteins	72% χ_1, 62% χ_{12}, buried 82% χ_1, 72% χ_{12} buried non-self: 55– 72% correct χ_1, 64–81% on correct backbones for these proteins
Kono, Doi [170, 192]	SIDEMOD	Tuffery library + additions	vdW + HB	Neural network	21 self-bb proteins + 2 non-self proteins	87.7% χ_1 for buried
Laughton [310]		Database search for local environment similarity	Repulsive only vdW as (r-r0)**2, r < r0	Monte Carlo search on rotamers	8 self-bb proteins, 1 non-self protein	72.8% χ_1, 79.6% χ_1 buried; 64.1% on non-self test

Tab. 5.3 (continued)

Authors	Availability	Rotamers & Flexibility	Potential energy	Search method	Test set + criteria	Claims
Leach, Lemon [311]		Desmet/Lavery backbone-independent rotamer library [181]	AMBER force field	DEE + A* Algorithm	8 self-bb proteins	65–73% of all dihedral angles correct; Provides global minimum energy + states of system within specified energy range of global minimum
Lee, Subbiah [190, 312]	GeneMine/ Look	10° increments on χ	vdW + 3-fold alkane potential	Simulated annealing	9 self-bb proteins	80–90% of interior residues; 70% of all χ_1
Levitt [177]	Segmod module in Look	Segment matching to PDB	ENCAD force field, including vdW + Hbonds	Monte Carlo simulation to choose segments from PDB to model each side chain, followed by averaging; energy minimization with ENCAD	8 proteins rebuilt from Cα positions	80% correct
Mendes et al. [154]		Tuffery214 + Gaussian distribution for flexibility around rotamer angles	GROMOS, distance-dependent dielectric	Mean-field approximation	20 self-bb proteins	87.7% χ_1, 85.7% χ_{12}; core better than SCWRL, surface not

Ogata & Umeyama [313, 314]	FAMS	Side chains from 72 proteins used as spare parts for side chains with similar arrangements of backbone atoms in local environment, as determined from principal component analysis	None	Starting from center of mass, each residue is placed in most likely conformation in turn; a steric conflict restarts the procedure beginning with the conflicting residue; in homology modeling, χ angles are borrowed from template for geometrically similar amino acids.	15 self-bb proteins	65% χ_1 on 15 self proteins
Samudrala & Moult [193]	RAMP	Canonical dihedral angles, 60°, 180°, −60°	Database atom-atom distances	Top 5 interactions with mainchain saved; then by best interactions in pairs	15 self-bb proteins	75% χ_1
Schiffer et al. [174]		Ponder & Richards	AMBER + surface-area solvation term	Molten zone around each mutated side chain is minimized starting from Ponder & Richards rotamers	1 non-self bb protein	solvation important for external residues

Tab. 5.3 (continued)

Authors	Availability	Rotamers & Flexibility	Potential energy	Search method	Test set + criteria	Claims
Shenkin et al. [315]		Ponder & Richards, extended through χ_4	$E = B - \varepsilon \log P$ where B is the total number of bad contacts, and log P is log P&R rotamer probabilities	Monte Carlo simulated annealing	49 self-bb proteins	74% χ_1
Snow, Amzel [316]		15° increments in dihedral angles	Molecular mechanics energy	"Coupled perturbation procedure": Potential energy surface of side chain determined first with nearby side chains truncated to alanine; then model is energy minimized with all atoms present	6 side-chains in 1 non-self antibody	Side chains correctly modeled
Summers, Karplus [155, 156]		Rigid rotation by 10°	CHARMM	Copy template dihedrals by maximal atomic occupancy, search by 10° for rest + many checks	1 non-self protein	92% of C-terminal lobe of rhizopuspepsin from penicillopepsin

Sutcliffe, Blundell [13]	COMPOSER	Secondary-structure dependent rotamers	None	Build from structurally aligned positions in set of homologous proteins	1 non-self protein	No results given
Tanimura, Kidera, Nakamura [317]		Backbone-independent rotamer libraries from 49 proteins	AMBER, $\varepsilon = 2r$	Dead-end elimination	11 self-bb proteins	68% correct χ_1, 80% buried χ_1 sc-mc and sc-sc are consistent with each other to favor low energy rotamers; principle of "minimal frustration" and Go's (1983) consistency principle
Tuffery, Etchebest, Hazout, Lavery [165, 191]	SMD	Tuffery library	"Flex" force field: vdw + elec ($\varepsilon = 3$), torsions	Sparse-matrix-drive algorithm, clustering interacting rotamers + genetic algorithm	15 self-bb proteins	percents not published
Vasquez [189]		Backbone-independent rotamer library calculated at modes of distributions (not means), similar to Tuffery; minimization in each step	vdW 12-6 and 9-6 + torsional PE for continuum searches	Mean-field, heat bath	60 self-bb proteins	80% χ_1, 70% χ_{12}; only slight improvement for flexible rotamers

Tab. 5.3 (continued)

Authors	Availability	Rotamers & Flexibility	Potential energy	Search method	Test set + criteria	Claims
Wendoloski, Salemme [318]		Borrowed from similar backbone segments of 5–9 aa in PDB	Simple bump check	Placed in order of highest probability first according to segment matching + steric checks	3 self-bb proteins	59, 53, 70% χ_1 54, 39, 59% χ_2
Wilson et al. [175]		Ponder & Richards + more Met, Asn	AMBER all atom + Eisenberg-McLachlan solvation term	Exhaustive search for small number of residues in active site		Designed enzyme with altered specificity

Methods for side-chain prediction are listed. When publicly available programs can be identified, they have been listed by name. Some methods have been implemented in publicly available programs, but no specific software or scripts are available.

Abbreviations used in table (see also caption to Table 5.2): HB – hydrogen bonding potential energy term; self-bb – prediction of side-chain conformation on to template backbone of modeled sequence; non-self – prediction of side-chain conformation onto backbone borrowed from a homologous protein structure.

mechanics potentials, interactions between atoms connected by 3 covalent bonds (atoms i and $i + 3$ in a chain) are not usually treated by van der Waals terms, but rather in torsion terms of the form [172],

$$E(\tau) = \sum_m K_m \cos(m\tau + a_m) \qquad (5.2)$$

where the sum over m may include one-fold, two-fold, three-fold, four-fold, and six-fold cosine terms. The K_m and a_m are constants specific for each dihedral angle and each term in the sum. These torsion terms are included in some side-chain prediction methods, but ignored in others [173].

Electrostatic interactions in the form of a Coulomb potential have been included in methods that rely on full molecular mechanics potentials, usually with a distance-dependent dielectric, $\varepsilon(r) = r$:

$$E = \frac{q_i q_j}{\varepsilon(r)r} \qquad (5.3)$$

Solvent interactions are also usually ignored, since these can be difficult or expensive to model properly (for exceptions see [174, 175]).

SCWRL uses an alternative strategy, based on a probabilistic potential based on the backbone-dependent rotamer library. There are two terms: the internal side-chain energy and the local side-chain-backbone interaction are modeled with an energy term proportional to $-\ln p_{rot}$ where p_{rot} is the probability of the rotamer for the particular side-chain type and backbone conformation and a simple truncated linear steric term that models the repulsive interactions between atoms [93, 94].

Side-chain conformation prediction is a combinatorial problem, since there are on the order of n_{rot}^N possible conformations, where n_{rot} is the average number of rotamers per side chain and N is the number of side chains. But in fact the space of conformations is much smaller than that, since side chains can only interact with a small number of neighbors, and in most cases clusters of interacting side chains can be isolated and each cluster can be solved separately [93, 165]. Also, many rotamers have prohibitively large interactions with the backbone and are at the outset unlikely to be part of the final predicted conformation. These can be eliminated from the search early on.

Many standard search methods have been used in side-chain conformation prediction, including Monte Carlo simulation [176–178], simulated annealing [179], self-consistent mean field calculations [154, 173, 180], and neural networks [170]. Self-consistent mean field calculations represent each side chain as a set of conformations, each with its own probability. Each rotamer of each side chain has a certain probability, $p(r_i)$. The total energy is a weighted sum of the interactions with the backbone and interactions of side chains with each other:

$$E_{tot} = \sum_{i=1}^{N} \sum_{r_i=1}^{n_{rot}(i)} p(r_i) E_{bb}(r_i) + \sum_{i=1}^{N-1} \sum_{r_i=1}^{n_{rot}(i)} \sum_{j=i+1}^{N} \sum_{r_j=1}^{n_{rot}(j)} p(r_i) p(r_j) E_{sc}(r_i, r_j) \qquad (5.4)$$

In this equation, $p(r_i)$ is the density or probability of rotamer r_i of residue i, $E_{bb}(r_i)$ is the energy of interaction of this rotamer with the backbone, and $E_{sc}(r_i, r_j)$ is the interaction energy (van der Waals, electrostatic) of rotamer r_i of residue i with rotamer r_j of residue j. Some initial probabilities are chosen for the p's in Eq. (5.4), and the energies calculated. New probabilities $p'(r_i)$ can then be calculated with a Boltzmann distribution based on the energies of each side chain and the probabilities of the previous step:

$$E(r_i) = E_{bb}(r_i) + \sum_{j=1, j \neq i}^{N} \sum_{r_j=1}^{n_{rot}(j)} p(r_j) E_{sc}(r_i, r_j)$$

$$p'(r_i) = \frac{\exp(-E(r_i)/kT)}{\sum_{r_i=1}^{n_{rot}(i)} \exp(-E(r_i)/kT)} \qquad (5.5)$$

Alternating steps of new energies and new probabilities can be calculated from the expressions in Eq. (5.5) until the changes in probabilities and energies in each step become smaller than some tolerance.

The dead-end elimination algorithm is a method for pruning the number of rotamers used in a combinatorial search by removing rotamers that can not be part of the global minimum energy conformation [181–185]. This method can be used for any search problem that can be expressed as a sum of single-side-chain terms and pairwise interactions. Goldstein's improvement on the original DEE can be expressed as follows [183]. If the total energy for all side chains is expressed as the sum of singlet and pairwise energies,

$$E = \sum_{i=1}^{N} E_{bb}(r_i) + \sum_{i=1}^{N-1} \sum_{j>i}^{N} E_{sc}(r_i, r_j) \qquad (5.6)$$

then a rotamer r_i can be eliminated from the search if there is another rotamer s_i for the same side chain that satisfies the following equation:

$$E_{bb}(r_i) - E_{bb}(s_i) + \sum_{j=1, j \neq i}^{N} \min_{r_j} \{ E_{sc}(r_i, r_j) - E_{sc}(s_i, r_j) \} > 0 \qquad (5.7)$$

In words, rotamer r_i of residue i can be eliminated from the search if another rotamer of residue i, s_i, always has a lower interaction energy with all other side chains regardless of which rotamer is chosen for the other side chains. More powerful versions have been developed that eliminate certain pairs of rotamers from the search [183, 185]. DEE-based methods have also proved very useful in protein design, where there is variation of residue type as well as conformation at each position of the protein [186–188].

Our side-chain prediction program SCWRL [93, 94] begins by determining the ϕ and ψ angles of the input backbone conformation, and then placing side chains initially in the most favored position according to the backbone-dependent rotamer library [166, 168, 169]. The χ_1 side-chain dihedrals vary as a function of the backbone conformation for each rotamer type, in contrast to the use of backbone-independent rotamer libraries. Once the side chains are placed, steric clashes are calculated and side chains are grouped into clusters of clashing residues. If any rotamer for any side chain in one of the original clusters can produce steric overlaps with other side chains not in the clusters, then these side chains are added to the clusters. The rotamers of these side chains are searched for interactions with other side chains, and so on. Once the clusters are determined, they are solved using a branch-and-bound algorithm. It should also be noted that SCWRL uses what is essentially a dead-end elimination step, since side chains which never get added to the clusters are not searched, since they are already in their optimum rotamer with respect to the backbone. Any other rotamer choice would increase the energy, consistent with Eq. (5.7) above.

If the clusters are too large to be solved quickly, the clusters are divided by finding a residue which when removed will divide the cluster into two parts with the smallest number of combinations for solving each subcluster. For instance if a cluster of N residues can be broken into two non-interacting parts by removing one residue (called the keystone residue), then the number of combinations required to solve each group separately once for each rotamer of the keystone residue, is approximately $n_{rot}(n_{rot}^a + n_{rot}^b)$ where $N = a + b$ and n_{rot} is the average number of rotamers for each side chain. If $a \approx b$, then the number of combinations is approximately the square root of the number of combinations of the undivided cluster.

In most methods, the search is over a well-defined set of rotamers for each residue. As described above, these represent local minima on the side-chain conformational potential energy map. In several methods, however, non-rotamer positions are sampled. Summers and Karplus used CHARMM to calculate potential energy maps for side chains based on 10° grids [155, 156]. Dunbrack and Karplus used CHARMM to minimize the energy of rotamers from canonical starting conformations (−60°, 180°, and +60°) [166]. Vasquez also used energy minimization [189], Lee and Subbiah used a search over 10° increments in dihedral angles with a simple van der Waals term and a threefold alkane potential on side-chain dihedrals [190]. Very recently Mendes et al. [154, 180] used a mean-field method to sample from Gaussian distributions about the conformations in the rotamer library of Tuffery et al. [165]. On a test-set of 20 proteins, they claim a higher rate of prediction of core residues than SCWRL but a lower rate for surface-exposed residues.

5.3.3.4 **Available programs for side-chain prediction**

While many methods for side-chain prediction have been presented over the years, only a small number of programs are publicly available at this time. Information on obtaining these programs is given in Table 5.6. The available programs include SMD [165, 191], Sidemod [192], Confmat [173], Torso [176], GeneMine [190], RAMP [193], and SCWRL [93, 94]. We found most of these programs easy to obtain, compile, and execute. One complication for using Torso and Confmat in modeling situations is that the programs do not maintain the residue numbering of the input file, but instead renumber the residues from 1 to N, the number of residues in the input file. This complicates the use of these programs for homology modeling. Also, only SCWRL provides a method for constraining the conformations of certain side chains, usually those that are preserved in amino acid type between the target sequence and template structure used for modeling.

Modeling programs which include backbone modeling also produce side-chain conformation predictions, although in most cases their side-chain prediction rates have not been studied. These programs include WhatIf, which uses a segment library (5–7 amino acids long) to determine a population of side-chain conformations for the central residue with a similar backbone conformation over the segment [194]. WhatIf's side-chain modeling is available on the web (see Table 5.6). MODELLER [16, 17] uses constraints based on the known side-chain conformations in the template structure. As such, it is not appropriate for modeling side chains on a backbone with no prior information on side-chain conformation. SwissModel [151] uses a backbone-independent rotamer library followed by energy minimization. ICM [147] uses a statistical potential energy function to sample side-chain conformations.

5.3.4
Methods for complete modeling

Homology modeling is a complex process. Automated protocols that begin with a sequence and produce a complete model are few, and the resulting models should be examined with great care (as of course should all models). But these methods usually allow for (and indeed recommend) some manual intervention in the choice of template structure or structures, and in the sequence alignment. In these steps, manual intervention is likely to have important consequences. Later stages of modeling (actual building of the structure) are more easily automated, and there are not usually obvious manual adjustments to make.

There are several publicly available programs available for homology modeling that are intended to make complete models from input sequences. These are MODELLER, developed by Andrej Šali and colleagues [16, 17, 41, 195], SwissModel, developed by Manuel Peitsch and colleagues [151, 196, 197], RAMP developed by Ram Samudrala and John Moult [193,

198, 199], and COMPOSER by Blundell and colleagues [13, 14]. Some details are summarized in Table 5.4. Program availability is given in Table 5.6. We described each program in turn.

5.3.5.1 Modeller

Modeller takes as input a protein sequence and a sequence alignment to the sequence(s) of known structure(s), and produces a comparative model. The program uses the input structure(s) to construct constraints on atomic distances, dihedral angles, and so forth, that when combined with statistical distributions derived from many homologous structure pairs in the Protein Data Bank, form a conditional probability distribution function for the degrees of freedom of the protein. For instance, a probability function for the backbone dihedrals of a particular residue to be built in the model can be derived by combining information in the known structure (given the alignment) and information about the amino acid type's Ramachandran distribution in the PDB. The number of restraints is very large; for a protein of 100 residues there may be as many as 20 000 restraints. The restraints are combined with the CHARMM force field to form a function to be optimized. This function is optimized using conjugate gradient minimization and molecular dynamics with simulated annealing.

Modeller provides some help with homologue identification and sequence alignment, but the full PDB should be searched outside of Modeller unless a very recent update of the program and database is obtained. Additional sequences of unknown structure can also be added to the information used by Modeller, although these must be obtained with other programs such as BLAST and FASTA. Modeller can combine the sequences and structures into a complete alignment which can then be examined using molecular graphics programs and edited manually.

5.3.5.2 SwissModel

SwissModel is intended to be a complete modeling procedure accessible via a web server that accepts the sequence to be modeled, and then delivers the model by electronic mail [151, 196]. In contrast to Modeller, SwissModel follows the standard protocol of homologue identification, sequence alignment, determining the core backbone, and modeling loops and side chains. SwissModel will search a sequence database of proteins in the PDB with BLAST, and will attempt to build a model for any PDB hits with p-values less than 10^{-5} and at least 30% sequence identity to the target (for a description of p-values and BLAST, see Chapter 2 of this volume). SwissModel allows for user intervention by specifying the template(s) and alignments to be used. If more than one structure is found, they will be superimposed on the template structure closest in sequence identity to the target. The structural superposition is accomplished by aligning the structures according to the initial sequence alignment, and then minimizing Cα-Cα distances be-

Tab. 5.4
Programs for complete homology modeling (alignment, loops, side chains)

Authors	Availability	Potential energy	Search method
Sutcliffe, Blundell [13, 14]	COMPOSER		Averaging several structures; picking fragments from each most similar to average
Sali, Blundell, Sanchez [16, 17, 195]	MODELLER	CHARMM + constraints from statistical analysis	Satisfaction of spatial restraints (prob. distribution functions) by minimization and simulated annealing
Yang, Honig [204]	PrISM		Multiple structures are aligned, and the most appropriate template is used for each segment of the target to be built. Loops are built *ab initio* and side chains are built using the template or based on mainchain torsions and a neural network algorithm.
Li, Tejero, Bruccoleri, Montelione [18, 19]	CONGEN + 2 set-up programs	CHARMM; $\varepsilon = r$	Distance constraints derived from known structure and alignment; random subset is used as input to CONGEN simulated annealing/ restrained molecular dynamics simulations
Peitsch, Guex [151, 196]	SwissModel	GROMOS	Alignment determines core backbone, while loops are built with database method. Side chains are built using the template side chain where possible, and GROMOS minimization otherwise.
Havel, Snow [15, 210]	DISGEO/Consensus (MSI, Inc.)	None	Distance geometry; constraints based on alignment and multiple structures
Samudrala, Moult [193, 198, 199, 209]	RAMP	Atom distance energy function from statistical analysis	Graph theory method that selects compatible conformations from multiple templates, multiple database loops, and side-chain rotamers

tween the proteins. This is in contrast to methods that compare *internal* distances in one structure with internal distances in the other, such as DALI [88] and CE [89].

SwissModel determines the core backbone from the alignment of the target sequence to the template sequence(s) by averaging the structures according to their local degree of sequence identity with the target sequence. The program builds new segments of backbone for loop regions by a database scan of the PDB using anchors of 4 Cα atoms on each end. This method is used to build only the Cα atoms, and the backbone is completed with a search of pentapeptide segments in the PDB that fit the Cα trace of the loop. Side chains are now built for those residues without information in the template structure by using the most common (backbone-independent) rotamer for that residue type. If a side chain can not be placed without steric overlaps, another rotamer is used. Some additional refinement is performed with energy minimization with the GROMOS [200] program.

SwissModel has been extensively tested by building over 1200 models of proteins of known structure, with template-target sequence identities ranging from 25 to 95%. It was found that 30% of models with sequence identity less than 30% could be built with backbone RMSD less than 3 Å. This rose rapidly with sequence identity, such that over 80% of models with sequence identity better than 50% could be built with RMSD's better than 3 Å. Assessments of side chain and loop modeling quality however have not been published.

5.3.5.3 CONGEN with Homology-derived constraints

Montelione et al. have used CONGEN [95, 201] to produce comparative models [18, 19, 202, 203]. Their method derives distance restraints from the template structure and the sequence alignment of the target and template for atoms they consider "homologous." These atoms include all backbone atoms and side-chain atoms of the same chemical type and hybridization state (i.e., sp^3, sp^2). In the vicinity of an insertion or deletion in the alignment (± 3 amino acids), no homologous atoms are defined so that loop regions are free to move. From the very large number of constraints generated by this procedure, a small fraction ($< 2\%$) are chosen randomly and used with restrained molecular dynamics and simulated annealing in the CONGEN program. CONGEN's side-chain loop and search routines can also be used. The method has only been tested on 4 proteins [18, 19], and so it is difficult to assess the accuracy.

5.3.5.4 PrISM

Yang and Honig have developed the PrISM package of programs that performs structure alignment, PDB sequence search, fold recognition, sequence alignment, secondary structure prediction, and homology modeling [204–208]. Its methods for homology modeling [204], in contrast to structure alignment [206–208], have not been published in detail. If multiple structures are available, PrISM aligns them and builds a composite template by selecting each secondary structure from the most appropriate template. Loop modeling is performed *ab initio* and side-chain dihedrals are taken either from the template or predicted based on mainchain torsion angles and a neural network algorithm.

5.3.5.5 **COMPOSER**

COMPOSER by Blundell and colleagues [13, 14] exploits the use of multiple template structures for building homology models. If a target sequence is related to more than one template (of different sequence) then all templates are used to provide an "average" framework for building the structure. Modeling of loops and side-chains is accomplished by borrowing from one of the template structure where possible, and by database methods in other cases.

5.3.5.6 **RAMP**

Samudrala and Moult described a method for "handling context sensitivity" of protein structure prediction, that is, simultaneous loop and side-chain modeling, using a graph theory method [198, 209] and an all-atom distance-dependent statistical potential energy function [199]. Their program RAMP is listed in Table 5.6.

5.3.5.7 **DISGEO/Consensus**

Havel and coworkers [15, 210] have described a method of producing comparative models by distance geometry based on restricted ranges for certain $C\alpha$-$C\alpha$ distances, dihedrals, and mainchain–side-chain distances. This method has become part of the InsightII package from Molecular Simulations, Inc.

5.4
Results

5.4.1
Range of targets

A very large number of homology models have been built over the years by many authors. Targets have included antibodies [211–216] and many proteins involved in human biology and medicine [19, 31, 217–256].

```
           * *
SecStr : EEEE EEE   EEEEEEEE   EEEEEEE   EEEE
1pso    1:VDEQPLENYLDMEYFGTIGIGTPAQDFTVVFDTGSSNLWVPSVYCSSLACTNHNRFNPED: 60
hbace1 62:EMVDNLRGKSGQGYYVEMTVGSPPQTLNILVDTGSSNFAVGAAPH----PFLHRYYQRQL:117
mbace1 62:EMVDNLRGKSGQGYYVEMTVGSPPQTLNILVDTGSSNFAVGAAPH----PFLHRYYQRQL:117
rbace1 62:EMVDNLRGKSGQGYYVEMTVGSPPQTLNILVDTGSSNFAVGAAPH----PFLHRYYQRQL:117
bbace1   :             LNILVDTGSSNFAVGAAPH----PFLHRYYQRQL:
sbace    :                                        YYDSEK:
zbace    :DMINNLKGDSGRGYYMQMIIGTPGQTLNILVDTGSSNFAVAAAAH----PYITHYFNRAL:
hbace2 79:AMVDNLQGDSGRGYYLEMLIGTPPQKLQILVDTGSSNFAVAGTPH----SYIDTYFDTER:134
mbace2 75:AMVDNLQGDSGRGYYLEMLIGTPPQKVQILVDTGSSNFAVAGAPH----SYIDTYFDSES:130


              **                                      *
SecStr :  EEEEE EEEEE    EEEEEEEEEEEE    EEEEEEEEEEEE  HHHHH
1pso   61:SSTYQSTSE-TVSITYGTGSMTGILGYDTVQVG----GISDTNQIFGLSETEPGSFLYYAP:116
hbace1 118:SSTYRDLRK-GVYVPYTQGKWEGELGTDLVSIPH-GPNVTV-RANIAAITESDKFFINGSN:175
mbace1 118:SSTYRDLRK-GVYVPYTQGKWEGELGTDLVSIPH-GPNVTV-RANIAAITESDKFFINGSN:175
rbace1 118:SSTYRDLRK-SVYVPYTQGKWEGELGTDLVSIPH-GPNVTV-RANIAAITESDKFFINGSN:175
bbace1   :SSTYRDLRK-GVYVPYTQGKWEGELGTDLVSIPH-GPNVTV-RANIAAITESDKF------:
sbace    :SSSVINSGIPDVDIEYTEGFWKGPLVTDLVSIPEAGLTEQV-RVDIVKITSSKKFFINGSG:
zbace    :SSTYQSTER-AVAVKYTQGEWEGELGTDLITIP
hbace2 135:SSTYRSKGF-DVTVKYTQGSWTGFVGEDLVTIPK-GFNTSF-LVNIATIFESENFFLPGIK:192
mbace2 131:SSTYHSKGF-DVTVKYTQGSWTGFVGEDLVTIPK-GFNSSF-LVNIATIFESENFFLPGIK:188


             *
SecStr :  EEEE        HHHHHHHH     EEEEE
1pso   117:FDGILGLAYPSISS--S-GATPVFDNIWNQGLVSQDLFSVYLSADD----------QSGS:163
hbace1 176:WEGILGLAYAEIARPDD-SLEPFFDSLVKQTHVP-NLFSLQLCGAGFPLNQSEVLASVGG:233
mbace1 176:WEGILGLAYAEIARPDD-SLEPFFDSLVKQTHIP-NIFSLQLCGAGFPLNQTEALASVGG:233
rbace1 176:WEGILGLAYAEIARPDD-SLEPFFDSLVKQTHIP-NIFSLQLCGAGFPLNQTEALASVGG:233
sbace    :WQGIIGLGYDELVRPNNPKVKSFMTSVIENTSVR-NVFSIQLCAA----NTMNFSDVTTG:
hbace2 193:WNGILGLAYATLAKPSS-SLETFFDSLVTQANIP-NVFSMQMCGAGLPVAGS---GTNGG:247
mbace2 189:WNGILGLAYAALAKPSS-SLETFFDSLVAQAKIP-DIFSMQMCGAGLPVAGS---GTNGG:243


                 *                                      *
SecStr :EEEE        EEEE     EEEEEEEEE EEEE     EEEEEEEE
1pso   164:VVIFGGIDSSYYTGSLNWVPVTVEGYWQITVDSITMNGEAIAC----AEGCQAIVDTGTS:219
hbace1 234:SMIIGGIDHSLYTGSLWYTPIRREWYYEVIIVRVEINGQDLKMDCKEYNYDKSIVDSGTT:293
mbace1 234:SMIIGGIDHSLYTGSLWYTPIRREWYYEVIIVRVEINGQDLKMDCKEYNYDKSIVDSGTT:293
rbace1 234:SMIIGGIDHSLYTGSLWYTPIRREWYYEVIIVRVEINGQDLKMDCKEYNYDKSIVDSGTT:293
bbace1 143:---IGGIDHSLYMGSLWYTPIRREWYYEVIIVRVRINGQDLKMDCKEYNYDKSIVDSGTT:200
sbace    :SLVFGDYDRT--DGTIFRTRIVHEWYYEVIVLGMKV    CREFNNDKSIVDSGTT:
hbace2 248:SLVLGGIEPSLYKGDIWYTPIKEEWYYQIEILKLEIGGQSLNLDCREYNADKAIVDSGTT:307
mbace2 244:SLVLGGIEPSLYKGDIWYTPIKEEWYYQIEILKLEIGGQNLNLDCREYNADKAIVDSGTT:303
rbace2   :               ILKLEIGGQSLNLDCREYNADKAIVDSGTT:


           +
SecStr :EEE HHHHHHHHHHH  EE          EE EEEEEE    EEEEE
1pso   220:LLTGPTSPIANIQSDIGASENSGD-------MVVSCSAISSLPDIVFTIN--------:260
hbace1 294:NLRLPKKVFEAAVKSIKAASSTEKFPDGFWLGEQLVCWQAGTTPWNIFPVISLYLMGEV:352
mbace1 294:NLRLPKKVFEAAVKSIKAASSTEKFPDGFWLGEQLVCWQAGTTPWNIFPVISLYLMGEV:352
rbace1 294:NLRLPKKVFEAAVKSIKAASSTEKFPDGFWLGEQLVCWQAGTTPWNIFPVISLYLMGEV:352
bbace1 201:NLRLPKKVFEAAVKSIKAASSTEKFPDGFWLGEQLVCWQAGTTPWNIFPVISLYLMGEV:259
sbace    :NLHLPEKVFN
hbace2 308:LLRLPQKVFDAVVEAVARASLIPEFSDGFWTGSQLACWTNSETPWSYFPKISIYLRDEN:366
mbace2 304:LLRLPQKVFDAVVEAVARTSLIPEFSDGFWTGAQLACWTNSETPWAYFPKISIYLRDEN:362
          LRLPQKVFDAVVEAVARTSLIPEFSDGFWTGAQLACWTNSETPWAYFPKISIYLRDEN:
```

Fig. 5.3
Multiple sequence alignment of BACE family members and human pepsin (Protein Data Bank entry 1pso, used as the basis of homology modeling). Identical residues are shaded gray, highly conserved residues in yellow, and active-site residues in the model in contact with the peptide substrate are indicated by an asterisk (*). BACE residue Arg 296, which is most often serine or threonine at this position in aspartyl proteases, is shown with a plus sign (+). The secondary structure of the human pepsin template structure is given to show that gaps in the alignment occur between elements of secondary structure (H is α-helix; E is β-strand). In sequence names, h = human; m = mouse; r = rat; b = bovine; z = zebrafish; s = sea squirt (*Halocynthia roretzi*). Fragment sequences of bbace1, rbace2, bbace2, and sbace were derived from EST sequences.

Two databases of homology models are available on the Internet – Mod-Base and SwissModel. ModBase [257], produced by Andrej Šali and colleagues at Rockefeller University, contains homology models built with the program MODELLER [16] for as many sequences in several complete genomes that can be modeled with structural homologues of better than 30% sequence identity. The models include evaluations for predicting their accuracy.

5.4.2
Example: amyloid precursor protein β-secretase

Alzheimer's disease is characterized by plaques in the brain consisting primarily of the 40–42 amino acid amyloid β-peptide (Aβ) [258]. Aβ derives from proteolysis of the amyloid precursor protein (APP) by the β and γ secretases to create the N and C-termini of the peptide respectively [259]. The β-secretase has recently been identified as a 501 amino-acid transmembrane protein by several research groups [260–263]. The enzyme, variously named BACE, memapsin2, and Asp2, is an aspartic protease related to pepsin, cathepsin D, and renin, with all the properties expected of the β-secretase.

Because BACE is a potential target for drug design to treat or prevent Alzheimer's disease, we recently built models of the BACE extracellular domain based on crystal structures of two aspartyl proteases. As described above, perhaps the most important step in homology modeling is to gather available experimental data on the protein's function. Even before the identification of BACE, the specificity of the β-secretase had already been studied extensively [264]. The sequence of APP at the protease cleavage site is EVKM-DAEF, where the protease breaks the peptide bond between M and D. These residues are labeled P4-P3-P2-P1-P1'-P2'-P3'-P4' in standard protease nomenclature. The protease also cleaves at another position within the Aβ peptide, at Glu 11, with sequence DSGY-EVHH. It is clear that there is at least some preference for a negatively charged residue at the P1' position, an unusual feature in aspartic proteases. This led us to hypothesize that the substrate binding site was very likely to contain a positively charged residue.

BACE also cleaves at the same location of APP with the so-called "Swedish mutation", KM → NL at P2-P1, found in families with early-onset Alzheimer's disease. This indicates that a hydrophobic residue seems to be preferred at the P1 position and a hydrophilic residue at P2. One other piece of data was a mutant of the APP substrate of M → V at position P1, which greatly reduced activity of BACE for the substrate [264, 265].

The second step was to gather *orthologous* sequences of BACE to define the family of sequences to which BACE belongs. We searched both the non-redundant protein sequence database at NCBI as well as the EST databases. The resulting alignment of BACE orthologues is shown in Figure 5.3

(s. p. 189). In mammals, there are two closely related sequences, now called BACE1 and BACE2, that appear to have arisen from a recent duplication. In zebrafish and seq squirts, there is a single BACE sequence in EST databases that shares feature of both BACE1 and BACE2. We include BACE1 and BACE2 in the alignment of orthologues, even though these sequences exist in the same species, because they are very closely related and their specific functions have not yet been determined.

To build the model we used PSI-BLAST to search the non-redundant protein sequence database and to produce a position-specific scoring matrix for BACE. This matrix was used to search the Protein Data Bank to identify homologues of known structure that could be used as templates to build a model of BACE. This process identified many aspartyl proteases. After looking at the alignments, we decided to use a 2.0 Å resolution crystal structure of human pepsin in a complex with pepstatin (Protein Databank entry 1PSO) [266]. This structure had no insertions and deletions in close proximity to the enzyme active site, while most other aspartyl proteases had insertions in the loops covering the active site. We were also interested in building in a model of the substrate sequence from APP. The crystal structures of most aspartyl proteases in the PDB contain a peptide-like inhibitor in the active site. To make these small molecules inhibitors rather than substrates, the peptide bond at the cleavage site is replaced with non-hydrolysable linkages. This is almost always accomplished with an extra atom or two in the backbone of the inhibitor. The only exception to this is a reduced peptide inhibitor in a 1.8 Å crystal structure of rhizopuspepsin (PDB entry 3APR) [267]. In this inhibitor, the carbonyl group of the scissile bond has been replaced with a methylene group ($-CH_2-$). We chose this substrate to model the APP substrate since the backbone of the substrate to be modeled could be borrowed directly from the substrate in the rhizopuspepsin structure. To model the interaction of the substrate with the enzyme, we superimposed the structure of the reduced peptide inhibitor onto the peptide portion of the pepstatin inhibitor in 1PSO. The backbone atoms of the peptide inhibitor before the scissile bond superimpose with the pepstatin backbone with an RMSD of 0.16 Å.

In Figure 5.4 we show the major proposed interactions of the APP substrate with the enzyme. For the wild-type peptide in Figure 5.4A with sequence EVKMDA, the P1' aspartic acid makes a salt-bridge with Arg296 of the enzyme. This is the hypothesized positive charge in the BACE active site. It is conserved in the closely related human transmembrane protein BACE2 [268], which indicates that this enzyme is likely to have similar substrate specificity to BACE. In other aspartic proteases, this residue is an Ala, Ser, Thr, Val, Leu, or Ile residue. In the model, the P2 lysine also makes a salt-bridge with the enzyme at Asp379. There are several hydrophobic contacts between the P1 methionine and enzyme residues Leu91, Tyr132, and Ile179, while the P3 valine interacts with Phe170.

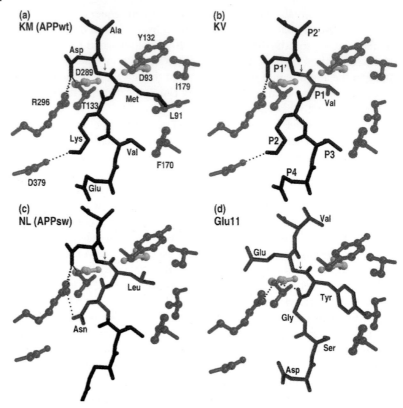

Fig. 5.4
Models of β-secretase with APP-derived substrates. (**A**) Wild-type substrate residues 592–597 of APP (GenBank accession CAA312830.1) with sequence EVKM*DA (P4-P3-P2-P1-P1'-P2') with cleavage site marked with *. (**B**) Mutant substrate with Val (red) substitute for P1 Met (sequence EVKV*DA). (**C**) Swedish mutation substrate with P2-P1 sequence of Asn-Leu (red) in place of wild-type Lys-Met (sequence EVNL*DA). (**D**) Alternative β-secretase cleavage site at APP residue Glu607 (P1') (sequence DSGY*EV). The figure was generated with MOLSCRIPT [287], Raster3D [288], and the GIMP [289].

We also modeled the mutant substrate EVKVDA shown in Figure 5.4B. The differences in interaction between the side chain and the enzyme explain why this substrate is cleaved less well by BACE [264]. While the $C_{\gamma 1}$ of Val and C_γ of Met are in the same location ($\chi_1 = -65°$), the $C_{\gamma 2}$ atom of Val contacts the catalytic Asp93 C_γ, $O_{\delta 1}$, and $O_{\delta 2}$ atoms. This is likely to interfere with the catalytic activity. Also, many of the hydrophobic contacts between P1 Met and the enzyme are missing in P1 Val. In Figure 5.4C a model of the "Swedish mutation" KM → NL at P2, P1, which is associated with early-onset Alzheimer's disease [269], shows that the Asn can hydrogen bond to

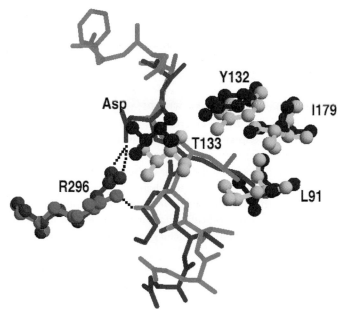

Fig. 5.5

Superposition of BACE crystal structure with homology model of BACE. The crystal structure of BACE (PDB entry 1FKN) was superimposed on the model using the MIDAS program [290]. The model peptide is shown in green stick figure and the crystal structure peptide is shown in red.

The crystal structure side-chains in contact with the peptide are shown in blue ball-and-stick figures, while the model side-chains are shown in yellow. Crystal structure Arg296 is shown in red and model structure Arg296 is shown in green, with a hydrogen bond depicted by dotted lines to Asp at P1′.

Arg296, which also hydrogen bonds to the P1′ Asp. The leucine also makes good hydrophobic contacts with Leu91, Tyr132, and Ile179. In Figure 5.4D, the minor cleavage site for β-secretase, at Glu11 of the Aβ peptide, is modeled as the substrate. The P1′ Glu is within hydrogen bonding distance of Arg296, although the hydrogen bond geometry is poor. This may explain why the Glu11 site is cleaved by BACE, but only as a minor component of BACE activity on APP.

The crystal structure of BACE was recently determined by Tang and colleagues [270]. In Figure 5.5, we show the substrate binding site from a structure alignment of the model and the crystal structure. A number of the substrate-specificity-determining residues were correctly determined both in location and conformation. In this crystal structure, an inhibitor is bound in the active site. This inhibitor has sequence EVNL-AAEF, where the L-A sequence has a non-peptide backbone. The P1′ position is occupied by an alanine rather than a negatively charged residue, as in both APP substrate positions. In our model, Arg296 forms a salt-bridge to Asp at P1′. This arginine is in nearly the exact same position in the crystal structure, but in a different rotamer and making a hydrogen bond to substrate Asn at P2. In the natural substrate, P2 is Lys and therefore very unlikely to make a hydrogen bond to Arg296. We believe that with a conformational change, Arg296 is easily able to form a salt bridge with P1′ negatively charged residues and is thus one of the prime determinants of specificity of BACE for APP. It is notable that because the crystal structure did not contain the native substrate sequence this interaction of Arg296 with P1′ was not discussed by the crystallographers, even though it is the only evident positively charged residue able to interact with P1′ side chains.

It is clear from this example that choosing templates and modeling goals is a process that takes some care and understanding of biological function. Automated procedures are less than satisfactory for the purposes described here.

5.5
Strengths and limitations

The strengths of homology modeling are based on the insights provided for protein function, structure, and evolution which would not be available in the absence of an experimental structure. In many situations, a model built by homology is sufficient for interpreting a great deal of experimental information and will provide enough information for designing new experiments. Homology modeling may also provide functional information beyond the identification of homologous sequences to the target. That is, a model may serve to distinguish orthologous and paralogous relationships.

The limitations are due to decreasing accuracy as the evolutionary distance between target and template increases. Alignment becomes more uncertain, insertions and deletions more frequent, and even secondary structural units may be of different lengths, numbers, and positions in very remote homologues. Predicting the locations of secondary structure units that are not present in the template structure is a difficult problem and there has been little attention paid to it.

The limitations of homology modeling also arise when we have insufficient information to build a model for an entire protein. For instance, we may be able to model one or more domains of a multi-domain protein or a multisubunit complex, but it may not be possible to predict the relative organization of the domains or subunits within the full protein. This remains a challenge for further research. And we are of course limited by structures present in the PDB, which are almost exclusively soluble proteins. Up to 30% of some genomes are membrane proteins, which are at present difficult to model because of the small number of membrane proteins of known structure. The recent structure of the G-protein-coupled receptor rhodopsin [271] creates an opportunity to model many of these membrane proteins more accurately than the previously available structure of bacteriorhodopsin [272, 273].

Another problem is the quality of data in sequencing and structure determination. There are substantial errors in determining protein sequences from genome sequences, either because of errors in the DNA sequence or in locating exons in eukaryotic DNA [274]. Over 50% of X-ray structures are solved at relatively low resolution, levels of greater than 2.0 Å. Despite progress in determining protein structures by NMR, these structures are of lower resolution than high quality X-ray structures. While high throughput structure determination will be of great value to modeling by homology, one concern is the quality of structure determination when the function of the proteins being determined is unknown.

5.6
Validation

Validation for homology modeling is available in two distinct ways: 1) the prediction rates for each method based on the prediction of known structures given information from other structures; 2) criteria used to judge each model individually. As shown in Tables 5.2–5.4, most structure prediction method papers have included predictions of known structures, serving as test sets of their accuracy. However, in many cases the number of test cases is inadequate. It is also very easy to select test structures that behave particularly well for a given method, and many methods do not stand up to scrutiny of large test sets performed by other researchers. Test sets vary in number of test cases as well as whether predictions of loops or side chains are performed by building replacements on the template structure scaffold, or in real homology modeling situations where the loops/side-chains are built on non-self scaffolds. The realistic case is more difficult to perform in a comprehensive way, since it requires many sequence-structure alignments to provide the input information on which models are to be built. Another problem is that each method is judged using widely varying criteria, and so no head-to-head comparison is possible from the published papers. The problem of biased test sets and subsequent development of larger benchmarks has a long history in the secondary structure prediction field [275, 276], but testing of loop and side-chain methods in this way has been much more limited.

5.6.1
Side-chain prediction accuracy

To remedy this situation for protein side-chain prediction, we developed a benchmark and tested several publicly available side-chain prediction programs. In the original publications, these methods were tested on very different test sets and using varying criteria for a "correct" prediction. In some cases, the test sets are rather small. To see why, recall that the standard deviation of the mean of a sample is the population standard deviation divided by the square root of the number of data points. If we treat the fraction of correct χ_1 rotamers for a single protein as a data point, we observe that the standard deviation of the data is approximately 0.06 (6%) on a large test set. If a test set consists of 10 proteins, then the standard deviation of the mean is $0.06/\sqrt{10} = 0.02$. The 95% confidence interval is therefore the mean ± 0.04, or an eight percentage point spread. Larger test sets are therefore called for.

We performed a test on 180 monomeric (at least within each asymmetric unit of the crystal structure) proteins with resolution better than 1.8 Å and less than 30% sequence identity between any two proteins in the list. The same criteria were used to assess each of the methods: a χ angle prediction within 40° of the crystal structure conformation. Pro residues were judged correct if the χ_1 angle was within 20° of the crystal structure conformation. All residues other than Gly and Ala were included. The results for the 18 amino acid types and for all residues together are shown in Table 5.5. In this table, "bbdep" means the prediction that would be made simply by choosing the most common rotamer in the backbone-dependent rotamer library, that is, the rotamer determined by ϕ, ψ for each residue. "bbind" indicates a prediction based on simply choosing the most common rotamer for each side chain in the backbone-independent rotamer library of Dunbrack. In this prediction, all non-Pro side chains are given a χ_1 dihedral of approximately $-60°$, except for Val which is most commonly near 180° (trans), and Ser and Thr which are most commonly near $+60°$. The most interesting result is that some of the methods currently available provide *worse* predictions than simply choosing the most common rotamer in the backbone-dependent rotamer library, a prediction that proceeds with simply looking up a conformation in a table. The bbdep library on its own can predict 73% of χ_1 rotamers correctly. SCWRL, Confmat, and Torso perform better than this prediction by optimizing rotamers based on interactions between side chains and between side chains and the backbone.

5.6.2
The CASP meetings

Another forum for testing homology modeling methods has been the ongoing series of CASP meetings (for "Critical Assessment of Protein Struc-

Tab. 5.5
Side chain program comparison

Residue	Number	BBIND	BBDEP	CCNFMAT	RAMP	SCWRL	SIDEMOD	SMD	TORSO
ILE	2127	75.9	87.6	91.1	54.1	91.3	89.7	66.3	90.7
VAL	2779	70.7	85.9	88.7	72.3	88.8	87.6	65.3	88.4
THR	2439	48.0	84.5	78.5	80.5	87.1	69.0	65.1	80.6
PRO	1996	60.4	76.8	51.1	*	74.8	39.2	*	52.9
PHE	1644	53.5	74.9	90.4	70.3	91.2	87.7	68.1	91.8
TYR	1626	53.0	74.2	88.0	67.3	90.0	81.6	65.0	89.8
LEU	3396	66.1	73.3	86.7	68.5	87.1	78.3	65.2	87.9
ASP	2483	50.2	71.3	75.8	62.4	81.1	58.2	57.9	70.0
HIS	895	54.3	70.8	82.8	63.4	83.1	74.4	59.5	84.1
CYS	678	57.8	70.5	85.3	60.5	83.0	78.0	81.5	78.5
TRP	668	46.8	69.9	83.9	60.5	86.1	76.1	57.9	84.0
ASN	2027	52.6	67.7	74.3	55.9	77.6	62.2	53.2	74.7
LYS	2311	56.4	67.4	71.4	58.9	73.2	57.1	57.0	73.1
GLN	1562	60.0	65.9	73.8	56.8	73.6	63.9	54.2	73.3
ARG	1885	57.1	65.4	75.4	59.5	74.7	64.0	54.7	69.0
MET	788	59.5	65.0	74.5	59.3	75.5	70.8	53.0	77.0
SER	2661	44.8	62.4	57.4	56.8	64.2	38.6	41.4	60.1
GLU	2255	55.0	62.0	67.3	52.6	68.7	56.8	51.3	66.8
ALL	34220	57.4	72.6	77.2	63.0	80.5	66.6	61.5	76.8

Percentage χ_1 dihedral angle prediction within 40° of crystal structure (20° for Pro) for predictions on 180 proteins of known structure, with resolution better than or equal to 1.8 Å. Residue types are sorted in increasing difficulty of prediction, according to backbone-dependent rotamer library prediction ("BBDEP"). "BBIND" column lists backbone-independent rotamer library prediction. The programs tested were CONFMAT [173], RAMP [193], SCWRL [93, 94], Sidemod [192], SMD [165, 191], and TORSO [319].

* RAMP and SMD do not predict proline conformations.

ture Prediction") organized by John Moult and colleagues [96–98, 277–280]. In the spring and summer before each meeting held in December 1994, 1996, 1998, and 2000, sequences of proteins whose structure was under active experimental determination by NMR or X-ray crystallography were distributed via the Internet. Anyone could submit structure predictions at various levels of detail (secondary structure predictions, sequence alignments to structures, and full 3-dimensional coordinates) before specific expiration dates for each target sequence. The models were evaluated via a number of computer programs written for the purpose, and then assessed by experts in each field, including comparative modeling, fold recognition, and *ab initio* structure prediction. The organizers invited predictors whose predictions were outstanding to present their methods and results at the meeting, and to described their work in a special issue of the journal Proteins, published in the following year.

Ordinarily when protein structure prediction methods are developed, they are tested on sets of protein structures where the answer is known. Unfortunately, it is easy to select targets, even subconsciously, for which a particular method under development may work well. Also, it is easy to optimize parameters for a small test set that do not work as well for larger test sets. While the number of prediction targets in CASP is limited to numbers on the order of 10–20 per category, these numbers are still higher than many of the test sets used in testing new methods under development, as shown in Tables 5.2 and 5.3.

5.6.3
Protein health

A number of programs have been developed to ascertain the quality of experimentally determined structures and these can be used to determine whether a protein model obeys appropriate stereochemical rules. The two most popular programs are ProCheck [281–283] and WhatCheck [284]. These programs check bond lengths and angles, dihedral angles, planarity of sp^2 groups, non-bonded atomic distances, disulfide bonds, and other characteristics of protein structures. One of the more useful checks is to see whether backbone geometries are in acceptable regions of the Ramachandran map. Backbone conformations in the forbidden regions are very likely to be incorrect. We have also developed a program, *bbdep*, that assesses whether side-chain rotamers in a structure are low-energy or high-energy conformations. The program reports unusual rotamers for the ϕ, ψ positions of each residue, as well as unusual $\chi_1, \chi_2, \chi_2, \chi_4$ rotamer combinations, and unusual dihedral angles.

It should be noted once again that correct geometry is no guarantee of correct structure prediction. In some cases, it may be better to tolerate a few steric conflicts or bad dihedral angles, rather than to minimize the structure's energy. While the geometry may look better, the final structure may be further away from the true structure (if it were known) than the un-minimized structure.

The availability of these programs is listed in Table 5.6.

5.7
Availability

Many programs are publicly available for the various steps in homology modeling and for evaluating and comparing structures, and for sequence alignment. I define "publicly available" as those programs which can be downloaded or used on a webserver for free or for a nominal charge to academic and non-profit research groups. There are commercial programs that cost on the order of $2000–$3000 per year or more (such as InsightII and ICM), and these are not included in this list. I have included as many programs as I could find by searching the Internet and contacting authors of papers that present methods for loop, side-chain, or full comparative modeling. The websites or electronic mail addresses for publicly available programs are listed in Table 5.6.

5.8
Appendix

5.8.1
Backbone conformations

In this Appendix, we describe some of the basic structural properties of proteins. Proteins are heteropolymers of 20 amino acid types with a backbone consisting of –NH-C(H)R-CO– (except proline) where R is some functional group defined for each amino acid type. A fragment of 3 amino acids is shown in Figure 5.6. As with all organic molecules, there are characteristic bond lengths and angles associated with the types of chemical bonds formed, and proteins have been found to conform to the properties of small organic molecules as determined by very high resolution structures of small peptides [285]. The structures of proteins can therefore be defined approximately by a set of *dihedral angles* that determine the orientation of chemical groups in the peptide chain. As shown in Figure 5.7, a dihedral angle is defined for a set of 4 bonded atoms, say A-B-C-D, as the angle between the planes A-B-C and B-C-D. Looking down the B-C bond (atom B nearest the viewer, C further away), a positive dihedral angle is determined by rotation of bond A-B (or rather plane A-B-C) in a counter-clockwise

Tab. 5.6
Web-sites and publicly available programs

	Down-load	Web-server	Internet address
Structure alignment			
CE	X	X	http://cl.sdsc.edu/ce.html
Dali	X	X	http://www2.ebi.ac.uk/dali/
VAST		X	http://www.ncbi.nlm.nih.gov:80/ Structure/VAST/vast.shtml
MINAREA	X		http://www.cmpharm.ucsf.edu/cohen/ pub/minarea.html
LOCK	X	X	http://gene.stanford.edu/lock/
SARF2		X	http://genomic.sanger.ac.uk/123D/ sarf2.html
SSAP	X		http://www.biochem.ucl.ac.uk/ ~orengo/ssap.html
CATH			http://www.biochem.ucl.ac.uk/bsm/ cath/
STAMP		X	http://www.hgmp.mrc.ac.uk/ Registered/Option/stamp.html
COMPARER	X	X	http://www-cryst.bioc.cam.ac.uk/ ~robert/cpgs/COMPARER/ comparer.html
Loop libraries			
Protein Loop Classification (Oliva and Sternberg)		X	http://www.bmm.icnet.uk/loop/ index.html
Sloop (Burke, Deane, Blundell)		X	http://www-cryst.bioc.cam.ac.uk/ ~sloop/
Lessel & Schomburg	X		ftp://ftp.uni-koeln.de/institute/ biochemie/pub/loop_db
Loop prediction methods			
BRAGI	X		ftp://ftp.gbf.de/pub/Bragi/
BTPRED		X	http://www.biochem.ucl.ac.uk/bsm/ btpred
CODA		X	http://www-cryst.bioc.cam.ac.uk/ ~charlotte/Coda/coda.html
RAMP	X		http://www.ram.org/computing/ramp/ ramp.html
CONGEN	X		http://www.congenomics.com/congen/ congen_toc.html
Drawbridge	X		http://www.cmpharm.ucsf.edu/cohen/ pub/
Confmat	X		Contact: koehl@allegro.stanford.edu
Protein Loop Classification (Oliva et al.)		X	http://www.bmm.icnet.uk/loop/
Swiss-PdbViewer	X	X	http://www.expasy.ch/spdbv
Rotamer libraries			
Backbone-dependent rotamer library (Dunbrack)	X		http://www.fccc.edu/research/labs/ dunbrack/sidechain.html

Tab. 5.6 (continued)

	Down-load	Web-server	Internet address
Lovell	X		http://kinemage.biochem.duke.edu/ website/rotamer.htm
DeMaeyer			http://www.fccc.edu/research/labs/ dunbrack/sidechain/demaeyer.rot
Tuffery			http://condor.urbb.jussieu.fr/ rotamer.html
Ponder & Richards	X		http://www.fccc.edu/research/labs/ dunbrack/sidechain/ ponder_richards.rot
Side-chain prediction methods			
Confmat	X		Contact: Patrice Koehl, koehl@allegro.stanford.edu
FAMS		X	http://physchem.pharm.kitasato-u.ac.jp/ FAMS/fams.html
RAMP	X		http://www.ram.org/computing/ramp/ ramp.html
SCWRL	X		http://www.fccc.edu/research/labs/ dunbrack/scwrl
Segmod/CARA (Genemine/Look3)	X		http://www.bioinformatics.ucla.edu/ ~genemine
Sidemod	X		http://www.rtc.riken.go.jp/~hkono/ SideChain
SMD	X		http://condor.urbb.jussieu.fr/Smd.html
torso	X	X	http://www2.ebi.ac.uk/dali/maxsprout/ maxsprout.html
WhatIf	X	X	http://www.cmbi.kun.nl/whatif
General Modeling Programs			
COMPOSER	X		http://www-cryst.bioc.cam.ac.uk/
MODELLER	X		http://guitar.rockefeller.edu/modeller/ modeller.html
PrISM	X		http://www.columbia.edu/~ay1/
CONGEN (Montelione)	X		http://www-nmr.cabm.rutgers.edu/ software/html/nmr_software.html
RAMP	X		http://www.ram.org/computing/ramp/ ramp.html
SwissModel		X	http://www.expasy.ch/swissmod/ SWISSMODEL.html
WhatIf	X	X	http://www.cmbi.kun.nl/whatif
Protein health			
ProCheck	X		http://www.biochem.ucl.ac.uk/~roman/ procheck/procheck.html
WhatCheck	X		http://www.cmbi.kun.nl/whatif/ whatcheck/
Promotif	X		http://www.biochem.ucl.ac.uk/bsm/ promotif/promotif.html
BBDEP	X		http://www.fccc.edu/research/labs/ dunbrack/sidechain.html

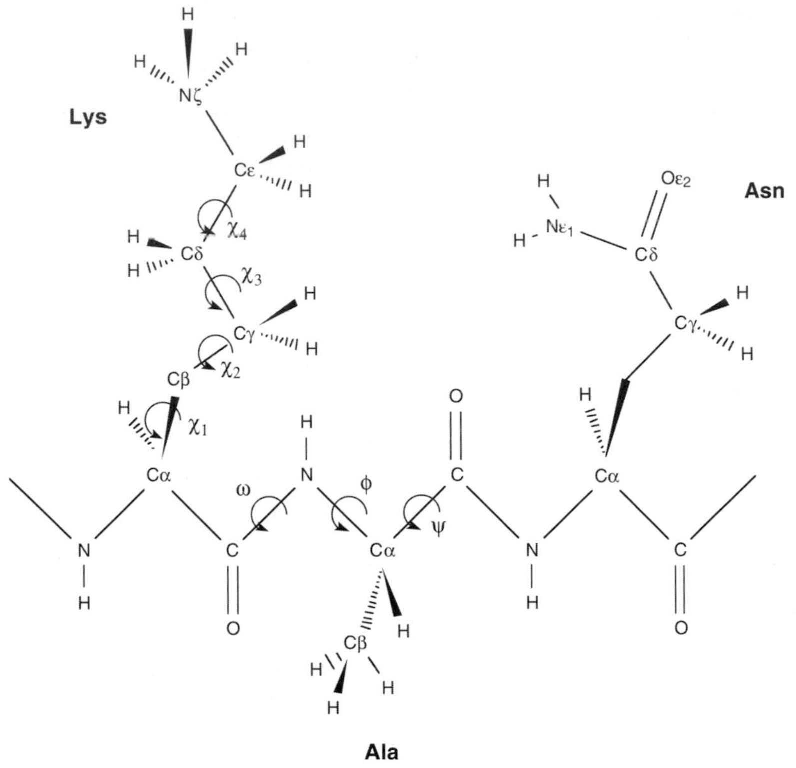

Fig. 5.6
Diagram of tripeptide with sequence Lys-Ala-Asn.
Backbone dihedrals ϕ, ψ, ω of the central Ala residue are
indicated. Side-chain dihedrals χ_1, χ_2, χ_3, and χ_4, of Lys are
also indicated.

direction about the B-C axis. Alternatively, a positive dihedral is determined
by rotation of bond C-D in a clockwise direction about the B-C axis.

Along the protein backbone, the dihedral angles ω, ϕ, and ψ are used
to describe the conformation, as shown in Figure 5.6. These are defined
as $\omega_i = C\alpha_{i-1}\text{-}C_{i-1}\text{-}N_i\text{-}C\alpha_i$, $\phi_i = C_{i-1}\text{-}N_i\text{-}C\alpha_i\text{-}C_i$, $\psi_i = N_i\text{-}C\alpha_i\text{-}C_i\text{-}N_{i+1}$. Because
the peptide bond between C and N has partial double-bond character, the
amide group is nearly flat. That is, the atoms N, Cα, H, C_{i-1}, O_{i-1}, and
Cα_{i-1} all lie in a plane. ω is therefore always close to 180° (*trans*) or 0° (*cis*).
For all residues except proline, ω is nearly always *trans* (less than 0.1% are
cis). Approximately 5% of proline peptide bonds are *cis*.

Only certain ϕ, ψ conformation pairs are observed. This is usually described in terms of the Ramachandran diagram, as shown in Figure 5.8,
with ϕ as the horizontal axis and ψ as the vertical axis. In Figure 5.8, the
experimentally determined backbone dihedral angles for the alanine, glycine, and proline residues of 699 proteins are depicted. There are three basic

Fig. 5.7
Dihedral angle definition. The
dihedral angle of atoms A,B,C,D is
shown as (**A**) a positive angle by
counter-clockwise rotation of the
B-A bond about the B-C bond;
(**B**) a negative angle by clockwise
rotation of B-A; (**C**) a positive
angle by clockwise rotation of the
C-D bond; (**D**) a negative angle by
counterclockwise rotation of C-D.

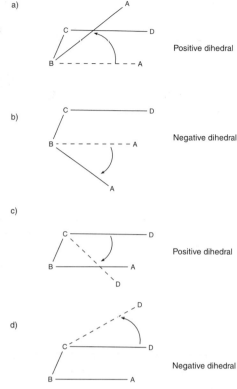

factors that determine the broad features of backbone conformations observed in proteins as shown in Figure 5.8: (1) steric interactions along the
protein backbone that prevent the existence of some conformations; (2)
electrostatic interactions in the formation of secondary structures by hydrogen bonding of backbone NH and CO groups; and (3) backbone-side-chain
interactions, both steric and electrostatic (and covalent in the case of proline), that account for the variation in Ramachandran distributions among
the 20 amino acids. We describe each of these in turn.

5.8.1.1 Steric interactions

Both the ϕ and ψ dihedrals are composed of 4 atoms with hybridization sp^2-
sp^2-sp^3-sp^2. The energy of a conformation with either ϕ or ψ approaching $0°$
is up to 5 kcal/mol higher than the energy minimum because of steric
interactions of the terminal atoms. In the case of ϕ these are the carbonyl
carbon of succeeding residues. In the case of ψ, these are the nitrogen
atoms of succeeding residues. The Ramachandran distribution for alanine
clearly demonstrates that these conformations ($\phi \sim 0°$ or $\psi \sim 0°$) are rare in

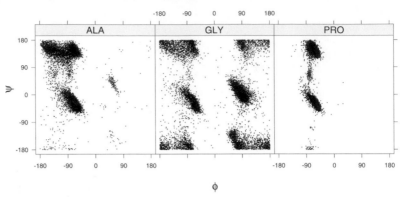

Fig. 5.8
Ramachandran plot. The density of ϕ, ψ values for alanines, glycines, and prolines in 699 proteins of known structure. Proteins were selected from the PDB with resolution better than or equal to 1.8 Å and less than 50% sequence identity among all pairs in the list using the *culledpdb* algorithm (http://www.fccc.edu/research/labs/dunbrack/culledpdb.html).

the database. We must also consider the presence of the β carbon in all non-glycine amino acids. The dihedral C_{i-1}-N_i-$C\alpha_i$-$C\beta_i$, is equal to $\phi_i - 120°$. This dihedral is near 0° when $\phi = +120°$. There are very few alanines (or other residues with β carbons) in the database with this conformation. The dihedrals $C\beta_i$-$C\alpha_i$-C_i-N_{i+1} and $C\beta_i$-$C\alpha_i$-C_i-O_i are equal to $\psi_i + 120°$ and $\psi_i - 60°$ respectively. These dihedrals are 0° when ψ_i is $-120°$ and $+60°$. Conformations with ψ_i near these values are also clearly relatively rare for residues with $C\beta$ atoms. Because glycine lacks $C\beta$ (its side chain is simply a hydrogen atom), glycine can not incur these steric penalties. In Figure 5.8B, the Ramachandran conformations of glycine residues in the same 699 proteins are plotted.

Another important effect is the steric interactions of non-hydrogen atoms separated by 4 covalent bonds. Let us label them A-B-C-D-E. There are two dihedrals that determine the separation in space of atoms A and E in this chain: $\theta_1 = $ A-B-C-D and $\theta_2 = $ B-C-D-E. When both θ_1 and θ_2 are 0°, atoms A and E are very close together. The distance between A and E falls off quickly when both θ_1 and θ_2 increase, or when they both decrease. The distance falls off much more slowly when θ_1 increases while θ_2 decreases or *vice versa*. In Figure 5.9, the distance between atoms A and E is shown as a function of the two dihedrals θ_1 and θ_2, assuming the bond lengths are 1.55 Å and the bond angles are 115° (the average of sp^2 and sp^3 angles). To determine the effect of this interaction on Ramachandran distributions we have to consider all chains of 5 atoms that vary with ϕ and/or ψ (along with the dihedrals A-B-C-D and B-C-D-E connecting them). These are shown in Table 5.7. The reader can confirm that these interactions reduce the

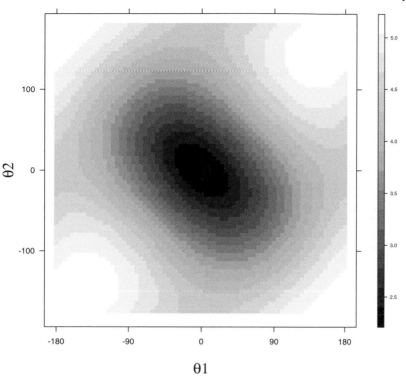

Fig. 5.9

Distance between atoms A and E in a five atom chain, A-B-C-D-E as a function of the A-B-C-D (θ_1) and B-C-D-E (θ_2) dihedrals.

The A–E distance was calculated by assuming the bond lengths are 1.55 Å and the bond angles are 115° (the average of sp^2 and sp^3 angles).

probability of certain Ramachandran conformations as seen in Figure 5.8A. Because glycine lacks Cβ it can attain backbone conformations that other residue types can not – those listed in Table 5.7 that include Cβ. Because proline contains a five-membered ring consisting of N-Cα-Cβ-Cγ-Cδ atoms, the ϕ dihedral is highly constrained to be near $\phi = -60 \pm 30°$. The Ramachandran distribution for proline is shown in Figure 5.8C.

5.8.1.2 Electrostatic interactions

Because the carbonyl and amide NH groups are highly polar, they form hydrogen bonds easily with water molecules. To bury these groups away from solvent (on the interior of a protein structure), requires that they are almost always involved in hydrogen bonds with other portions of the backbone or with side-chain groups. An efficient way of fulfilling this obligation occurs in the regular secondary structures of α-helices and β-sheets. In the

Tab. 5.7
Steric interactions along backbone that determine Ramachandran distributions

A	B	C	D	E	θ_1	θ_2	ω	ϕ	ψ
$C\alpha_{i-1}$	C_{i-1}	N_i	$C\alpha_i$	C_i	ω	ϕ	0	0	
$C\alpha_{i-1}$	C_{i-1}	N_i	$C\alpha_i$	$C\beta_i$	ω	$\phi - 120°$	0	120	
O_{i-1}	C_{i-1}	N_i	$C\alpha_i$	C_i	$\omega - 180°$	ϕ	180	0	
O_{i-1}	C_{i-1}	N_i	$C\alpha_i$	$C\beta_i$	$\omega - 180°$	$\phi - 120°$	180	120	
C_{i-1}	N_i	$C\alpha_i$	C_i	N_{i+1}	ϕ	ψ	0	0	
C_{i-1}	N_i	$C\alpha_i$	C_i	O_i	γ	$\psi - 180°$	0	180	
N_i	$C\alpha_i$	C_i	N_{i+1}	$C\alpha_{i+1}$	ψ	ω	0		0
$C\beta_i$	$C\alpha_i$	C_i	N_{i+1}	$C\alpha_{i+1}$	$\psi + 120°$	ω	0		-120

Possible steric interactions of 2 atoms (A and E) on the ends of a
5-atom chain (A-B-C-D-E). The dihedral A-B-C-D is denoted θ_1 and the
dihedral B-C-D-E is denoted θ_2. These dihedrals for each set of atoms
are expressed as functions of backbone dihedrals ϕ, ψ, and ω. The
values of ϕ, ψ, and ω when θ_1 and θ_2 are zero are given in the last
three columns. Large steric interactions will occur when θ_1 and θ_2
are both 0°. This interaction will fall off quickly as both dihedrals
decrease or both dihedrals increase, but much more slowly as one
increases from 0° and the other decreases (see Figure 5.7).

case of α-helices, hydrogen bonds are formed between the C=O of amino
acid i and the amide HN group of amino acid $i + 4$ when the backbone dihedral angles are close to $(-45°, -55°)$. Most α-helices are at least 2 turns
long, and often longer. β-sheets are secondary structures consisting of two
or more parallel or anti-parallel backbone segments that form hydrogen
bonds between them with their backbone NH and C=O groups. For several
residue types, we have drawn Ramachandran diagrams for residues in α-
helices, β-sheets, and coil regions in separate plots in Figure 5.10 to demonstrate that regular secondary structures populate only certain portions of
the map, and that coil residues occur in many areas.

5.8.1.3 Backbone-side-chain interactions

We describe the effect of backbone conformations on side-chain conformations below, but first we show that non-Gly, non-Pro amino acids do
not have identical Ramachandran distributions, due to interactions between
the side chain and the backbone. In Figure 5.10B and 5.10C we show the
Asn and Val Ramachandran distributions in each secondary structure to
demonstrate this effect. Asn in particular has high content of residues with
$\phi > 0°$ in coil because of backbone–side-chain hydrogen bonds. Val has very
few residues with $\phi > 0°$ because of steric clashes of the two $C\gamma$ atoms with
the backbone.

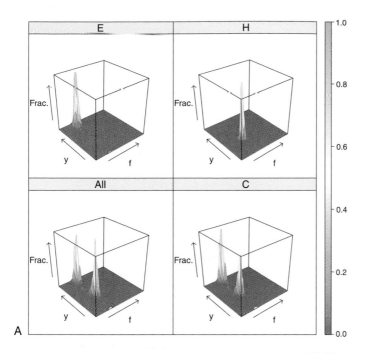

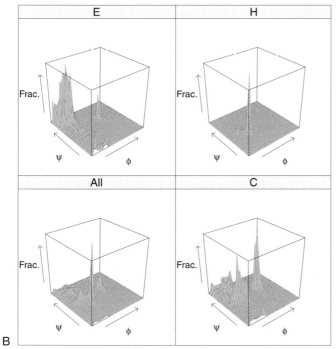

Fig. 5.10
Ramachandran distributions of A) Alanine; B) Asparagine; C) Valine separated by secondary structure type. Data was taken from the same 699 proteins of known structure as described in the caption to Figure 5.8. Secondary structure was determined with the STRIDE program [291]. H denotes Helix, E denotes Sheet, and C denotes Coil.

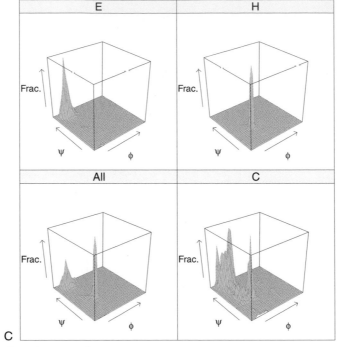

Fig. 5.10 (continued)

5.8.2
Side-chain conformational analysis

5.8.2.1 Backbone-conformation-independent interactions

Side chains can also be described by their dihedral angles, and these are denoted χ_1, χ_2, χ_3, and χ_4. Except for glycine, whose side chain is only a single hydrogen atom bonded to the α-carbon of the backbone, and alanine, whose side chain is only a single methyl group bonded to the α-carbon of the backbone, all other side chains have a χ_1 dihedral angle defined as N-Cα-Cβ-X where X is either Cγ, or Cγ1 (Val, Ile), Oγ (Ser), Oγ1 (Thr), or Sγ (Cys). Longer side chains have additional χ dihedrals, as shown in Figure 5.6. In Figure 5.11 we show the distribution of χ_1 dihedrals and χ_1, χ_2. for Lys. As we expect from organic chemistry, the dihedrals are clustered near $+60°$, $180°$, and $-60°$. These conformations are denoted as g^+, t, and g^- respectively, where g represents "gauche" and t represents trans. It should be noted that some authors reverse the g^+ and g^- definitions.

The rotamers are not at all evenly distributed among the possible staggered dihedrals. The reason for this can be seen in the Newman diagram in

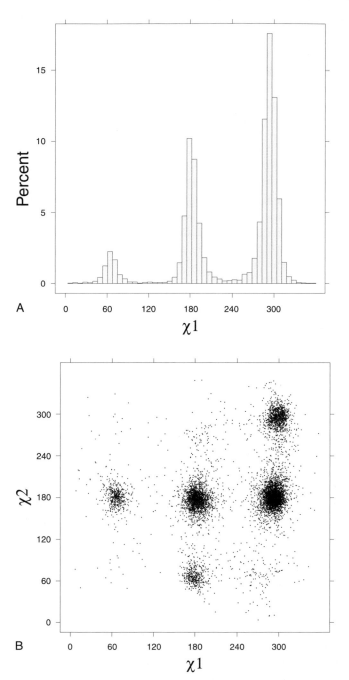

Fig. 5.11
Distribution of A) χ_1 and B) $\chi_1-\chi_2$ side-chain dihedrals for Lysine. Lysines in the same 699 proteins as described in the caption to Figure 5.8 were used to produce the plot.

Figure 5.12. First, the gauche interaction (dihedral $= +60°$ or $-60°$) of the χ_1 dihedral, N-Cα-Cβ-Cγ, increases the energy by about 0.9 kcal/mol per gauche interaction [286]. The t and g^- rotamers each have one gauche interaction with the backbone (backbone C and N respectively), while the g^+ has gauche interactions with *both* the backbone C and N atoms. Its energy is therefore about 0.9 kcal/mol higher than the other two rotamers. This is a backbone-conformation independent interaction; that is, it does not depend on backbone dihedrals ϕ and ψ. Second, both gauche and 1–5 interactions (as shown in Figure 5.9) affect the distribution of χ_1–χ_2 pairs of dihedrals. As discussed earlier for interactions along the backbone, the distance between atoms A and E of a 5-atom chain remain close to one another if the dihedrals are of opposite sign. For χ_1–χ_2 dihedral pairs, this means that when χ_1 is $+60°$ and χ_2 is $-60°$, or vice versa, the energy rises because of the steric interaction of the nitrogen and Cδ atoms in the 5-atom chain N-Cα-Cβ-Cγ-Cδ. But this kind of interaction also occurs between Cδ and the carbonyl carbon. This occurs when χ_1 is $+60°$ and χ_2 is $+60°$ and when χ_1 is $180°$ and χ_2 is $-60°$. The effects of these interactions can be observed in the χ_1–χ_2 distribution for lysine in Figure 5.11B.

5.8.2.2 Backbone-conformation-dependent interactions

One of the prime determinants of χ_1 rotamer choice is steric interaction of the γ heavy atom with the backbone in a manner that is dependent on the conformation of the backbone, that is, the dihedrals ϕ and ψ. In Figure 5.13, the proportions of the g^+, t, and g^- rotamers are shown on the Ramachandran map for Lys, Phe, Asp, and Val. The patterns of low energy and high energy rotamers are immediately apparent, with some regularities (and some differences) between the different amino acid types. The reason for strong backbone dependence can be observed in Figure 5.12B. The possible interactions are listed in Table 5.8. Steric interactions can occur between heavy atoms at positions i and $i+4$ when the two dihedrals connecting them take on values in the gray region of Figure 5.9. We count from the γ carbon to atoms C_{i-1}, O_i, and N_{i+1} of the backbone, and determine the disallowed combinations of dihedrals in these 5-atom chains, C_{i-1}-N_i-Cα-Cβ-Cγ, O_i-C_i-Cα-Cβ-Cγ, and N_{i+1}-C_i-Cα-Cβ-Cγ respectively. The second dihedral in each of these sequences is either χ_1 or $\chi_1 - 120°$ (see Figure 5.6). Since the χ_1 dihedral is likely to be in a staggered position ($60°$, $180°$, $-60°$) relative to backbone N and C, we need to determine the range of the dihedrals C_{i-1}-N_i-Cα-Cβ, O_i-C_i-Cα-Cβ, and N_{i+1}-C_i-Cα-Cβ allowed. For instance, if χ_1 is $-60°$, Figure 5.9 indicates that C_{i-1}-N_i-Cα-Cβ is disallowed when it takes on values of $+30° \pm 50°$. Since this dihedral is equal to $\phi - 120°$, the g^+ rotamer is disallowed when $\phi = +150° \pm 50°$. This is evident in the plots of Figure 5.13, where the g^- rotamer is uncommon on the far left of each plot ($\phi \sim -180°$). By the same reasoning, the g^+ rotamer is not allowed when $\phi = +90° \pm 50°$.

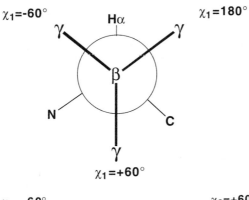

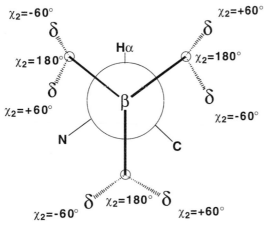

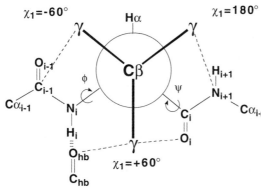

Fig. 5.12
Newman diagrams for protein backbone-side-chain interactions. (**A**) Backbone-independent interactions, looking down the Cα-Cβ bond from Cβ; certain combinations of χ_1 and χ_2 are forbidden because of close interactions between Cδ and backbone N and C of the same residue. In the lower diagram, the $\chi_2 = 180°$ rotamer is pointing towards the viewer. The four χ_1, χ_2 combinations which

are forbidden are those with Cδ (denoted "δ") are closest to the backbone N and C atoms. (**B**) Backbone-dependent interactions, looking down the Cα-Cβ bond from Cβ; certain combinations of χ_1 and ϕ and certain combinations of χ_1 and ψ are essentially forbidden because of interactions between Cγ and backbone atoms C_{i-1}, N_{i+1}, O_i, and a hydrogen bond acceptor to HN_i.

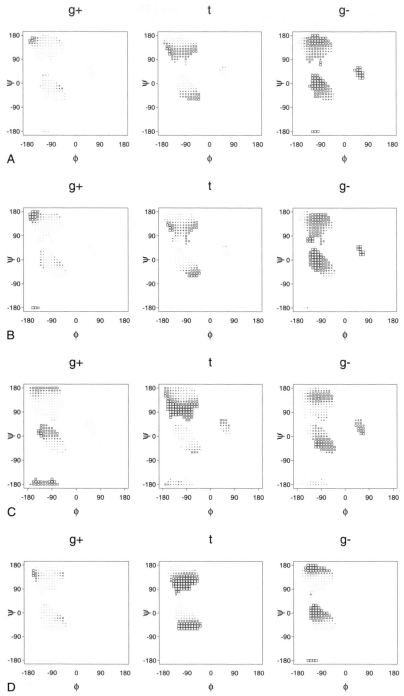

Fig. 5.13
Distribution of χ_1 side-chain rotamers as a function of backbone dihedrals ϕ and ψ for (**A**) Lys; (**B**) Phe; (**C**) Asp; (**D**) Val. Each square represents the fraction of each rotamer type in $10°$ ranges of ϕ or ψ in a sample of 699 proteins as described in Figure 5.8. Only ranges of ϕ or ψ for which there were at least 20 residues available in this sample of proteins are plotted.

Tab. 5.8
Steric interactions along backbone that determine Ramachandran distributions

A	B	C	D	E	θ_1	θ_1 value	θ_2	θ_2 value	χ_1	ϕ	ψ
$C\gamma$	$C\beta$	$C\alpha$	N_i	C_{i-1}	χ_1	$+60°$	ϕ	$-30°$	$+60°$	$-30°$	
$C\gamma$	$C\beta$	$C\alpha$	N_i	$O..HN_i$	χ_1	$+60°$	$\phi-180°$	$-30°$	$+60°$	$150°$	
$C\gamma$	$C\beta$	$C\alpha$	N_i	C_{i-1}	χ_1	$-60°$	ϕ	$+30°$	$-60°$	$+30°$	
$C\gamma$	$C\beta$	$C\alpha$	N_i	$O..HN_i$	χ_1	$-60°$	$\phi-180°$	$+30°$	$-60°$	$-150°$	
$C\gamma$	$C\beta$	$C\alpha$	C_i	O_i	$\chi_1-120°$	$+60°$	$\psi-60°$	$-30°$	$180°$		$+30°$
$C\gamma$	$C\beta$	$C\alpha$	C_i	N_{i+1}	$\chi_1-120°$	$+60°$	$\psi+120°$	$-30°$	$180°$		$-150°$
$C\gamma$	$C\beta$	$C\alpha$	C_i	O_i	$\chi_1-120°$	$-60°$	$\psi-60°$	$+30°$	$+60°$		$+90°$
$C\gamma$	$C\beta$	$C\alpha$	C_i	N_{i+1}	$\chi_1-120°$	$-60°$	$\psi+120°$	$+30°$	$+60°$		$-90°$
$C\gamma2$	$C\beta$	$C\alpha$	N_i	C_{i-1}	$\chi_1+120°$	$+60°$	ϕ	$-30°$	$-60°$	$-30°$	
$C\gamma2$	$C\beta$	$C\alpha$	N_i	$O..HN_i$	$\chi_1+120°$	$+60°$	$\phi-180°$	$-30°$	$-60°$	$150°$	
$C\gamma2$	$C\beta$	$C\alpha$	N_i	C_{i-1}	$\chi_1+120°$	$-60°$	ϕ	$+30°$	$180°$	$+30°$	
$C\gamma2$	$C\beta$	$C\alpha$	N_i	$O..HN_i$	$\chi_1+120°$	$-60°$	$\phi-180°$	$+30°$	$180°$	$-150°$	
$C\gamma2$	$C\beta$	$C\alpha$	C_i	O_i	χ_1	$+60°$	$\psi-60°$	$-30°$	$+60°$		$+30°$
$C\gamma2$	$C\beta$	$C\alpha$	C_i	N_{i+1}	χ_1	$+60°$	$\psi+120°$	$-30°$	$+60°$		$-150°$
$C\gamma2$	$C\beta$	$C\alpha$	C_i	O_i	χ_1	$-60°$	$\psi-60°$	$+30°$	$-60°$		$+90°$
$C\gamma2$	$C\beta$	$C\alpha$	C_i	N_{i+1}	χ_1	$-60°$	$\psi+120°$	$+30°$	$-60°$		$-90°$

Possible steric interactions of 2 atoms (A and E) on the ends of a 5-atom chain (A-B-C-D-E). The dihedral A-B-C-D is denoted θ_1 and the dihedral B-C-D-E is denoted θ_2. These dihedrals for each set of atoms are expressed as functions of side-chain dihedral χ_1 and the backbone dihedrals ϕ and ψ. Because χ_1 must be near one of the canonical staggered dihedral values ($+60°$, $180°$, and $-60°$), steric interactions can occur when the value of θ_1 is $+60°$ or $-60°$, and θ_2 is $-30° \pm 50°$

or $+30° \pm 50°$ respectively (see Figure 5.7). For the side-chain types with one γ heavy atom, the entries listed for $C\gamma$ apply. For side-chains with two γ heavy atoms (Val, Thr, Ile), interactions listed for $C\gamma$ and $C\gamma2$ apply. Because of the definition of χ_1 for Ile and Thr, χ_1 for these residues is equal to $\chi_1 + 120°$ as listed in the Table. So for instance the $-60°$ rotamer in the Table is equivalent to the $+60°$ rotamer for Ile and Thr.

The dihedral O_i-C_i-$C\alpha$-$C\beta$ is equal to $\psi - 60°$. $\chi_1 - 120°$ is $-60°$ for the g^+ rotamer, and therefore the g^+ rotamer is forbidden when $\psi - 60°$ is $+30° \pm 50°$, or when $\psi = +90° \pm 50°$, because of a steric interaction between $C\gamma$ and O_i. The interaction with backbone atom N_{i+1} occurs $180°$ away from this, or when ψ is $-90° \pm 50°$. Both of these interactions are evident in the low proportions of the g^+ rotamers in these regions in Figure 5.13. Finally, the same interactions occur for the t rotamer ($\chi_1 - 120° = +60°$) when $\psi - 60°$ is $-30° \pm 50°$ or when $\psi = +30° \pm 50°$ and again $180°$ away when $\psi = -150° \pm 50°$. Because valine, isoleucine, and threonine have two γ heavy atoms, interactions with both atoms must be considered in this analysis. It is clear in Figure 5.13C that valine is highly restricted by the backbone conformation, since in most regions of the Ramachandran map two out of three χ_1 rotamers are disallowed, leaving only one likely conformation.

5.9
Acknowledgments

This work was funded in part by grant CA06927 from the National Institutes of Health, a grant from the American Cancer Society, and an appropriation from the Commonwealth of Pennsylvania. I thank the Research Computing Services of Fox Chase Cancer Center for computer support.

References

1 F. C. BERNSTEIN, T. F. KOETZLE, G. J. B. WILLIAMS, E. F. J. MEYER, M. D. BRICE, J. R. RODGERS, O. KENNARD, T. SHIMANOUCHI, M. TASUMI. The Protein Data Bank: A computer-based archival file for macromolecular structures. *J. Mol. Biol.* **1977**, 112, 535–542.

2 H. M. BERMAN, J. WESTBROOK, Z. FENG, G. GILLILAND, T. N. BHAT, H. WEISSIG, I. N. SHINDYALOV, P. E. BOURNE. The Protein Data Bank. *Nucleic Acids Res.* **2000**, 28, 235–242.

3 A. G. MURZIN, S. E. BRENNER, T. HUBBARD, C. CHOTHIA. SCOP: a structural classification of proteins database for the investigation of sequences and structures. *J. Mol. Biol.* **1995**, 247, 536–540.

4 C. A. ORENGO, A. D. MICHIE, S. JONES, D. T. JONES, M. B. SWINDELLS, J. M. THORNTON. CATH – a hierarchic classification of protein domain structures. *Structure.* **1997**, 5, 1093–1108.

5 L. HOLM, C. SANDER. The FSSP database: fold classification based on structure-structure alignment of proteins. *Nucleic Acids Res.* **1996**, 24, 206–209.

6 R. L. DORIT, L. SCHOENBACH, W. GILBERT. How big is the universe of exons? *Science.* **1990**, 250, 1377–1382.

7 L. HOLM, C. SANDER. Mapping the protein universe. *Science.* **1996**, 273, 595–603.

8 M. F. PERUTZ, J. C. KENDREW, H. C. WATSON. Structure and function of haemoglobin. *J. Mol. Biol.* **1965**, 13, 669–678.

9 W. J. BROWNE, A. C. NORTH, D. C. PHILLIPS. A possible three-dimensional structure of bovine alpha-lactalbumin based on that of hen's egg-white lysozyme. *J. Mol. Biol.* **1969**, 42, 65–86.

10 J. GREER. Model for haptoglobin heavy chain based upon structural homology. *Proc Natl Acad Sci U S A.* **1980**, 77, 3393–3397.

11 J. GREER. Comparative modeling methods: application to the family of the mammalian serine proteases. *Proteins.* **1990**, 7, 317–334.

12 T. L. BLUNDELL, B. L. SIBANDA, M. J. E. STERNBERG, J. M. THORNTON. Knowledge-based prediction of protein structures and the design of novel molecules. *Nature.* **1987**, 326, 347–352.

13 M. J. SUTCLIFFE, I. HANEEF, D. CARNEY, T. L. BLUNDELL. Knowledge based modeling of homologous proteins, part I: three-dimensional frameworks derived from the simultaneous superposition of multiple structures. *Prot. Eng.* **1987**, 5, 377–384.

14 M. J. SUTCLIFFE, F. R. HAYES, T. L. BLUNDELL. Knowledge based modeling of homologous proteins, Part II: Rules for the conformations of substituted sidechains. *Prot. Eng.* **1987**, 1, 385–392.

15 T. F. HAVEL, M. E. SNOW. A new method for building protein conformations from sequence alignments with homologues of known structure. *J. Mol. Biol.* **1991**, 217, 1–7.

16 A. SALI, T. L. BLUNDELL. Comparative protein modelling by satisfaction of spatial restraints. *J. Mol. Biol.* **1993**, 234, 779–815.

17 R. SANCHEZ, A. SALI. Evaluation of comparative protein structure modeling by MODELLER-3. *Proteins.* **1997**, Suppl, 50–58.

18 H. LI, R. TEJERO, D. MONLEON, D. BASSOLINO-KLIMAS, C. ABATE-SHEN, R. E. BRUCCOLERI, G. T. MONTELIONE. Homology modeling using simulated annealing of restrained molecular dynamics and conformational search calculations with CONGEN: application in predicting the three-dimensional structure of murine homeodomain Msx-1. *Protein Sci.* **1997**, 6, 956–970.

19 P. V. SAHASRABUDHE, R. TEJERO, S. KITAO, Y. FURUICHI, G. T. MONTELIONE. Homology modeling of an RNP domain from a human RNA-binding protein: Homology-constrained energy optimization provides a criterion for distinguishing potential sequence alignments. *Proteins.* **1998**, 33, 558–566.

20 O. LICHTARGE, H. R. BOURNE, F. E. COHEN. An evolutionary trace method defines binding surfaces common to protein families. *J. Mol. Biol.* **1996**, 257, 342–358.

21 J. S. FETROW, A. GODZIK, J. SKOLNICK. Functional Analysis of the Escherichia coli Genome Using the Sequence-to-Structure-

to-Function Paradigm: Identification of Proteins Exhibiting the Glutaredoxin/Thioredoxin Disulfide Oxidoreductase Activity. *J. Mol. Biol.* **1998**, 282, 703–711.

22 N. BLOMBERG, R. R. GABDOULLINE, M. NILGES, R. C. WADE. Classification of protein sequences by homology modeling and quantitative analysis of electrostatic similarity. *Proteins.* **1999**, 37, 379–387.

23 D. DEVOS, A. VALENCIA. Practical limits of function prediction. *Proteins.* **2000**, 41, 98–107.

24 J. C. WHISSTOCK, J. A. IRVING, S. P. BOTTOMLEY, R. N. PIKE, A. M. LESK. Serpins in the Caenorhabditis elegans genome. *Proteins.* **1999**, 36, 31–41.

25 R. LANDGRAF, D. FISCHER, D. EISENBERG. Analysis of heregulin symmetry by weighted evolutionary tracing. *Protein Eng.* **1999**, 12, 943–951.

26 D. R. CAFFREY, L. A. O'NEILL, D. C. SHIELDS. A method to predict residues conferring functional differences between related proteins: application to MAP kinase pathways. *Protein Sci.* **2000**, 9, 655–670.

27 B. ZOLLER, B. DAHLBACK. Linkage between inherited resistance to activated protein C and factor V gene mutation in venous thrombosis . *Lancet.* **1994**, 343, 1536–1538.

28 J. P. KRAUS, M. JANOSIK, V. KOZICH, R. MANDELL, V. SHIH, M. P. SPERANDEO, G. SEBASTIO, R. DE FRANCHIS, G. ANDRIA, L. A. KLUIJTMANS, H. BLOM, G. H. BOERS, R. B. GORDON, P. KAMOUN, M. Y. TSAI, W. D. KRUGER, H. G. KOCH, T. OHURA, M. GAUSTADNES. Cystathionine beta-synthase mutations in homocystinuria. *Hum Mutat.* **1999**, 13, 362–375.

29 F. J. COUCH, B. L. WEBER. Mutations and polymorphisms in the familial early-onset breast cancer (BRCA1) gene. Breast Cancer Information Core. *Hum Mutat.* **1996**, 8, 8–18.

30 I. T. WEBER, M. MILLER, M. JASKOLSKI, J. LEIS, A. M. SKALKA, A. WLODAWER. Molecular modeling of the HIV-1 protease and its substrate binding site. *Science.* **1989**, 243, 928–931.

31 I. T. WEBER. Evaluation of homology modeling of HIV protease. *Proteins.* **1990**, 7, 172–184.

32 C. S. RING, E. SUN, J. H. MCKERROW, G. K. LEE, P. J. ROSENTHAL, I. D. KUNTZ, F. E. COHEN. Structure-based inhibitor design by using protein models for the development of anti-parasitic agents. *Proc. Natl. Acad. Sci. USA.* **1993**, 90, 3583–3587.

33 S. F. ALTSCHUL, T. L. MADDEN, A. A. SCHÄFFER, J. ZHANG, Z. ZHANG, W. MILLER, D. J. LIPMAN. Gapped BLAST and PSI-BLAST: a new generation of database programs. *Nucleic Acids Res.* **1997**, 25, 3389–3402.

34 J. M. SAUDER, J. W. ARTHUR, R. L. DUNBRACK, JR. Large-scale comparison of protein sequence alignment algorithms with structure alignments. *Proteins.* **2000**, 40, 6–22.

35 D. CHRISTENDAT, A. YEE, A. DHARAMSI, Y. KLUGER, A. SAVCHENKO, J. R. CORT, V. BOOTH, C. D. MACKERETH, V. SARIDAKIS, I. EKIEL, G. KOZLOV, K. L. MAXWELL, N. WU, L. P. McINTOSH, K. GEHRING, M. A. KENNEDY, A. R. DAVIDSON, E. F. PAI, M. GERSTEIN, A. M. EDWARDS, C. H. ARROWSMITH. Structural proteomics of an archaeon. *Nat Struct Biol.* **2000**, 7, 903–909.

36 M. Huynen, T. Doerks, F. Eisenhaber, C. Orengo, S. Sunyaev, Y. Yuan, P. Bork. Homology-based fold predictions for *Mycoplasma genitalium* proteins. *J. Mol. Biol.* **1998**, 280, 323–326.

37 Y. I. Wolf, S. E. Brenner, P. A. Bash, E. V. Koonin. Distribution of protein folds in the three superkingdoms of life. *Genome Res.* **1999**, 9, 17–26.

38 B. Rost. Protein structures sustain evolutionary drift. *Fold. Des.* **1997**, 2, S19–S24.

39 C. Chothia, A. M. Lesk. The relation between the divergence of sequence and structure in proteins. *EMBO J.* **1986**, 5, 823–826.

40 M. C. Peitsch, M. R. Wilkins, L. Tonella, J. C. Sanchez, R. D. Appel, D. F. Hochstrasser. Large-scale protein modelling and integration with the SWISS-PROT and SWISS-2DPAGE databases: the example of Escherichia coli. *Electrophoresis.* **1997**, 18, 498–501.

41 R. Sanchez, A. Sali. Large-scale protein structure modeling of the Saccharomyces cerevisiae genome. *Proc Natl Acad Sci USA.* **1998**, 95, 13597–13602.

42 C. Dodge, R. Schneider, C. Sander. The HSSP database of protein structure-sequence alignments and family profiles. *Nucleic Acids Res.* **1998**, 26, 313–315.

43 D. L. Wheeler, C. Chappey, A. E. Lash, D. D. Leipe, T. L. Madden, G. D. Schuler, T. A. Tatusova, B. A. Rapp. Database resources of the National Center for Biotechnology Information. *Nucleic Acids Res.* **2000**, 28, 10–14.

44 A. Bairoch, R. Apweiler. The SWISS-PROT protein sequence database and its supplement TrEMBL in 2000. *Nucleic Acids Res.* **2000**, 28, 45–48.

45 J. D. Thompson, D. G. Higgins, T. J. Gibson. CLUSTAL W: improving the sensitivity of progressive multiple sequence alignment through sequence weighting, position-specific gap penalties and weight matrix choice. *Nucleic Acids Res.* **1994**, 22, 4673–4680.

46 O. Gotoh. Significant improvement in accuracy of multiple sequence alignments by iterative refinement as assessed by reference to structural alignments. *J. Mol. Biol.* **1996**, 264, 823–838.

47 F. Corpet. Multiple sequence alignment with hierarchical clustering. *Nucleic Acids Res.* **1988**, 16, 10881–10890.

48 J. U. Bowie, N. D. Clarke, C. O. Pabo, R. T. Sauer. Identification of protein folds: Matching hydrophobicity patterns of sequence sets with solvent accessibility patterns of known structures. *Proteins.* **1990**, 7, 257–264.

49 R. Lüthy, J. U. Bowie, D. Eisenberg. Assessment of protein models with three-dimensional profiles. *Nature.* **1992**, 36, 83–85.

50 M. J. Sippl, S. Weitckus. Detection of native-like models for amino acid sequences of unknown three-dimensional structure in a data base of known protein conformations. *Proteins.* **1992**, 13, 258–271.

51 D. T. Jones, W. R. Taylor, J. M. Thornton. A new approach to protein fold recognition. *Nature.* **1992**, 358, 86–89.

52 C. Ouzounis, C. Sander, M. Scharf, R. Schneider. Prediction of protein structure by evaluation of sequence-structure fitness: Aligning sequences to contact profiles derived from three-dimensional structures. *J. Mol. Biol.* **1993**, 232, 805–825.

53 S. H. BRYANT, C. E. LAWRENCE. An empirical energy function for threading protein sequence through the folding motif. *Proteins.* **1993**, 16, 92–112.

54 R. ABAGYAN, D. FRISHMAN, P. ARGOS. Recognition of distantly related proteins through energy calculations. *Proteins.* **1994**, 19, 132–140.

55 S. H. BRYANT, S. F. ALTSCHUL. Statistics of sequence-structure threading. *Curr. Opin. Struct. Biol.* **1995**, 5, 236–244.

56 D. T. JONES, R. T. MILLER, J. M. THORNTON. Successful protein fold recognition by optimal sequence threading validated by rigorous blind testing. *Proteins.* **1995**, 23, 387–397.

57 D. R. WESTHEAD, V. P. COLLURA, M. D. ELDRIDGE, M. A. FIRTH, J. LI, C. W. MURRAY. Protein fold recognition by threading: comparison of algorithms and analysis of results. *Prot. Eng.* **1995**, 8, 1197–1204.

58 Y. MATSUO, K. NISHIKAWA. Assessment of a protein fold recognition method that takes into account four physicochemical properties: Sidechain packing, solvation, hydrogen-bonding, and local conformation. *Proteins.* **1995**, 23, 370–375.

59 M. WILMANNS, D. EISENBERG. Inverse protein folding by the residue pair preference profile method: estimating the correctness of alignments of structurally compatibile sequences. *Prot. Eng.* **1995**, 8, 627–639.

60 T. MADEJ, J.-F. GIBRAT, S. H. BRYANT. Threading a database of protein cores. *Proteins.* **1995**, 23, 356–369.

61 J. SELBIG. Contact pattern-induced pair potentials for protein fold recognition. *Prot. Eng.* **1995**, 8, 339–351.

62 R. B. RUSSELL, R. R. COPLEY, G. J. BARTON. Protein fold recognition by mapping predicted secondary structures. *J. Mol. Biol.* **1996**, 259, 349–365.

63 L. MIRNY, E. DOMANY. Protein fold recognition and dynamics in the space of contact maps. *Proteins.* **1996**, 26, 391–410.

64 R. T. MILLER, D. T. JONES, J. M. THORNTON. Protein fold recognition by sequence threading: tools and assessment techniques. *FASEB J.* **1996**, 10, 171–178.

65 T. R. DEFAY, F. E. COHEN. Multiple sequence information for threading algorithms. *J. Mol. Biol.* **1996**, 262, 314–323.

66 N. N. ALEXANDROV, R. NUSSINOV, R. M. ZIMMER. Fast protein fold recognition via sequence to structure alignment and contact capacity potentials. *Pac. Symp. Biocomput.* **1996**, 53–72.

67 G. M. CRIPPEN. Failures of inverse folding and threading with gapped alignment. *Proteins.* **1996**, 26, 167–171.

68 D. FISCHER, D. RICE, J. U. BOWIE, D. EISENBERG. Assigning amino acid sequences to 3-dimensional protein folds. *FASEB J.* **1996**, 10, 126–136.

69 R. H. LATHROP, T. F. SMITH. Global optimum protein threading with gapped alignment and empirical pair score functions. *J. Mol. Biol.* **1996**, 255, 641–665.

70 D. FISCHER, D. EISENBERG. Protein fold recognition using sequence-derived predictions. *Prot. Science.* **1996**, 5, 947–955.

71 J. U. BOWIE, K. ZHANG, M. WILMANNS, D. EISENBERG. Three-dimensional profiles for measuring compatibility of amino acid sequences with three-dimensional structure. *Meth. Enz.* **1996**, 266, 598–616.

72 W. ZHENG, S. J. CHO, I. I. VAISMAN, A. TROPSHA, *A new approach to protein fold recognition based on Delaunay tessellation of protein structure*, in *Pacific Symposium on Biocomputing '97*, R. B. ALTMAN, A. K. DUNKER, L. HUNTER, T. E. KLEIN, Editors. 1996, World Scientific: Singapore. p. 486–497.

73 N. N. ALEXANDROV. SARFing the PDB. *Prot. Eng.* **1996**, 9, 727–732.

74 D. T. JONES, J. M. THORNTON. Potential energy functions for threading. *Curr. Opin. Struct. Biol.* **1996**, 6, 210–216.

75 W. R. TAYLOR. Multiple sequence threading: An analysis of alignment quality and stability. *J. Mol. Biol.* **1997**, 269, 902–943.

76 V. DiFRANCESCO, J. GARNIER, P. J. MUNSON. Protein topology recognition from secondary structure sequences: Application of the hidden Markov models to the alpha class proteins. *J. Mol. Biol.* **1997**, 267, 446–463.

77 D. FISCHER, A. ELOFSSON, D. RICE, D. EISENBERG. Assessing the performance of fold recognition methods by means of a comprehensive benchmark. *Pac Symp Biocomput.* **1996**, 300–318.

78 R. B. RUSSELL, M. A. S. SAQI, P. A. BATES, R. A. SAYLE, M. J. E. STERNBERG. Recognition of analogous and homologous protein folds: assessment of prediction success and associated alignment accuracy using empirical substitution matrices. *Prot. Eng.* **1998**, 11, 1–9.

79 X. DE LA CRUZ, J. M. THORNTON. Factors limiting the performance of prediction-based fold recognition methods. *Prot. Science.* **1999**, 8, 750–759.

80 M. BROWN, R. HUGHEY, A. KROGH, I. S. MIAN, K. SJOLANDER, D. HAUSSLER. Using Dirichlet mixture priors to derive hidden Markov models for protein families. *Ismb.* **1993**, 1, 47–55.

81 P. BALDI, Y. CHAUVIN, T. HUNKAPILLER, M. A. McCLURE. Hidden Markov models of biological primary sequence information. *Proc Natl Acad Sci U S A.* **1994**, 91, 1059–1063.

82 S. R. EDDY, G. MITCHISON, R. DURBIN. Maximum discrimination hidden Markov models of sequence consensus. *J Comput Biol.* **1995**, 2, 9–23.

83 T. L. BAILEY, M. GRIBSKOV. The megaprior heuristic for discovering protein sequence patterns. *Ismb.* **1996**, 4, 15–24.

84 K. SJÖLANDER, K. KARPLUS, M. BROWN, R. HUGHEY, A. KROGH, I. S. MIAN, D. HAUSSLER. Dirichlet mixtures: a method for improved detection of weak but significant protein sequence homology. *Comput Appl Biosci.* **1996**, 12, 327–345.

85 S. R. EDDY. Hidden Markov models. *Curr Opin Struct Biol.* **1996**, 6, 361–365.

86 K. KARPLUS, K. SJOLANDER, C. BARRETT, M. CLINE, D. HAUSSLER, R. HUGHEY, L. HOLM, C. SANDER. Predicting protein structure using hidden Markov models. *Proteins*. **1997**, Suppl, 134–139.

87 K. KARPLUS, C. BARRETT, R. HUGHEY. Hidden Markov models for detecting remote protein homologies. *Bioinformatics*. **1998**, 14, 846–856.

88 L. HOLM, C. SANDER. Dali: a network tool for protein structure comparison. *Trends Biochem Sci*. **1995**, 20, 478–480.

89 I. N. SHINDYALOV, P. E. BOURNE. Protein structure alignment by incremental combinatorial extension (CE) of the optimal path. *Prot. Eng*. **1998**, 11, 739–747.

90 R. L. STRAUSBERG, K. H. BUETOW, M. R. EMMERT-BUCK, R. D. KLAUSNER. The cancer genome anatomy project: building an annotated gene index. *Trends Genet*. **2000**, 16, 103–106.

91 A. HAMOSH, A. F. SCOTT, J. AMBERGER, D. VALLE, V. A. McKUSICK. Online Mendelian Inheritance in Man (OMIM). *Hum. Mutat*. **2000**, 15, 57–61.

92 M. KRAWCZAK, E. V. BALL, I. FENTON, P. D. STENSON, S. ABEYSINGHE, N. THOMAS, D. N. COOPER. Human gene mutation database – a biomedical information and research resource. *Hum. Mutat*. **2000**, 15, 45–51.

93 M. J. BOWER, F. E. COHEN, R. L. DUNBRACK, JR. Prediction of protein side-chain rotamers from a backbone-dependent rotamer library: a new homology modeling tool. *J Mol Biol*. **1997**, 267, 1268–1282.

94 R. L. DUNBRACK, JR. Comparative modeling of CASP3 targets using PSI-BLAST and SCWRL. *Proteins Suppl*. **1999**, Suppl. 3, 81–87.

95 R. E. BRUCCOLERI, M. KARPLUS. Prediction of the folding of short polypeptide segments by uniform conformational sampling. *Biopolymers*. **1987**, 26, 137–168.

96 J. MOULT. The current state of the art in protein structure prediction. *Curr. Opin. Biotechnology*. **1996**, 7, 422–427.

97 J. MOULT, T. HUBBARD, S. H. BRYANT, K. FIDELIS, J. T. PEDERSEN. Critical assessment of methods of protein structure prediction (CASP): round II. *Proteins*. **1997**, Suppl, 2–6.

98 J. MOULT, T. HUBBARD, K. FIDELIS, J. T. PEDERSEN. Critical assessment of methods of protein structure prediction (CASP): round III. *Proteins*. **1999**, Suppl, 2–6.

99 P. KOEHL, M. LEVITT. A brighter future for protein structure prediction. *Nature Struct. Biol*. **1999**, 6, 108–111.

100 N. N. ALEXANDROV, R. LUETHY. Alignment algorithm for homology modeling and threading. *Protein Sci*. **1998**, 7, 254–258.

101 L. WESSON, D. EISENBERG. Atomic solvation parameters applied to molecular dynamics of proteins in solution. *Prot. Science*. **1992**, 1, 227–235.

102 A. JANARDHAN, S. VAJDA. Selecting near-native conformations in homology modeling: the role of molecular mechanics and solvation terms. *Protein Sci*. **1998**, 7, 1772–1780.

103 R. J. PETRELLA, T. LAZARIDIS, M. KARPLUS. Protein sidechain conformer prediction: a test of the energy function [published erratum appears in Fold Des 1998;3(6):588]. *Fold Des*. **1998**, 3, 353–377.

104 C. S. RAPP, R. A. FRIESNER. Prediction of loop geometries using a generalized born model of solvation effects. *Proteins.* **1999**, 35, 173–183.

105 T. LAZARIDIS, M. KARPLUS. Effective energy function for proteins in solution. *Proteins.* **1999**, 35, 133–152.

106 B. L. SIBANDA, J. M. THORNTON. Beta-hairpin families in globular proteins. *Nature.* **1985**, 316, 170–174.

107 C. M. WILMOT, J. M. THORNTON. Analysis and prediction of the different types of beta-turn in proteins. *J. Mol. Biol.* **1988**, 203, 221–232.

108 C. M. WILMOT, J. M. THORNTON. Beta-turns and their distortions: a proposed new nomenclature. *Protein Eng.* **1990**, 3, 479–493.

109 B. L. SIBANDA, J. M. THORNTON. Conformation of beta hairpins in protein structures: classification and diversity in homologous structures. *Methods Enzymol.* **1991**, 202, 59–82.

110 A. J. SHEPHERD, D. GORSE, J. M. THORNTON. Prediction of the location and type of beta-turns in proteins using neural networks. *Protein Sci.* **1999**, 8, 1045–1055.

111 L. E. DONATE, S. D. RUFINO, L. H. J. CANARD, T. L. BLUNDELL. Conformational analysis and clustering of short and medium size loops connecting regular secondary structures: A database for modeling and prediction. *Prot. Science.* **1996**, 5, 2600–2616.

112 V. GEETHA, P. J. MUNSON. Linkers of secondary structures in proteins. *Prot. Science.* **1997**, 6, 2538–2547.

113 J. F. LESZCZYNSKI, G. D. ROSE. Loops in globular proteins: a novel category of secondary structure. *Science.* **1986**, 234, 849–855.

114 B. OLIVA, P. A. BATES, E. QUEROL, F. X. AVILÉS, M. J. E. STERNBERG. An automated classification of the structure of protein loops. *J. Mol. Biol.* **1997**, 266, 814–830.

115 M. J. ROOMAN, S. J. WODAK, J. M. THORNTON. Amino acid sequence templates derived from recurrent turn motifs in proteins: critical evaluation of their predictive power. *Protein Eng.* **1989**, 3, 23–27.

116 J. KWASIGROCH, J. CHOMILIER, J. MORNON. A global taxonomy of loops in globular proteins. *J. Mol. Biol.* **1996**, 259, 855–872.

117 R. T. WINTJENS, M. J. ROOMAN, S. J. WODAK. Automatic classification and analysis of alpha alpha-turn motifs in proteins. *J Mol Biol.* **1996**, 255, 235–253.

118 A. C. MARTIN, K. TODA, H. J. STIRK, J. M. THORNTON. Long loops in proteins. *Protein Eng.* **1995**, 8, 1093–1101.

119 C. S. RING, D. G. KNELLER, R. LANGRIDGE, F. E. COHEN. Taxonomy and conformational analysis of loops in proteins [published erratum appears in Journal of Molecular Biology 1992 Oct 5;227(3):977]. *J. Mol. Biol.* **1992**, 224, 685–699.

120 M. B. SWINDELLS, M. W. MacARTHUR, J. M. THORNTON. Intrinsic ϕ, ψ propensities of amino acids, derived from the coil regions of known structures. *Nature Struct. Biol.* **1995**, 2, 596–603.

121 J. S. EVANS, S. I. CHAN, W. A. GODDARD, 3rd. Prediction of polyelectrolyte polypeptide structures using Monte Carlo conformational search methods with implicit solvation modeling. *Protein Sci.* **1995**, 4, 2019–2031.

122 T. A. JONES, S. THIRUP. Using known substructures in protein model building and crystallography. *EMBO J.* **1986**, 5, 819–822.

123 C. CHOTHIA, A. M. LESK, A. TRAMONTANO, M. LEVITT, S. J. SMITH-GILL, G. AIR, S. SHERIFF, E. A. PADLAN, D. DAVIES, W. R. TULIP, et al. Conformations of immunoglobulin hypervariable regions. *Nature.* **1989**, 342, 877–883.

124 N. L. SUMMERS, M. KARPLUS. Modeling of globular proteins: A distance-based search procedure for the construction of insertion/deletion regions and Prōnon-Pro mutations. *J. Mol. Biol.* **1990**, 216, 991–1016.

125 M. J. ROOMAN, S. J. WODAK. Weak correlation between predictive power of individual sequence patterns and overall prediction accuracy in proteins. *Proteins.* **1991**, 9, 69–78.

126 K. FIDELIS, P. S. STERN, D. BACON, J. MOULT. Comparison of systematic search and database methods for constructing segments of protein structures. *Prot. Eng.* **1994**, 7, 953–960.

127 P. KOEHL, M. DELARUE. A self consistent mean field approach to simultaneous gap closure and sidechain positioning in homology modeling. *Nature Struct. Biol.* **1995**, 2, 163–170.

128 T. FECHTELER, U. DENGLER, D. SCHOMBURG. Prediction of protein three-dimensional structures in insertion and deletion regions: a procedure for searching data bases of representative protein fragments using geometric scoring criteria. *J Mol Biol.* **1995**, 253, 114–131.

129 S. D. RUFINO, L. E. DONATE, L. H. J. CANARD, T. L. BLUNDELL. Predicting the conformational class of short and medium size loops connecting regular secondary structures: Application to comparative modeling. *J. Mol. Biol.* **1997**, 267, 352–367.

130 H. W. T. V. VLIJMEN, M. KARPLUS. PDB-based protein loop prediction: Parameters for selection and methods for optimization. *J. Mol. Biol.* **1997**, 267, 975–1001.

131 P. A. BATES, R. M. JACKSON, M. J. STERNBERG. Model building by comparison: a combination of expert knowledge and computer automation. *Proteins.* **1997**, Suppl, 59–67.

132 U. LESSEL, D. SCHOMBURG. Importance of anchor group positioning in protein loop prediction. *Proteins.* **1999**, 37, 56–64.

133 J. WOJCIK, J. P. MORNON, J. CHOMILIER. New efficient statistical sequence-dependent structure prediction of short to medium-sized protein loops based on an exhaustive loop classification. *J Mol Biol.* **1999**, 289, 1469–1490.

134 U. LESSEL, D. SCHOMBURG. Creation and characterization of a new, non-redundant fragment data bank. *Protein Eng.* **1997**, 10, 659–664.

135 J. MOULT, M. N. G. JAMES. An algorithm for determining the conformation of polypeptide segments in proteins by systematic search. *Proteins.* **1986**, 1, 146–163.

136 R. E. BRUCCOLERI, E. HABER, J. NOVOTNY. Structure of antibody hypervariable loops reproduced by a conformational search algorithm [published erratum appears in Nature 1988 Nov 17;336(6196):266]. *Nature.* **1988**, 335, 564–568.

137 D. Bassolino-Klimas, R. E. Bruccoleri, S. Subramaniam. Modeling the antigen combining site of an anti-dinitrophenyl antibody, ANO2. *Protein Sci.* **1992**, 1, 1465–1476.

138 B. R. Brooks, R. E. Bruccoleri, B. D. Olafson, D. J. States, S. Swaminathan, M. Karplus. CHARMM: A program for macromolecular energy, minimization, and dynamics calculations. *J. Comput. Chem.* **1983**, 4, 187–217.

139 Q. Zheng, R. Rosenfeld, S. Vajda, C. DeLisi. Determining protein loop conformation using scaling-relaxation techniques. *Prot. Science.* **1993**, 2, 1242–1248.

140 Q. Zheng, R. Rosenfeld, C. DeLisi, D. J. Kyle. Multiple copy sampling in protein loop modeling: computational efficiency and sensitivity to dihedral angle perturbations. *Prot. Science.* **1994**, 3, 493–506.

141 Q. Zheng, D. J. Kyle. Accuracy and reliability of the scaling-relaxation method for loop closure: An evaluation based on extensive and multiple copy conformational samplings. *Proteins.* **1996**, 24, 209–217.

142 D. Rosenbach, R. Rosenfeld. Simultaneous modeling of multiple loops in proteins. *Prot. Science.* **1995**, 4, 496–505.

143 P. S. Shenkin, D. L. Yarmush, R. M. Fine, H. J. Wang, C. Levinthal. Predicting antibody hypervariable loop conformation. I. Ensembles of random conformations for ringlike structures. *Biopolymers.* **1987**, 26, 2053–2085.

144 R. M. Fine, H. Wang, P. S. Shenkin, D. L. Yarmush, C. Levinthal. Predicting antibody hypervariable loop conformations. II: Minimization and molecular dynamics studies of MCPC603 from many randomly generated loop conformations. *Proteins.* **1986**, 1, 342–362.

145 R. Sowdhamini, S. D. Rufino, T. L. Blundell. A database of globular protein structural domains: clustering of representative family members into similar folds. *Fold. Des.* **1996**, 1, 209–220.

146 E. O. Purisima, H. A. Scheraga. An approach to the multiple-minima problem by relaxing dimensionality. *Proc. Natl. Acad. Sci. USA.* **1986**, 83, 2782–2786.

147 T. Cardozo, M. Totrov, R. Abagyan. Homology modeling by the ICM method. *Proteins.* **1995**, 23, 403–414.

148 U. Rao, M. M. Teeter. Improvement of turn structure prediction by molecular dynamics: a case study of alpha 1-purothionin. *Protein Eng.* **1993**, 6, 837–847.

149 R. Abagyan, M. Totrov. Biased probability Monte Carlo conformational searches and electrostatic calculations for peptides and proteins. *J Mol Biol.* **1994**, 235, 983–1002.

150 J. S. Evans, A. M. Mathiowetz, S. I. Chan, W. A. Goddard, 3rd. De novo prediction of polypeptide conformations using dihedral probability grid Monte Carlo methodology. *Protein Sci.* **1995**, 4, 1203–1216.

151 N. Guex, M. C. Peitsch. SWISS-MODEL and the Swiss-PdbViewer: an environment for comparative protein modeling. *Electrophoresis.* **1997**, 18, 2714–2723.

152 C. M. Deane, T. L. Blundell. A novel exhaustive search algorithm for predicting the conformation of polypeptide segments in proteins. *Proteins.* **2000**, 40, 135–144.

153 M. LEVITT, M. GERSTEIN, E. HUANG, S. SUBBIAH, J. TSAI. Protein folding: the endgame. *Annu Rev Biochem.* **1997**, 66, 549–579.

154 J. MENDES, A. M. BAPTISTA, M. A. CARRONDO, C. M. SOARES. Improved modeling of side-chains in proteins with rotamer-based methods: a flexible rotamer model. *Proteins.* **1999**, 37, 530–543.

155 N. L. SUMMERS, W. D. CARLSON, M. KARPLUS. Analysis of sidechain orientations in homologous proteins. *J. Mol. Biol.* **1987**, 196, 175–198.

156 N. L. SUMMERS, M. KARPLUS. Construction of side-chains in homology modelling: Application to the C-terminal lobe of rhizopuspepsin. *J. Mol. Biol.* **1989**, 210, 785–811.

157 V. SASISEKHARAN, P. K. PONNUSWAMY. Backbone and sidechain conformations of amino acids and amino acid residues in peptides. *Biopolymers.* **1970**, 9, 1249–1256.

158 V. SASISEKHARAN, P. K. PONNUSWAMY. Studies on the conformation of amino acids. X. Conformations of norvalyl, leucyl, aromatic side groups in a dipeptide unit. *Biopolymers.* **1971**, 10, 583–592.

159 M. KARPLUS, R. G. PARR. An approach to the internal rotation problem. *J. Chem. Phys.* **1963**, 38, 1547–1552.

160 J. JANIN, S. WODAK, M. LEVITT, B. MAIGRET. Conformations of amino acid side-chains in proteins. *J. Mol. Biol.* **1978**, 125, 357–386.

161 T. N. BHAT, V. SASISEKHARAN, M. VIJAYAN. An analysis of sidechain conformation in proteins. *Int. J. Peptide Protein Res.* **1979**, 13, 170–184.

162 E. BENEDETTI, G. MORELLI, G. NEMETHY, H. A. SCHERAGA. Statistical and energetic analysis of sidechain conformations in oligopeptides. *Int. J. Peptide Protein Res.* **1983**, 22, 1–15.

163 J. W. PONDER, F. M. RICHARDS. Tertiary templates for proteins: Use of packing criteria in the enumeration of allowed sequences for different structural classes. *J. Mol. Biol.* **1987**, 193, 775–792.

164 M. J. McGREGOR, S. A. ISLAM, M. J. E. STERNBERG. Analysis of the relationship between sidechain conformation and secondary structure in globular proteins. *J. Mol. Biol.* **1987**, 198, 295–310.

165 P. TUFFERY, C. ETCHEBEST, S. HAZOUT, R. LAVERY. A new approach to the rapid determination of protein side chain conformations. *J. Biomol. Struct. Dynam.* **1991**, 8, 1267–1289.

166 R. L. DUNBRACK, JR., M. KARPLUS. Backbone-dependent rotamer library for proteins. Application to side-chain prediction. *J Mol Biol.* **1993**, 230, 543–574.

167 H. SCHRAUBER, F. EISENHABER, P. ARGOS. Rotamers: To be or not to be? An analysis of amino acid sidechain conformations in globular proteins. *J. Mol. Biol.* **1993**, 230, 592–612.

168 R. L. DUNBRACK, JR., M. KARPLUS. Conformational analysis of the backbone-dependent rotamer preferences of protein sidechains. *Nature Struct. Biol.* **1994**, 1, 334–340.

169 R. L. DUNBRACK, JR., F. E. COHEN. Bayesian statistical analysis of protein sidechain rotamer preferences. *Prot. Science.* **1997**, 6, 1661–1681.

170 H. Kono, J. Doi. Energy minimization method using automata network for sequence and side-chain conformation prediction from given backbone geometry. *Proteins.* **1994**, 19, 244–255.

171 S. C. Lovell, J. M. Word, J. S. Richardson, D. C. Richardson. Asparagine and glutamine rotamers: B-factor cutoff and correction of amide flips yield distinct clustering. *Proc Natl Acad Sci U S A.* **1999**, 96, 400–405.

172 J. MacKerell, A. D., M. B. D. Bashford, J. Dunbrack, R. D., M. J. F. J. Evanseck, S. Fischer, J. Gao, H. Guo, S. Ha, D. Joseph-McCarthy, L. Kuchnir, K. Kuczera, F. T. K. Lau, C. Mattos, S. Michnick, T. Ngo, D. T. Nguyen, B. Prodhom, W. E. Reiher, III, B. Roux, M. Schlenkrich, J. Smith, R. Stote, J. Straub, M. Watanabe, J. Wiórkiewicz-K. All-atom empirical potential for molecular modeling and dynamics studies of proteins. *J. Phys. Chem.* **1998**, B102, 3586–3616.

173 P. Koehl, M. Delarue. Application of a self-consistent mean field theory to predict protein side-chains conformation and estimate their conformational entropy. *J Mol Biol.* **1994**, 239, 249–275.

174 C. A. Schiffer, J. W. Caldwell, P. A. Kollman, R. M. Stroud. Prediction of homologous protein structures based on conformational searches and energetics. *Proteins.* **1990**, 8, 30–43.

175 C. Wilson, L. Gregoret, D. Agard. Modeling sidechain conformation for homologous proteins using an energy-based rotamer search. *J. Mol. Biol.* **1993**, 229, 996–1006.

176 L. Holm, C. Sander. Fast and simple Monte Carlo algorithm for side chain optimization in proteins: Application to model building by homology. *Proteins.* **1992**, 14, 213–223.

177 M. Levitt. Accurate modeling of protein conformation by automatic segment matching. *J. Mol. Biol.* **1992**, 226, 507–533.

178 C. A. Laughton. Prediction of protein sidechain conformations from local three-dimensional homology relationships. *J. Mol. Biol.* **1994**, 235, 1088–1097.

179 J. K. Hwang, W. F. Liao. Side-chain prediction by neural networks and simulated annealing optimization. *Protein Eng.* **1995**, 8, 363–370.

180 J. Mendes, C. M. Soares, M. A. Carrondo. Improvement of side-chain modeling in proteins with the self-consistent mean field theory method based on an analysis of the factors influencing prediction. *Biopolymers.* **1999**, 50, 111–131.

181 J. Desmet, M. De Maeyer, B. Hazes, I. Lasters. The dead-end elimination theorem and its use in protein sidechain positioning. *Nature.* **1992**, 356, 539–542.

182 I. Lasters, J. Desmet. The fuzzy-end elimination theorem: Correctly implementing the sidechain placement algorithm based on the dead-end elimination theorem. *Prot. Eng.* **1993**, 6, 717–722.

183 R. F. Goldstein. Efficient rotamer elimination applied to protein side-chains and related spin glasses. *Biophys J.* **1994**, 66, 1335–1340.

184 I. Lasters, M. DeMaeyer, J. Desmet. Enhanced dead-end elimination in the search for the global minimum energy conformation of a collection of protein sidechains. *Prot. Eng.* **1995**, 8, 815–822.

185 J. Desmet, M. De Maeyer, I. Lasters. Theoretical and algorithmical optimization of the dead-end elimination theorem. *Pac Symp Biocomput.* **1997**, 122–133.

186 B. I. Dahiyat, C. A. Sarisky, S. L. Mayo. De novo protein design: towards fully automated sequence selection. *J Mol Biol.* **1997**, 273, 789–796.

187 D. B. Gordon, S. L. Mayo. Branch-and-terminate: a combinatorial optimization algorithm for protein design. *Structure Fold Des.* **1999**, 7, 1089–1098.

188 C. A. Voigt, D. B. Gordon, S. L. Mayo. Trading accuracy for speed: A quantitative comparison of search algorithms in protein sequence design. *J Mol Biol.* **2000**, 299, 789–803.

189 M. Vasquez. An evaluation of discrete and continuum search techniques for conformational analysis of side-chains in proteins. *Biopolymers.* **1995**, 36, 53–70.

190 C. Lee, S. Subbiah. Prediction of protein side-chain conformation by packing optimization. *J Mol Biol.* **1991**, 217, 373–388.

191 P. Tuffery, C. Etchebest, S. Hazout. Prediction of protein side chain conformations: a study on the influence of backbone accuracy on conformation stability in the rotamer space. *Protein Eng.* **1997**, 10, 361–372.

192 H. Kono, J. Doi. A new method for side-chain conformation prediction using a Hopfield network and reproduced rotamers. *J. Comp. Chem.* **1996**, 17, 1667–1683.

193 R. Samudrala, J. Moult. Determinants of side chain conformational preferences in protein structures. *Protein Eng.* **1998**, 11, 991–997.

194 G. Chinea, G. Padron, R. W. Hooft, C. Sander, G. Vriend. The use of position-specific rotamers in model building by homology. *Proteins.* **1995**, 23, 415–421.

195 A. Sali, J. P. Overington. Derivation of rules for comparative protein modeling from a database of protein structure alignments. *Prot. Science.* **1994**, 3, 1582–1596.

196 M. C. Peitsch. ProMod and Swiss-Model: Internet-based tools for automated comparative protein modelling. *Biochem Soc Trans.* **1996**, 24, 274–279.

197 M. C. Peitsch. Large scale protein modelling and model repository. *Ismb.* **1997**, 5, 234–236.

198 R. Samudrala, J. Moult. A graph-theoretic algorithm for comparative modeling of protein structure. *J Mol Biol.* **1998**, 279, 287–302.

199 R. Samudrala, J. Moult. An all-atom distance-dependent conditional probability discriminatory function for protein structure prediction. *J Mol Biol.* **1998**, 275, 895–916.

200 W. F. v. Gunsteren, P. H. Hünenberger, A. E. Mark, P. E. Smith, I. G. Tironi. Computer simulation of protein motion. *Comp. Phys. Comm.* **1995**, 91, 305–319.

201 R. E. BRUCCOLERI, M. KARPLUS. Conformational sampling using high-temperature molecular dynamics. *Biopolymers.* **1990**, 29, 1847–1862.

202 R. TEJERO, D. BASSOLINO-KLIMAS, R. E. BRUCCOLERI, G. T. MONTELIONE. Simulated annealing with restrained molecular dynamics using CONGEN: energy refinement of the NMR solution structures of epidermal and type-alpha transforming growth factors. *Protein Sci.* **1996**, 5, 578–592.

203 D. BASSOLINO-KLIMAS, R. TEJERO, S. R. KRYSTEK, W. J. METZLER, G. T. MONTELIONE, R. E. BRUCCOLERI. Simulated annealing with restrained molecular dynamics using a flexible restraint potential: theory and evaluation with simulated NMR constraints. *Protein Sci.* **1996**, 5, 593–603.

204 A. S. YANG, B. HONIG. Sequence to structure alignment in comparative modeling using PrISM. *Proteins.* **1999**, 37, 66–72.

205 B. HONIG. Protein folding: from the levinthal paradox to structure prediction. *J Mol Biol.* **1999**, 293, 283–293.

206 A. S. YANG, B. HONIG. An integrated approach to the analysis and modeling of protein sequences and structures. I. Protein structural alignment and a quantitative measure for protein structural distance. *J Mol Biol.* **2000**, 301, 665–678.

207 A. S. YANG, B. HONIG. An integrated approach to the analysis and modeling of protein sequences and structures. II. On the relationship between sequence and structural similarity for proteins that are not obviously related in sequence. *J Mol Biol.* **2000**, 301, 679–689.

208 A. S. YANG, B. HONIG. An integrated approach to the analysis and modeling of protein sequences and structures. III. A comparative study of sequence conservation in protein structural families using multiple structural alignments. *J Mol Biol.* **2000**, 301, 691–711.

209 R. SAMUDRALA, J. MOULT. Handling context-sensitivity in protein structures using graph theory: bona fide prediction. *Proteins.* **1997**, Suppl, 43–49.

210 T. F. HAVEL. Predicting the structure of the flavodoxin from *Escherichia coli* by homology modeling, distance geometry, and molecular dynamics. *Molecular Simulation.* **1993**, 10, 175–210.

211 M. T. MAS, K. C. SMITH, D. L. YARMUSH, K. AISAKA, R. M. FINE. Modeling the anti-CEA antibody combining site by homology and conformational search. *Proteins.* **1992**, 14, 483–498.

212 C. L. CASIPIT, R. TAL, V. WITTMAN, P. A. CHAVAILLAZ, K. ARBUTHNOTT, J. A. WEIDANZ, J. A. JIAO, H. C. WONG. Improving the binding affinity of an antibody using molecular modeling and site-directed mutagenesis. *Protein Sci.* **1998**, 7, 1671–1680.

213 S. DERET, L. DENOROY, M. LAMARINE, R. VIDAL, B. MOUGENOT, B. FRANGIONE, F. J. STEVENS, P. M. RONCO, P. AUCOUTURIER. Kappa light chain-associated Fanconi's syndrome: molecular analysis of monoclonal immunoglobulin light chains from patients with and without intracellular crystals. *Protein Eng.* **1999**, 12, 363–369.

214 S. PETIT, F. BRARD, G. COQUEREL, G. PEREZ, F. TRON. Structural models of antibody variable fragments: a method for investigating binding mechanisms. *J Comput Aided Mol Des.* **1998**, 12, 147–163.

215 R. SUENAGA, K. MITAMURA, N. I. ABDOU. V gene sequences of lupus-derived human IgM anti-ssDNA antibody: implication for the importance of the location of DNA-binding amino acids. *Clin Immunol Immunopathol.* **1998**, 86, 72–80.

216 J. E. McELVEEN, M. R. CLARK, S. J. SMITH, H. F. SEWELL, F. SHAKIB. Primary sequence and molecular model of the variable region of a mouse monoclonal anti-Der p 1 antibody showing a similar epitope specificity as human IgE. *Clin Exp Allergy.* **1998**, 28, 1427–1434.

217 B. L. SIBANDA, T. BLUNDELL, P. M. HOBART, M. FOGLIANO, J. S. BINDRA, B. W. DOMINY, J. M. CHIRGWIN. Computer graphics modeling of human renin. *FEBS Lett.* **1984**, 174, 102–111.

218 E. A. PADLAN, B. A. HELM. Modeling of the lectin-homology domains of the human and murine low-affinity Fc epsilon receptor (Fc epsilon RII/CD23). *Receptor.* **1993**, 3, 325–341.

219 J. A. FERNANDEZ, B. O. VILLOUTREIX, T. M. HACKENG, J. H. GRIFFIN, B. N. BOUMA. Analysis of protein S C4b-binding protein interactions by homology modeling and inhibitory antibodies. *Biochemistry.* **1994**, 33, 11073–11078.

220 M. T. PISABARRO, A. R. ORTIZ, L. SERRANO, R. C. WADE. Homology modeling of the Abl-SH3 domain. *Proteins.* **1994**, 20, 203–215.

221 L. LI, T. DARDEN, C. FOLEY, R. HISKEY, L. PEDERSEN. Homology modeling and molecular dynamics simulation of human prothrombin fragment 1. *Protein Sci.* **1995**, 4, 2341–2348.

222 V. ROSSI, C. GABORIAUD, M. LACROIX, J. ULRICH, J. C. FONTECILLA-CAMPS, J. GAGNON, G. J. ARLAUD. Structure of the catalytic region of human complement protease C1s: study by chemical cross-linking and three-dimensional homology modeling. *Biochemistry.* **1995**, 34, 7311–7321.

223 R. T. KROEMER, W. G. RICHARDS. Homology modeling study of the human interleukin-7 receptor complex. *Protein Eng.* **1996**, 9, 1135–1142.

224 R. T. KROEMER, S. W. DOUGHTY, A. J. ROBINSON, W. G. RICHARDS. Prediction of the three-dimensional structure of human interleukin-7 by homology modeling. *Protein Eng.* **1996**, 9, 493–498.

225 N. H. THOMA, P. F. LEADLAY. Homology modeling of human methylmalonyl-CoA mutase: a structural basis for point mutations causing methylmalonic aciduria. *Protein Sci.* **1996**, 5, 1922–1927.

226 S. MODI, M. J. PAINE, M. J. SUTCLIFFE, L. Y. LIAN, W. U. PRIMROSE, C. R. WOLF, G. C. ROBERTS. A model for human cytochrome P450 2D6 based on homology modeling and NMR studies of substrate binding. *Biochemistry.* **1996**, 35, 4540–4550.

227 G. Raddatz, H. Bisswanger. Receptor site and stereospecifity of dihydrolipoamide dehydrogenase for R- and S-lipoamide: a molecular modeling study. *J Biotechnol.* **1997**, 58, 89–100.

228 P. Prusis, H. B. Schioth, R. Muceniece, P. Herzyk, M. Afshar, R. E. Hubbard, J. E. Wikberg. Modeling of the three-dimensional structure of the human melanocortin 1 receptor, using an automated method and docking of a rigid cyclic melanocyte-stimulating hormone core peptide. *J Mol Graph Model.* **1997**, 15, 307–317, 334.

229 D. K. Pettit, T. P. Bonnert, J. Eisenman, S. Srinivasan, R. Paxton, C. Beers, D. Lynch, B. Miller, J. Yost, K. H. Grabstein, W. R. Gombotz. Structure-function studies of interleukin 15 using site-specific mutagenesis, polyethylene glycol conjugation, and homology modeling. *J Biol Chem.* **1997**, 272, 2312–2318.

230 B. K. Klein, Y. Feng, C. A. McWherter, W. F. Hood, K. Paik, J. P. McKearn. The receptor binding site of human interleukin-3 defined by mutagenesis and molecular modeling. *J Biol Chem.* **1997**, 272, 22630–22641.

231 D. F. Lewis, M. G. Bird, D. V. Parke. Molecular modelling of CYP2E1 enzymes from rat, mouse and man: an explanation for species differences in butadiene metabolism and potential carcinogenicity, and rationalization of CYP2E substrate specificity. *Toxicology.* **1997**, 118, 93–113.

232 M. Lacroix, V. Rossi, C. Gaboriaud, S. Chevallier, M. Jaquinod, N. M. Thielens, J. Gagnon, G. J. Arlaud. Structure and assembly of the catalytic region of human complement protease C1r: a three-dimensional model based on chemical cross-linking and homology modeling. *Biochemistry.* **1997**, 36, 6270–6282.

233 O. H. Kapp, J. Siemion, W. C. Eckelman, V. I. Cohen, R. C. Reba. Molecular modeling of the interaction of diagnostic radiopharmaceuticals with receptor proteins: m2 antagonist binding to the muscarinic m2 subtype receptor. *Recept Signal Transduct.* **1997**, 7, 177–201.

234 J. Bajorath, W. J. Metzler, P. S. Linsley. Molecular modeling of CD28 and three-dimensional analysis of residue conservation in the CD28/CD152 family. *J Mol Graph Model.* **1997**, 15, 135–139, 108–111.

235 C. L. Byington, R. L. Dunbrack, Jr., F. G. Whitby, F. E. Cohen, N. Agabian. Entamoeba histolytica: computer-assisted modeling of phosphofructokinase for the prediction of broad-spectrum antiparasitic agents. *Exp Parasitol.* **1997**, 87, 194–202.

236 J. P. Annereau, V. Stoven, F. Bontems, J. Barthe, G. Lenoir, S. Blanquet, J. Y. Lallemand. Insight into cystic fibrosis by structural modelling of CFTR first nucleotide binding fold (NBF1). *C R Acad Sci III.* **1997**, 320, 113–121.

237 M. L. Hoover, R. T. Marta. Molecular modelling of HLA-DQ suggests a mechanism of resistance in type 1 diabetes. *Scand J Immunol.* **1997**, 45, 193–202.

238 S. Miertus, J. Tomasi, G. Mazzanti, E. E. Chiellini, R. Solaro, E. Chiellini. Modelling of the 3-D structure of IFN-alpha-k and characterization of its surface molecular properties. *Int J Biol Macromol.* **1997**, 20, 85–95.

239 A. I. WACEY, M. KRAWCZAK, G. KEMBALL-COOK, D. N. COOPER. Homology modelling of the catalytic domain of early mammalian protein C: evolution of structural features. *Hum Genet.* **1997**, 101, 37–42.

240 T. PIIRONEN, B. O. VILLOUTREIX, C. BECKER, K. HOLLINGSWORTH, M. VIHINEN, D. BRIDON, X. QIU, J. RAPP, B. DOWELL, T. LOVGREN, K. PETTERSSON, H. LILJA. Determination and analysis of antigenic epitopes of prostate specific antigen (PSA) and human glandular kallikrein 2 (hK2) using synthetic peptides and computer modeling. *Protein Sci.* **1998**, 7, 259–269.

241 F. PEELMAN, N. VINAIMONT, A. VERHEE, B. VANLOO, J. L. VERSCHELDE, C. LABEUR, S. SEGURET-MACE, N. DUVERGER, G. HUTCHINSON, J. VANDEKERCKHOVE, J. TAVERNIER, M. ROSSENEU. A proposed architecture for lecithin cholesterol acyl transferase (LCAT): identification of the catalytic triad and molecular modeling. *Protein Sci.* **1998**, 7, 587–599.

242 S. MORO, A. H. LI, K. A. JACOBSON. Molecular modeling studies of human A3 adenosine antagonists: structural homology and receptor docking. *J Chem Inf Comput Sci.* **1998**, 38, 1239–1248.

243 M. LIU, M. E. MURPHY, A. R. THOMPSON. A domain mutations in 65 haemophilia A families and molecular modelling of dysfunctional factor VIII proteins. *Br J Haematol.* **1998**, 103, 1051–1060.

244 P. V. JENKINS, K. J. PASI, S. J. PERKINS. Molecular modeling of ligand and mutation sites of the type A domains of human von Willebrand factor and their relevance to von Willebrand's disease. *Blood.* **1998**, 91, 2032–2044.

245 R. G. EFREMOV, F. LEGRET, G. VERGOTEN, A. CAPRON, G. M. BAHR, A. S. ARSENIEV. Molecular modeling of HIV-1 coreceptor CCR5 and exploring of conformational space of its extracellular domain in molecular dynamics simulation. *J Biomol Struct Dyn.* **1998**, 16, 77–90.

246 D. YAMAMOTO, N. SHIOTA, S. TAKAI, T. ISHIDA, H. OKUNISHI, M. MIYAZAKI. Three-dimensional molecular modeling explains why catalytic function for angiotensin-I is different between human and rat chymases. *Biochem Biophys Res Commun.* **1998**, 242, 158–163.

247 I. STOILOV, A. N. AKARSU, I. ALOZIE, A. CHILD, M. BARSOUM-HOMSY, M. E. TURACLI, M. OR, R. A. LEWIS, N. OZDEMIR, G. BRICE, S. G. AKTAN, L. CHEVRETTE, M. COCA-PRADOS, M. SARFARAZI. Sequence analysis and homology modeling suggest that primary congenital glaucoma on 2p21 results from mutations disrupting either the hinge region or the conserved core structures of cytochrome P4501B1. *Am J Hum Genet.* **1998**, 62, 573–584.

248 J. S. SONG, H. PARK, H. J. HONG, M. H. YU, S. E. RYU. Homology modeling of the receptor binding domain of human thrombopoietin. *J Comput Aided Mol Des.* **1998**, 12, 419–424.

249 D. K. Smith, H. R. Treutlein. LIF receptor-gp130 interaction investigated by homology modeling: implications for LIF binding. *Protein Sci.* **1998**, 7, 886–896.

250 S. P. Scott, J. C. Tanaka. Use of homology modeling to predict residues involved in ligand recognition. *Methods Enzymol.* **1998**, 293, 620–647.

251 L. L. Johnson, D. A. Bornemeier, J. A. Janowicz, J. Chen, A. G. Pavlovsky, D. F. Ortwine. Effect of species differences on stromelysin-1 (MMP-3) inhibitor potency. An explanation of inhibitor selectivity using homology modeling and chimeric proteins. *J Biol Chem.* **1999**, 274, 24881–24887.

252 F. Fraternali, A. Pastore. Modularity and homology: modelling of the type II module family from titin. *J Mol Biol.* **1999**, 290, 581–593.

253 Y. T. Chang, G. H. Loew. Homology modeling and substrate binding study of human CYP4A11 enzyme. *Proteins.* **1999**, 34, 403–415.

254 R. J. Auchus, W. L. Miller. Molecular modeling of human P450c17 (17alpha-hydroxylase/17,20-lyase): insights into reaction mechanisms and effects of mutations. *Mol Endocrinol.* **1999**, 13, 1169–1182.

255 G. Wishart, D. H. Bremner, K. R. Sturrock. Molecular modelling of the 5-hydroxytryptamine receptors. *Receptors Channels.* **1999**, 6, 317–335.

256 N. Venkatesh, S. Krishnaswamy, S. Meuris, G. S. Murthy. Epitope analysis and molecular modeling reveal the topography of the C-terminal peptide of the beta-subunit of human chorionic gonadotropin. *Eur J Biochem.* **1999**, 265, 1061–1066.

257 R. Sanchez, U. Pieper, N. Mirkovic, P. I. de Bakker, E. Wittenstein, A. Sali. MODBASE, a database of annotated comparative protein structure models. *Nucleic Acids Res.* **2000**, 28, 250–253.

258 G. G. Glenner, C. W. Wong. Alzheimer's disease: initial report of the purification and characterization of a novel cerebrovascular amyloid protein. *Biochem Biophys Res Commun.* **1984**, 120, 885–890.

259 J. Kang, H. G. Lemaire, A. Unterbeck, J. M. Salbaum, C. L. Masters, G. H. Grzeschik, G. Multhaup, K. Beyreuther, B. Muller-Hill. The precursor of Alzheimer's disease amyloid A4 protein resembles a cell surface receptor. *Nature.* **1987**, 325, 733–736.

260 I. Hussain, D. Powell, D. R. Howlett, D. G. Tew, T. D. Meek, C. Chapman, I. S. Gloger, K. E. Murphy, C. D. Southan, D. M. Ryan, T. S. Smith, D. L. Simmons, F. S. Walsh, C. Dingwall, G. Christie. Identification of a novel aspartic protease (Asp2) as β-secretase. *Mol. Cell. Neurosci.* **1999**, 14, 419–427.

261 R. Vassar, B. D. Bennett, S. Babu-Khan, S. Kahn, E. A. Mendiaz, P. Denis, J.-C. Louis, F. Collins, J. Treanor, G. Rogers, M. Citron. β-secretase cleavage of Alzheimer's amyloid precursor protein by the transmembrane aspartic protease BACE. *Science.* **1999**, 286, 735–741.

262 S. SINHA, J. P. ANDERSON, R. BARBOUR, G. S. BASI, R. CACCAVELLO, D. DAVIS, M. DOAN, H. F. DOVEY, N. FRIGON, J. HONG, K. JACOBSON-CROAK, N. JEWETT, P. KEIM, J. KNOPS, I. LIEBERBURG, M. POWER, H. TAN, G. TATSUNO, J. TUNG, D. SCHENK, P. SEUBERT, S. M. SUOMENSAARI, S. WANG, D. WALKER, V. JOHN, et al. Purification and cloning of amyloid precursor protein beta-secretase from human brain. *Nature.* **1999**, 402, 537–540.

263 R. YAN, M. J. BIENKOWSKI, M. E. SHUCK, H. MIAO, M. C. TORY, A. M. PAULEY, J. R. BRASHIER, N. C. STRATMAN, W. R. MATHEWS, A. E. BUHL, D. B. CARTER, A. G. TOMASSELLI, L. A. PARODI, R. L. HEINRIKSON, M. E. GURNEY. Membrane-anchored aspartyl protease with Alzheimer's disease beta-secretase activity. *Nature.* **1999**, 402, 533–537.

264 M. CITRON, D. B. TEPLOW, D. J. SELKOE. Generation of amyloid beta protein from its precursor is sequence specific. *Neuron.* **1995**, 14, 661–670.

265 R. VASSAR, B. D. BENNETT, S. BABU-KHAN, S. KAHN, E. A. MENDIAZ, P. DENIS, D. B. TEPLOW, S. ROSS, P. AMARANTE, R. LOELOFF, Y. LUO, S. FISHER, J. FULLER, S. EDENSON, J. LILE, M. A. JAROSINSKI, A. L. BIERE, E. CURRAN, T. BURGESS, J. C. LOUIS, F. COLLINS, J. TREANOR, G. ROGERS, M. CITRON. beta-Secretase Cleavage of Alzheimer's Amyloid Precursor Protein by the Transmembrane Aspartic Protease BACE. *Science.* **1999**, 286, 735–741.

266 M. FUJINAGA, M. M. CHERNAIA, N. I. TARASOVA, S. C. MOSIMANN, M. N. G. JAMES. Crystal structure of human pepsin and its complex with pepstatin. *Prot. Science.* **1995**, 4, 960–972.

267 K. SUGUNA, E. A. PADLAN, C. W. SMITH, W. D. CARLSON, D. R. DAVIES. Binding of a reduced peptide inhibitor to the aspartic proteinase from Rhizopus chinensis: implications for a mechanism of action. *Proc. Natl. Acad. Sci. USA.* **1987**, 84, 7009–7013.

268 A. J. SAUNDERS, T.-W. KIM, R. E. TANZI. BACE maps to chromosome 11 and a BACE homolog, BACE2, reside in the obligate down syndrome region of chromosome 21. *Science.* **1999**, 286, 1255a.

269 X.-D. CAI, T. E. GOLDE, S. G. YOUNKIN. Release of excess amyloid β protein from a mutant amyloid β precursor. *Science.* **1993**, 259, 514–516.

270 L. HONG, G. KOELSCH, X. LIN, S. WU, S. TERZYAN, A. K. GHOSH, X. C. ZHANG, J. TANG. Structure of the protease domain of memapsin 2 (beta-secretase) complexed with inhibitor. *Science.* **2000**, 290, 150–153.

271 K. PALCZEWSKI, T. KUMASAKA, T. HORI, C. A. BEHNKE, H. MOTOSHIMA, B. A. FOX, I. LE TRONG, D. C. TELLER, T. OKADA, R. E. STENKAMP, M. YAMAMOTO, M. MIYANO. Crystal structure of rhodopsin: A G protein-coupled receptor. *Science.* **2000**, 289, 739–745.

272 H. LUECKE, B. SCHOBERT, H. T. RICHTER, J. P. CARTAILLER, J. K. LANYI. Structure of bacteriorhodopsin at 1.55 A resolution. *J Mol Biol.* **1999**, 291, 899–911.

273 R. HENDERSON, J. M. BALDWIN, T. A. CESKA, F. ZEMLIN, E. BECKMANN, K. H. DOWNING. Model for the structure of bacteriorhodopsin based on high-resolution electron cryo-microscopy. *J Mol Biol.* **1990**, 213, 899–929.

274 D. J. STATES, D. BOTSTEIN. Molecular sequence accuracy and the analysis of protein coding regions. *Proc Natl Acad Sci USA.* **1991**, 88, 5518–5522.

275 B. ROST, C. SANDER, R. SCHNEIDER. Redefining the goals of structure prediction. *J. Mol. Biol.* **1994**, 235, 13–26.

276 A. ZEMLA, C. VENCLOVAS, K. FIDELIS, B. ROST. A modified definition of Sov, a segment-based measure for protein secondary structure prediction assessment. *Proteins.* **1999**, 34, 220–223.

277 C. VENCLOVAS, A. ZEMLA, K. FIDELIS, J. MOULT. Criteria for evaluating protein structures derived from comparative modeling. *Proteins.* **1997**, Suppl, 7–13.

278 C. VENCLOVAS, A. ZEMLA, K. FIDELIS, J. MOULT. Some measures of comparative performance in the three CASPs. *Proteins.* **1999**, Suppl, 231–237.

279 A. ZEMLA, C. VENCLOVAS, J. MOULT, K. FIDELIS. Processing and analysis of CASP3 protein structure predictions. *Proteins.* **1999**, Suppl, 22–29.

280 J. MOULT. Predicting protein three-dimensional structure. *Curr Opin Biotechnol.* **1999**, 10, 583–588.

281 R. A. LASKOWSKI, D. S. MOSS, J. M. THORNTON. Main-chain bond lengths and bond angles in protein structures. *J Mol Biol.* **1993**, 231, 1049–1067.

282 A. L. MORRIS, M. W. MACARTHUR, E. G. HUTCHINSON, J. M. THORNTON. Stereochemical quality of protein structure coordinates. *Proteins.* **1992**, 12, 345–364.

283 R. A. LASKOWSKI, M. W. MACARTHUR, D. S. MOSS, J. M. THORNTON. PROCHECK: a program to check the stereochemical quality of protein structures. *J. Appl. Cryst.* **1993**, 26, 283–291.

284 G. VRIEND. WHAT IF: A molecular modeling and drug design program. *J. Mol. Graphics.* **1990**, 8, 52–56.

285 R. A. ENGH, R. HUBER. Accurate bond and angle parameterss for X-ray protein structure refinement. *Acta. Cryst.* **1991**, A47, 392–400.

286 K. B. WIBERG, M. A. MURCKO. Rotational barriers. 2. Energies of alkane rotamers. An examination of gauche interactions. *J. Amer. Chem. Soc.* **1988**, 110, 8029–8038.

287 P. J. KRAULIS. MolScript: A program to produce both detailed and schematic plots of protein structures. *J. Appl. Cryst.* **1991**, 24, 946–950.

288 E. A. MERRITT, D. J. BACON. Raster3D photorealistic molecular graphics. *Meth. Enz.* **1997**, 277, 505–524.

289 P. MATTIS, S. KIMBALL, *The GNU image manipulation program.* 1998, Gimp.org: Berkeley, CA.

290 T. E. FERRIN, C. C. HUANG, L. E. JARVIS, R. LANGRIDGE. The MIDAS display system. *J. Mol. Graphics.* **1988**, 6, 13–27.

291 D. FRISHMAN, P. ARGOS. Knowledge-based protein secondary structure assignment. *Proteins.* **1995**, 23, 566–579.

292 R. Abagyan, S. Batalov, T. Cardozo, M. Totrov, J. Webber, Y. Zhou. Homology modeling with internal coordinate mechanics: deformation zone mapping and improvements of models via conformational search. *Proteins*. **1997**, Suppl, 29–37.

293 K. C. Chou, L. Carlacci. Simulated annealing approach to the study of protein structures. *Protein Eng*. **1991**, 4, 661–667.

294 L. Carlacci, S. W. Englander. The loop problem in proteins: a Monte Carlo simulated annealing approach. *Biopolymers*. **1993**, 33, 1271–1286.

295 J. Higo, V. Collura, J. Garnier. Development of an extended simulated annealing method: application to the modeling of complementary determining regions of immunoglobulins. *Biopolymers*. **1992**, 32, 33–43.

296 V. Collura, J. Higo, J. Garnier. Modeling of protein loops by simulated annealing. *Protein Sci*. **1993**, 2, 1502–1510.

297 V. P. Collura, P. J. Greaney, B. Robson. A method for rapidly assessing and refining simple solvent treatments in molecular modelling. Example studies on the antigen-combining loop H2 from FAB fragment McPC603. *Protein Eng*. **1994**, 7, 221–233.

298 B. Robson, E. Platt. Refined models for computer calculations in protein engineering. Calibration and testing of atomic potential functions compatible with more efficient calculations. *J Mol Biol*. **1986**, 188, 259–281.

299 M. J. Dudek, H. A. Scheraga. Protein structure prediction using a combination of sequence homology and global enegy minimization I. Global energy minimization of surface loops. *J. Comp. Chem*. **1990**, 11, 121–151.

300 K. A. Palmer, H. A. Scheraga. Standard-geometry chains fitted to X-ray derived structures: Validation of the rigid-geometry approximation. I. Chain closure through a limited search of loop conformations. *J. Comp. Chem*. **1990**, 12, 505–526.

301 K. A. Palmer, H. A. Scheraga. Standard-geometry chains fitted to X-ray derived structures: Validation of the rigid-geometry approximation. II. Systematic searches for short loops in proteins: Applications to bovine pancreatic ribonuclease A and human lysozyme. *J. Comp. Chem*. **1990**, 13, 329–350.

302 C. S. Ring, F. E. Cohen. Modeling protein structures: construction and their applications. *FASEB J*. **1993**, 7, 783–790.

303 S. Sudarsanam, S. Srinivasan. Sequence-dependent conformational sampling using a database of phi(i)+1 and psi(i) angles for predicting polypeptide backbone conformations. *Protein Eng*. **1997**, 10, 1155–1162.

304 S. Sudarsanam, R. F. DuBose, C. J. March, S. Srinivasan. Modeling protein loops using a phi i + 1, psi i dimer database. *Protein Sci*. **1995**, 4, 1412–1420.

305 P. A. Bates, M. J. Sternberg. Model building by comparison at CASP3: Using expert knowledge and computer automation. *Proteins*. **1999**, 37, 47–54.

306 J. L. PELLEQUER, S. W. CHEN. Does conformational free energy distinguish loop conformations in proteins? *Biophys J.* **1997**, 73, 2359–2375.

307 M. DE MAEYER, J. DESMET, I. LASTERS. All in one: a highly detailed rotamer library improves both accuracy and speed in the modelling of sidechains by dead-end elimination. *Fold Des.* **1997**, 2, 53–66.

308 F. EISENMENGER, P. ARGOS, R. ABAGYAN. A method to configure protein side-chains from the main-chain trace in homology modelling. *J Mol Biol.* **1993**, 231, 849–860.

309 D. A. KELLER, M. SHIBATA, E. MARCUS, R. L. ORNSTEIN, R. REIN. Finding the global minimum: a fuzzy end elimination implementation. *Protein Eng.* **1995**, 8, 893–904.

310 C. A. LAUGHTON. Prediction of protein side-chain conformations from local three-dimensional homology relationships. *J Mol Biol.* **1994**, 235, 1088–1097.

311 A. R. LEACH, A. P. LEMON. Exploring the conformational space of protein side chains using dead-end elimination and the A* algorithm. *Proteins.* **1998**, 33, 227–239.

312 C. LEE. Testing homology modeling on mutant proteins: Predicting structural and thermodynamic effects in the Ala 98 → Val mutants of T4 lysozyme. *Fold. Des.* **1995**, 1, 1–12.

313 K. OGATA, H. UMEYAMA. Prediction of protein side-chain conformations by principal component analysis for fixed main-chain atoms. *Protein Eng.* **1997**, 10, 353–359.

314 K. OGATA, M. OHYA, H. UMEYAMA. Amino acid similarity matrix for homology modeling derived from structural alignment and optimized by the Monte Carlo method. *J Mol Graph Model.* **1998**, 16, 178–189, 254.

315 P. S. SHENKIN, H. FARID, J. S. FETROW. Prediction and evaluation of side-chain conformations for protein backbone structures. *Proteins.* **1996**, 26, 323–352.

316 M. E. SNOW, L. M. AMZEL. Calculating three-dimensional changes in protein structure due to amino-acid substitutions: the variable region of immunoglobulins. *Proteins.* **1986**, 1, 267–279.

317 R. TANIMURA, A. KIDERA, H. NAKAMURA. Determinants of protein side-chain packing. *Protein Sci.* **1994**, 3, 2358–2365.

318 J. J. WENDOLOSKI, F. R. SALEMME. PROBIT: A statistical approach to modeling proteins from partial coordinate data using substructure libraries. *J. Mol. Graphics.* **1992**, 10, 124–127.

319 L. HOLM, C. SANDER. Database algorithm for generating protein backbone and sidechain coordinates from a Cα trace: Application to model building and detection of coordinate errors. *J. Mol. Biol.* **1991**, 218, 183–194.

6

Protein Structure Prediction

Ralf Zimmer and Thomas Lengauer

Proteins fold into nice-looking, three-dimensional structures [1, 2]. These structures are an essential factor of the protein's function. Therefore, extensive experimental efforts have been invested to solve the structures for the most interesting, abundant, or medically, biologically, or pharmaceutically important proteins. Classically, X-ray crystallography has been the dominant technique for resolving protein structures. Here the protein has to be purified and crystallized, before a diffraction diagram can be measured by exposing the crystal to X-ray radiation. The diffraction diagram then has to be Fourier transformed to obtain the electron density map of the protein. The latter problem is aggravated by the fact that the diffraction diagram only reveals the intensity but not the phase of the interfering X-rays. In order to obtain the electron density from the diffraction diagram computer-based modeling methods have been established. More recently, NMR techniques have been developed and refined to solve larger and larger proteins structures in solution. These methods yield 2D data that have to be interpreted by a laborious computer-based process, in order to reveal the electron density [1]. Thus, for both approaches, not only are the experimental costs enormous but also the computer methods involved. The motivation to go through the time-consuming and expensive process of determining a detailed atomic structure of a bio-molecule ranges from the fundamental research interest to "see" the molecule under investigation and to structurally understand how it works, to being able to exploit the protein structure for rationally designing molecular interaction partners, which influence the protein by either inhibiting or activating it (see also Chapters 1, 7 and 8).

As part of the protein folding problem, determining the 3D structure of a protein from its 1D sequence is a grand challenge in molecular biology ("the second half of the genetic code"). Since Anfinsen's [3] refolding experiments it seems that the primary sequence often completely determines the tertiary structure according to the principles of thermodynamics. But also often protein folding appears to be dependent on a number of additional factors: post-translational editing, covalent modifications, interactions with other molecules and membranes, and on cofactors or chaperones necessary for correct folding.

Bioinformatics approaches towards solving protein structures are called protein structure prediction methods. They can roughly be categorized into three types according to the input data used and the available knowledge to be exploited: 2D prediction, *ab initio* structure prediction, and homology-based modeling. 2D or secondary structure prediction methods just assign secondary structures to the residues of a protein, i.e. they decide whether the residue occurs in a α-helix, a β-strand or neither. *Ab initio* methods predict structure from sequence without any additional information on the protein. They usually employ energy minimization or molecular dynamics methods (see Chapter 5 and Appendix). Homology-based prediction as a first step needs to construct a rough model for a protein sequence after the known structure of another protein (the so-called template) via a sequence-structure alignment, a so-called threading, as a basis for further homology-based modeling (see Chapter 5). Threading methods only predict the course of the protein backbone, whereas homology-based modeling methods compute a full-atom model of the protein 3D structure. Different structure prediction methods have different applications. E.g., protein threading can extend the current capabilities of sequence analysis methods (see Chapter 2) with respect to detecting remote homology and similarities in order to derive functional and structural features. Homology-based modeling aims at computing detailed 3D structural information to be used either in the understanding of the protein function or as a basis for rational drug design.

6.1
Overview

This Chapter concentrates on secondary structure prediction and on protein threading methods since, given current data and state-of-the-art of the field, these are the most promising approaches to tertiary structure prediction, fold recognition and remote homology detection on a large genomic scale. *Ab initio* methods are not covered here because, to date, they are less application relevant, with the exception of recent approaches combining local sequence–structure similarities with *ab initio* conformation sampling (see Section 6.2.5). Homology-based modeling based on an alignment or on a rough structural model is discussed in Chapter 5 of this volume.

Bioinformatics approaches to protein structure prediction and the identification of homology and similarities involve the solution of search problems. As the range of possible protein sequences and folding conformations is astronomical, appropriate pruning of the search space is inherently necessary. The so-called Levinthal paradox [4] states that even real proteins cannot try out all the possible conformations during the time they fold into their native structure. One possibility would be to concentrate on local sub-problems and assemble the subsolutions to form overall solutions. Fortunately, proteins are composed of highly regular recurring elements (secondary structures) such as helices or strands. The secondary structures, in turn, form certain supersecondary structure motifs. Unfortunately, building up a protein structure locally is not possible directly, as short peptide segments of a protein can fold into different structures in different protein environments. It is immediately apparent that the 3D structure brings amino acids that are distant in the sequence in close contact in the fold. Such long-range contacts often may overrule local conformational tendencies and have to be dealt with, e.g. via conformational search approaches [5].

For the detection of remote homologies it may be required to investigate complex evolutionary and functional relationships. *Homologues* are proteins that relate back to the same common ancestor in evolution. Evolutionary relationship often points to similar structure and function. Finding close homologues, such as protein family members or related (*paralogous*, see also Chapter 5) sequences from other species can be detected via similarities exhibited with standard sequence search methods. Such methods include sequence alignment [6–8] and heuristic statistical search methods such as BLAST [9–13] and FASTA [14–18]. Current profile and multiple alignment methods, HMMs [19–28] and iterative searches [29] allow to identify more distant and insecure homology (see also Chapter 2 on sequence analysis in this volume). Sequence, structure and function analysis problems are closely intertwined, thus, so are the associated prediction problems. Structure prediction problems range from aligning a sequence to a sequence of known structure across threading the sequence to structural environments without significant sequence similarity, across mapping the sequence to structural templates according to contacts between residues within protein structures [30], to mere functional similarity based on indicative functional motifs [31]. Typically, biological problems have fuzzy, non-local aspects which are invariably highly interconnected. This is also the case for protein structure prediction, which apparently looks like a simple optimization problem, but turns out to be a knowledge discovery task exploiting heterogeneous information with efficient algorithms in the most discriminative and critical way.

The practical problem of performing as detailed as possible a prediction for a given sequence usually involves a procedure of analysis steps (Figure 6.1), so that results from early steps can and should be used in subsequent steps, e.g. predicted secondary structure can help for selecting fold sets or improve sequence–structure alignments, multiple alignments can help to both improve secondary structure prediction and fold recognition, functional predictions can provide further evidence for top scoring threading models. In order to deal with various kinds of information in the individual steps, it is helpful to provide appropriate means to store features and results of prediction steps. One such approach to represent and visualize

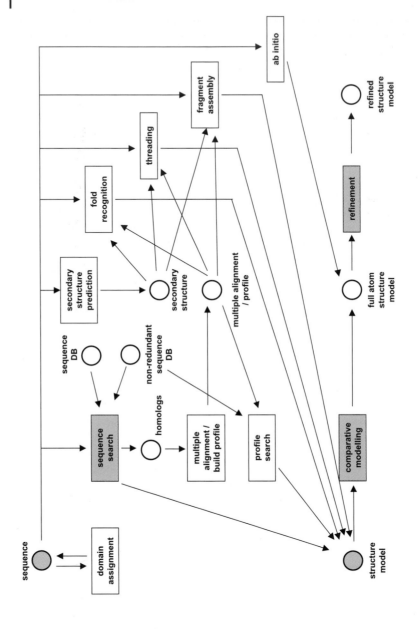

intermediate results and features in a uniform way and to exploit them efficiently in subsequent prediction programs is to use an extensible (XML-based) protein description language (ProML [32, 33]).

It is probably fair to conclude that most of the improvements on prediction performance are due to efficient use of more evolutionary information (homologs) in individual methods and/or due to clever processes of analyses [317] trying to mimic the approach of human experts, which is still the state-of-the-art [34] in protein structure prediction.

6.1.1
Definition of terms

Ab initio approaches try to perform predictions based on first principles and the input sequence alone. In their pure form, they do not use database-derived potentials, knowledge-based approaches or transfer of information from homologous proteins [35]. Recently, *ab initio* methods have been combined with fragment searching and threading to yield a new fragment assembly approach, which appears to be quite successful for small proteins [5, 36, 37]. In contrast to homology prediction and protein threading completely new folds that have never been seen before can be computed with *ab*

◄ **Fig. 6.1**
Overview of possible steps of a manual or automated hierarchical prediction process from a sequence to a (refined) model structure. Given a sequence a search against a (non-redundant) sequence database can be performed using appropriate searching methods (FASTA, BLAST, GAP-BLAST, etc). If a significant hit to a sequence of a protein with known structure is found, a structural model can be inferred. If a simple sequence search fails an iterative searching procedure with intermediate (probabilistic) sequence profiles can be applied to find more distant homologs. Again if significant hits can be identified they can be used for predicting a structural model. Otherwise, the sequence can be partitioned into its (predicted) domains and the analysis performed for the individual parts separately. This is especially important for the subsequent structure prediction steps. A secondary structure prediction yields additional information on the protein structure, which can also be used for fold recognition, full threading, combined threading *ab initio* prediction (fragment assembly), and even pure *ab initio* approaches. The used (non-redundant) databases of sequences and structures and representative sets of sequences, folds, classifications, have influence on the searching procedures and the statistical significance of the hits. The different procedures are described in the main text except for pure *ab initio* approaches and comparative modeling and structure refinement that are dealt with in Chapter 5 of this volume. What is omitted from the figure are functional predictions or knowledge as well as additional experimental information (e.g., on accessibility, secondary or domain structure, distance constraints, functional or biochemical network context), which can be used in various ways in sophisticated structure prediction methods either via validating predicted models or in the model generation process itself (see Section 6.4). Procedures represented by shaded boxes are discussed in other Chapters of this volume.

initio methods, in principle. However, an overall solution of the *ab initio* protein structure prediction problem seems to be still out of reach.

Threading aligns a protein sequence (the target) to the sequence of a known protein structure (the template) producing a sequence-to-structure alignment by optimizing some objective function. If this function contains only sequence related terms threading reduces to sequence alignment, if structural and structure environment information are also considered threading tries to optimize the compatibility of a given sequence with a given structure. If the scoring functions contains pairwise interactions between spatially distant amino acids the problem of computing an optimal sequence–structure alignment becomes NP complete [38].

There is a subproblem of threading called *fold recognition*. Here, we only ask for identifying a suitable template structure among a set of candidate structures. This problem is easier than threading, as it can be observed that the correct fold can be identified even on the basis of an inferior alignment. In addition, it is much easier to rate the quality of fold recognition, since the best fitting template structure can be identified more readily than scoring the quality of an alignment. This makes fold recognition a popular version of the problem for benchmarking purposes.

Homology modeling takes a sequence-to-structure alignment and computes a full atom model for the target sequence. It does so by modeling gaps in the alignment of target and template which most often represent loops on the outside of the target protein. In addition, gaps in the template have to be modeled and side chains that differ between template and target have to be inserted. Homology-based modeling ends with a refinement step and yields a full-atom model, in general (see Chapter 5 for details).

Homology-based protein structure prediction methods are the methods of choice, in practice. The reason is that the number of protein folds occurring in nature is much smaller than the number of protein sequences, as nature conserves successful structural architectures much more than the exact protein sequence. On the sequence level much variation is found in orthologous proteins from different organisms with same structure and function. Thus, if we are looking for a fold of a natural protein, it is much more economical to test the limited set of folds realized by nature than to search in an astronomically large space of potential protein structures.

Secondary structures [39, 40] are regular local structure elements identified in the 1950s [41, 42] which are characteristic abstractions for proteins and believed to be of great importance for the 3D folding. Secondary structure elements are defined via characteristic main chain torsion angles. Usually, three discrete states, α-helix (H), β-strand (E, extended), or other (L, loop) are distinguished [1] (see Chapter 5 for details).

The *fold* of a protein is usually represented by its secondary structure elements and their arrangement in 3D space. According to current fold classifications, i.e., SCOP [43] or CATH [44, 45], there are several hundreds of different folds. Classifying proteins into a hierarchy from class, fold, superfamily to family (SCOP [318]) or class, architecture, topology, and homology (CATH [319]), is not a trivial task, probably because the fold space of proteins is not easily partitioned into a hierarchy of distinct fold clusters.

3D structural coordinates: The protein is composed of a chain of amino acids, which in turn are composed of atoms. The position of these atoms can be inferred from experimental measurements, most prominently, X-ray diffraction patterns or NMR spectra. Therefore, a very detailed description of a protein is given by the exact (x, y, z) coordinates of all the protein's atoms. It has to be kept in mind, however, that structure coordinates are still an abstraction of the protein in its actual environment interacting with solvent and/or other macromolecules. This is due to experimental measurement errors, crystallization effects, and the static picture implicated by the coordinates. In addition, it is not necessarily true, that the most accurate or comprehensive information, say regarding protein function, can be inferred best from the most detailed structural representation.

Remote homology detection identifies, for a given sequence, one (*candidate recognition*) or all (*related pair recognition*) of the evolutionarily related proteins in a set of candidates. This may or may not include the computation of a detailed structural model. Interestingly, it is often possible to detect remote homology with the help of structural criteria without actually being able to compute a reasonable alignment and structural model.

6.1.2
What is covered in this chapter

This Chapter discusses protein structure prediction methods and their use for homology-based modeling (Chapter 5), docking and design of ligands (Chapter 7), structural genomics, drug target finding (see also Chapter 6 of Volume 2), and function assignments. As outlined in the introduction, this Chapter focuses on homology-based or knowledge-based methods rather than physics-based (*ab initio*) approaches.

Secondary structure prediction assigns a local helical (alpha) or extended (beta) structure to amino acid chains. The problem has been approached in an *ab initio* manner [46–48] for some time and many prediction methods have been developed based on properties of amino acids. Major advances have been achieved by employing homology-based approaches to 2D prediction. Such methods using sequence family information or consensus formation from several prediction methods are discussed in Section 6.4.1.

In general *3D protein structure prediction* is an unsolved grand challenge problem in computational biology. It has been studied for decades starting with the seminal experiments of Anfinsen in 1961 [3]. Turn-key solutions for the overall problem of computing the complete 3D structure from an amino acid sequence are not in sight for the foreseeable future. Due to the increasing numbers of sequence and structure data and the more and more complete coverage of sequence (see genome projects for bacteria [320] and higher organisms, e.g. *C. elegans*, *D. melanogaster*, *A. thaliana*, Rice, Rat, Mouse, Human [321]) and structure space (structural genomics projects in the US [322], in Europe [323], and in Japan [324]) homology-based methods are already quite successful. The different methods used for homology based tertiary structure prediction are described in the Methods, Section 6.4.

Related problems are *protein folding, molecular dynamics, sequence design*, and *protein design*. The *protein folding* problem involves predicting in more or less detail the sequence of steps leading from an unfolded protein chain to its final structure. A lot of theoretical work has been done in this field [4, 49–59]. Especially, lattice models of proteins [60–64] and the associated folding algorithms have been studied extensively, though the practical relevance of the results for genomics and pharmaceutical applications remains limited. All of these so-called lattice models have in common a highly simplified view of a protein as a chain inside a regular 3D lattice. Residues are abstracted to nodes in the chain that come to lie on vertices of the lattice. Only few residue properties are distinguished and modeled. The solution to the full atom protein folding problem would also solve the structure prediction problem. *Molecular dynamics* studies what happens to a protein structure under varying conditions, such as increasing temperature, unfolding, or interaction with other molecules (see Appendix). *Sequence design* asks for optimized sequences to fit a given structure or fold (it is also called the inverse folding problem), i.e. sequences which assume the given fold and have additional modified features such as a modified stability or active sites. *Protein design* involves rational engineering of a protein with a specific new fold or modifying an existing fold by changing certain amino acids to attain new or modified functions [65–69].

The rest of this Chapter is restricted to methods for homology-based protein structure prediction given a protein sequence that occurs in nature. The difference in dealing with natural and designed proteins is an essential one: natural sequences have evolved to fulfill a certain purpose and function. Evolution has challenged certain parts of the sequence and structure design and conserved the necessary ingredients. Traces of this evolutionary history can be detected and used for predictions within the territory of folded and functional proteins. Predictions can thus be made by similarity transfer: similarity of certain features can be analyzed with care to predict similarity of implied features. Such similarities are not likely to occur in the same fashion in artificially designed proteins.

Homology-based protein structure prediction in concert with experimental high-throughput methods for protein structure prediction within a global structural genomics effort might well effectively solve the prediction protein structure prediction problem before a satisfying solution to the protein folding problem is in sight. Evidence for this perspective is that progress in the protein structure prediction field is rapid and can be expected at an accelerating pace simply because of the amount of data gathered both on sequences and structures. This will be even more the case as dedicated structural genomics projects [325] have started with the goal of charting the space of naturally occurring protein folds. Together with improved methods of protein classification a detailed map of the protein structure space will become available, which allows to assign a structure to a given sequence on the basis of the wealth of structural folds and their relations as well as the associated sequences already assigned to those structural classes.

In the following, we first describe the input data on which the protein structure problem is based to generate the output data (Section 6.2). Section 6.3 reviews the principal methods of structure prediction. Section 6.4 discusses the results obtained by structure prediction methods in the context of various application areas. Ways to assess and validate prediction results and the strengths and limitations of the overall approach are dealt with in Sections 6.5 and 6.6.

6.2
Data

6.2.1
Input data

In the simplest case the problem instance is defined by the amino acid sequence (*the target sequence*) only.

More detailed instances of the problem contain the available folds to be used as structure templates and the set of all available resolved 3D structures; the available sequence database and the sequences that are evolutionarily related to the target sequence, possibly including a multiple alignment of this protein family. It can also be helpful to distinguish between orthologous and paralogous members of the protein family. Today, tools such as PSI-BLAST (see Chapter 2) are used to locate the involved protein family, to collect the members of the family, and to generate the associated multiple alignment and/or sequence profile. Similarly, for each of the structurally resolved proteins from the set of templates evolutionary and/or structurally related proteins can be collected in order to determine the respective protein neighborhood and multiple alignment.

In addition, the availability of experimental or predicted secondary structure information on the query sequence as well as other experimental constraints such as knowledge on disulfide bridges, on distances or distance intervals between residues and functional groups within proteins and on accessibility of side chains could have large impact on the prediction methods and results as well as on the confidence in the predicted model structure.

In general, providing the information accompanying the protein sequence or any of the fold templates is subject of dedicated research and the quality of the structure prediction heavily depends on the quality of these derived input data.

6.2.2
Output data

The simplest output is an alignment of the query sequence onto the best (closest relative) template structure. Such an alignment is the starting point for several analysis and modeling steps.

A comparison of the alignment score of the target sequence with (many) different template proteins can be the basis of rating the significance of the produced structure model. *P*-values are a version of this approach that also applies to sequence analysis methods (see Chapter 2 of this volume), for which is has been studied extensively [9, 11, 70, 71]. In addition, the resulting models can be subjected to a wide variety of plausibility tests in order to ascertain their protein-likeness. Many of these checks will be made on the full-atom model of the protein that results from the procedures described in Chapter 5 of this volume.

Additional output data could be predicted secondary structure and accessibility annotation for the query sequence (see Section 6.3.1, [72]), a multiple alignment of related sequences and structures, structural models, and variance and mutation analysis of closely related proteins.

The following paragraphs list additional data resources relevant for structure prediction approaches. For knowledge-based methods it is essential to make the most efficient use of the relevant information to come up with a plausible prediction. Doing so seems to be the most promising approach to the problem.

6.2.3
Additional input data

Most protein threading methods rest on the input data mentioned in Section 6.2.1. In order to improve protein structure prediction we can benefit significantly from a wide variety of additional information that is contained in biological databases. We already mentioned the protein family information that is contained in a growing number of databases which exist both on the sequence [73–76] and structure level [43–45, 77–82]. This information can be used to produce profiles for both the target protein sequence and its family and the template fold database. Another approach is to exploit conserved motifs [31, 83, 84] and functional patterns [85–87] for performing the structure prediction, either via a post-processing step eliminating fold candidates not compatible with necessary functional constraints, or via the selection of candidates or partial alignments on the basis of searching for motifs and patterns. This approach has been followed, for example in RDP (see Section 6.3.2, [32]).

An even more involved and innovative approach is to use knowledge or predictions of functional information on the protein and network information in the form of biochemical, e.g. metabolic and regulatory networks. Here the information that a set of proteins is contained in a certain pathway could provide additional constraints for candidate structures fitting into certain locations of that pathway, assuming that additional information on the functional and/or structural constraints of these pathway members is available. Again, such information can be used to select plausible folds to thread against, to filter conspicuous candidates from the ranked list of the prediction, or to use associated constraints in the prediction step itself [32]. In particular, networks generated from experimentally established (e.g. via large-scale yeast2hybrid screens [88–91]) and/or predicted interactions [92–96] of proteins can provide structural constraints to be of use for homology-based structure prediction.

In the next two subsections, we describe in more detail two of the main non-algorithmic ingredients of structure prediction methods, namely the structure classifications that represent something like the global view of protein structure space, and the scoring function rating a sequence–structure alignment. In many threading methods, these components are chosen by default and applied to all protein target sequences. However, a threading method can be improved by specially tailoring these components to the target sequence.

6.2.4
Structure comparison and classification

In several ways, protein structure prediction methods rely on classifications of the known protein structures that are based on systematic comparison of the 3D structures. First, from such classifications overall principles constituting different protein folds can be inferred, which can be of use in protein structure prediction algorithms and/or associated scoring systems. Second, representative subsets of folds can be constructed which are exploited as the universe of possible structural templates for sequences to be predicted, e.g. as a list of folds against which to thread the sequence. Third, such classifications can be used to discriminate between the available choices. I.e. it is not only necessary to identify a probably close fold but to identify those folds that are the most probable models for the target sequence among all available fold types. Fourth, structure classifications are used as a standard-of-truth in benchmarks for fold recognition methods. Finally, they aim at providing a complete overview of the protein structure universe.

Structure comparison methods are a way to compare three-dimensional structures. They are important for at least two reasons. First, they allow for inferring a similarity or distance measure to be used for the construction of structural classifications of proteins. Second, they can be used to assess the success of prediction procedures by measuring the deviation from a given standard-of-truth, usually given via the experimentally determined native protein structure. Formally, the problem of structure superposition is given as two sets of points in 3D space each connected as a linear chain. The objective is to provide a maximum number of point pairs, one from each of the two sets such that an optimal translation and rotation of one of the point sets (structural superposition) minimizes the rms (root mean square deviation) between the matched points. Obviously, there are two contrary criteria to be optimized: the rms to be minimized and the number of matched residues to be maximized. Clearly, a smaller number of residue pairs can be superposed with a smaller rms and, clearly, a larger number of equivalent residues with a certain rms is more indicative of significant overall structural similarity.

Analogously to the structure prediction problem, for the structure comparison problem the first step of matching points in both sets is the crucial and quality-determining part. In fact, the second step of finding an optimal superposition of a given set of matched pairs of points can be solved efficiently and optimally [97–99]. For the first step a broad spectrum of methods has been proposed over the years: simple alignment, matching of features [100, 101], compactness and recurrence criteria [102], secondary structure matching (with or without obeying the chain order) [103, 104], to geometric hashing techniques completely ignoring the chain information of the residues [105–107]. For assessing the quality of 3D structure superpositions, besides the rms, the so-called soap film method of comparing protein structures [108] has been proposed. This method measures the distance between structures via the size of the area between the associated 3D curves of the protein backbones. In addition, assessment procedures for homology models have been developed during the CASP experiments. These can be classified into sequence dependent and sequence independent variants [109–111]. The former use the predicted alignment to identify equivalent residue pairs and compute the rms of the associated optimal superposition of model and experimental structure. The latter determines an appropriate set of matched pairs, given the model and native structure coordinates. These pairs are used for the superposition. The latter approach is a bit more tolerant to alignment errors, especially when an overall correct fold has been identified via a (partially) faulty alignment.

As mentioned above, several of these approaches have been employed to generate exhaustive clusterings of the protein structures into structural classes available via the resulting databases (DDD/FSSP/Dali [102], 3Dee [79], JOY/HOMSTRAD/DDBASE [112–114], structure cores LPFC [115]), These superposition-based classifications are complemented with classifications based on other criteria (CATH) such as topology and orientation of secondary structure elements up to manual classification from visual inspection of the folds (SCOP). A recent systematic comparison of these classifications has been carried out by Hadley and Jones [116]. They suggest selecting a somewhat redundant larger set of threading templates that is compiled from clusters of all individual classifications and to base the evaluation on those examples on which all the classifications agree. Generally, the classifications agree on a large part (about two-thirds) of the protein structure data. There are some systematic inconsistencies among the different taxonomies due to the different approaches taken.

From classifications representative lists for threading as well as assessment criteria for prediction results can be obtained. In addition, involved prediction methods can take advantage of a global view of protein structure space to better discriminate between the available alternatives than can be done on alignment scores of a sequence to an individual fold alone. We expect more comprehensive global knowledge of protein structure space to significantly improve protein structure prediction accuracy, in the future.

Domain assignment: domains are basic building blocks of protein 3D structures. Several algorithms have been proposed to identify domains in a protein structure using criteria such as physico-chemical properties of amino acids in certain regions, e.g. hydrophobicity and accessibility, compactness of protein parts and their recurrence in other proteins [116–120].

For threading, often representative sets of protein structures are selected from domain fold libraries. Threading algorithms then align the sequence under consideration against a library of domains in order to detect the domain(s) constituting the unknown sequence. Due to the small signal-to-noise ratio, the performance could be increased by splitting the sequence into putative domains before threading them against the domain library. This can be done on the basis of the sequence via dedicated domain prediction methods [121–123] or by approximately matching the length of the respective structural domain in a straightforward manner.

6.2.5
Scoring functions and (empirical) energy potentials

A standard common assumption is that native structures of proteins are those conformations with minimal free energy [2, 57, 58]. Free energy cannot be determined exactly: many factors have to be taken into account, not all relevant factors are known, neither are their respective weightings. Threading has to estimate alignment conformations between sequence and structure and rate different alignments against each other. For this purpose it has to be taken into account that native structures and the model structures will be more or less different. Therefore, the energy evaluation via the scoring function has to keep a balance, in order to account for the expected differences of individual proteins and, at the same time, provide the necessary detail to achieve sufficient discrimination between different folds. In addition, almost always only partial protein models can be evaluated (e.g. due to incorporating the backbone conformation but no side-chain conformations, missing loop regions, or gaps). Usually the model and the template structure are different enough to prohibit the application of detailed atom-level physico-chemical potentials. Therefore, knowledge-based, i.e. database-derived potentials on different levels of abstraction have been successfully computed via database analysis and inverse Boltzmann statistics [30, 124] and applied to threading optimizations.

Inverse Boltzmann statistics rely on the relationship between the energy of a molecular state and the probability with which it is observed. This relationship is formulated in the Boltzmann law. Usually, the law is applied to derive the probability of observing a molecular state from its given energy. Here, we apply the law in the inverse direction. I.e., we count the frequency of observing a molecular state (or partial state) and derive from that a pseudo-energy of this state. We call the partial molecular states whose frequencies we are counting *events*. Event classes can be freely defined. We will give a few examples below. We collect the statistics on events by scanning through a representative set of protein structures. Using the Boltzmann law we deduce the contribution that this event makes to the total pseudo-energy of the protein structure. Contributions of all observed events in a protein structure are added up to yield the pseudo-energy of the protein structure.

This procedure is applied ubiquitously in protein structure prediction. Despite the caveats that the Boltzmann law is only applicable in thermal equilibrium and that it cannot readily be applied additively to partial molecular states scoring functions based on inverse Boltzmann statistics have proved very valuable in protein structure prediction.

6.2.5.1 **Pair-interaction potentials**

Essentially there are two types of potentials that cater to two different sets of optimization methods. The more accurate potential functions score pair-interactions between amino acid residues in the protein (see Figure 6.2a). The residue pairs are grouped, for instance by the type of the two residues, their distance in the protein sequence, and their distance in the protein structure. E.g. residue A and L occurring at a distance of 6–15 residues apart in the protein sequence and at a distance between 3 Å and 6 Å in space may form such a group. The overall score of a sequence–structure alignment is the sum of all contributions determined by the inverse Boltzmann law of pairs of amino acids observed in the structure of the target sequence that is obtained by aligning the target sequence to the template structure as the alignment prescribes (see Figure 6.2). This kind of potential has been introduced by Sippl [30, 125].

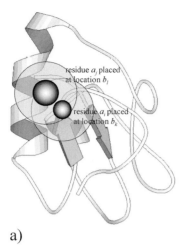 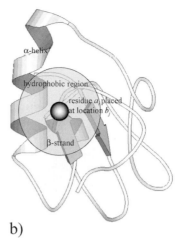

a)

b)

Fig. 6.2
Pair potentials vs contact capacity potentials a) a pair potential attributes an energetic contribution to each pair of amino acids a_i and a_j that are mapped by a sequence-structure alignment onto locations of the amino acids b_k and b_l of the protein template structure. The contribution depends on the types of residues a_i and a_j as well as on geometric features of their locations b_k and b_l, such as their distance in space or in sequence. Often, only certain pairs of amino acid residues are scored, such as those that are contacting or close in space. The total energy is computed as the sum of all pair contributions. b) a contact capacity potential rates the energetic contribution of an amino acid a_i from the protein target

sequence A that is mapped onto a location b_j inside the protein template structure B. The energetic contribution is determined by the type of residue a_i and the chemical features close to the location b_j inside the protein template structure B. In the case of the figure, a_i comes to lie a the tip of a strand and next to a helix inside the protein core in a hydrophobic region. These pieces of information can be assembled in a fingerprint that characterizes the environment of location b_j. Note that the contact capacity potential is derived from the chemical makeup of the template protein exclusively. In both variations of potentials, the contribution of residue a_i as mapped onto location b_j is determined via inverse Boltzmann statistics.

Today, there is quite a variety of such potentials [125–132]. The performance of potentials for different tasks in structure prediction varies. Reviews and comparisons of database and molecular mechanics force fields as well an discussion of their controversial relationship to free energy can be found in [133–135].

The involved alignment optimization problem, i.e. the problem of finding a sequence–structure alignment with minimum pair interaction score is NP-complete [38].

6.2.5.2 Contact capacity potentials

Because of the computational problems a different set of potential functions has been proposed that allows for more efficient optimization algorithms.

c) Superposition of the backbones of elastase (red) and trypsin (blue)

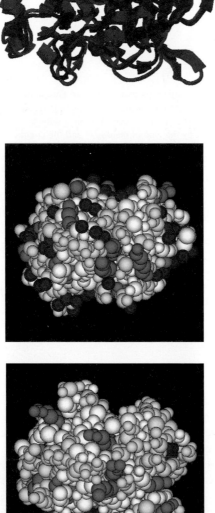

a) Elastase

b) Trypsin

Fig. 6.3

This figure shows that two proteins that have very similar structures can have widely differing chemical makeup. a) shows a full-atom model of porcine elastase (1est.pdb, 240 amino acids). The atoms are colored according to the chemical composition of the respective residue (red = positively charged, blue = negatively charged, yellow = polar, white = nonpolar). b) shows the same picture of bovine trypsin (1tpp.pdb, 223 amino acids). Elastase and trypsin are both serine proteases and have a sequence identity of about 36% and 39%, respectively. c) shows a superposition of the backbones of elastase and trypsin, revealing that their structures are practically identical (222 superposed residues with an rms of 2.6.9). Still, the chemical composition revealed by the coloration of a) and b) is quite different.

Such potentials are sometimes called profiles, as the structure is abstracted into a linear representation (the profile) of positional score values for the 20 amino acids at the respective position. A profile does not explicitly score the energy of pair-interactions in the resulting protein structure. Rather the idea of the profile is based on the hypothesis, that the protein target sequence inherits not only the backbone conformation from the template structure but also the chemical composition. Then, e.g. we can assume hydrophobic patches in the target protein where they occur in the template protein, H-bond donor regions are preserved as are polar regions etc. (see Figure 6.2b) If we assume this hypothesis, then we have to score not pairs of amino acids but single amino acids, each against a given and constant chemical environment. Aligning, say, a residue a_i in the target sequence against a residue b_j in the template sequence would be scored according to how often a residue of type a_i is found within the chemical environment that the template protein structure presents for residue b_j (see Figure 6.2, page 252). The alignment optimization problem now becomes much simpler. Essentially, the sequence alignment methods described in Chapter 2 can be used. We only have to replace the amino-acid substitution matrix used for scoring in sequence alignment by the threading profile. Thus alignment optimization can be done with dynamic programming in polynomial time (see Section 6.3.2). In practice, in this case several optimal alignments can be computed per second on a workstation.

Of course, the hypothesis that the chemical composition be identical in the template and the target protein, is often false (see Figure 6.3). Thus one has to apply both kinds of potential functions in concert to make suitable structure predictions (see Section 6.3).

Bowie et al. first introduced profile scoring functions for protein threading [136]. They consider the type of residue and geometric aspects of its environment such as inside/outside or the secondary structure element in which the residue occurs. Later different variants of this scheme have been developed. They essentially differ in how they describe the chemical environment of the residue in the template structure. Contact capacity potentials (CCPs) are a variant used in the threader 123D (Alexandrov et al [137]). CCPs essentially count contacts to neighboring residues, irrespective of the type of residue. The 2D structure environment is considered in a version of CCPs and contacts are distinguished with respect to whether they involve residues that are close (short range) or far apart (long range) in the protein sequence. Effectively, contact capacity potentials aim at approximating generalized hydrophobicity measures and secondary structure preferences, at the same time.

6.2.5.3 Contacts in protein structures

Many potentials of both types rely on a particular definition of residue-residue contact. Starting with Sippl [30, 125] many authors have used the distance between residue pairs to classify contacts. Distances below a certain threshold (e.g. less than 7 Å for the relevant C$_\beta$ atoms) constitute a contact. This definition is debatable and often contradicts the actual geometry of a protein structure. E.g. a third residue can be located between both residues even though they are at close distance. We have introduced a more appropriate notion of residue-residue contact [138]. Here, residues are assumed to be in (direct) contact only if they share a contacting face of a certain minimum area. These *direct* contacts and the associated contact areas are determined via a so-called Voronoi tessellation of the protein. A Voronoi tessellation is a decomposition of the space occupied by the protein into so-called Voronoi cells, one cell for each atom. The Voronoi cell for an atom is composed of those points in space that are located closer to the particular atom than to all other atoms. There is analogous variant of Voronoi tessellation that is based on complete residues instead of single atom. An alternative tesselation on the residue level can be obtained by using the union of atom cells belonging to a residue as the representative for the residue cell. The potentials resulting from Voronoi tessellations have been demonstrated to improve the performance of fold recognition methods [138].

6.2.5.4 Calibration of the scoring function

The definition and calibration of appropriate scoring contributions capturing relevant features of protein structure and sequence-structure compatibility has started with Margaret Dayhoff's derivation of her famous substitution matrices [139]. Various other attempts have been made to define better scoring terms for sequence compatibility [140–144], secondary structure preference and contact potentials (see above). Classification and calibration techniques such as linear programming, supervised learning, support vector machines have been used to systematically optimize substitution matrices [145], the discrimination of scores [146] and the relative weights of different individually optimized scoring contributions among each other [147–149]. E.g. the fold recognition performance of the ToPLign 123D* threader using calibrated weights of 18 scoring contributions (sequence, secondary structure, gap insertion and extension, local and global contact capacity, each weighted individually in three secondary structure elements) has been improved by more than 40% as compared to the original 123D.

6.3
Methods

This section gives an overview of secondary structure and tertiary structure prediction methods for proteins. Methods for (secondary) structure prediction of other biomolecules, such as RNA are not discussed here.

6.3.1
Secondary structure prediction

Secondary structure prediction assigns a local structure to amino acids chains. Usually, local structure is characterized by one of three discrete states, helix (H), strand (E for extended), or other (C for coil). This measure of evaluation is also called three-state (Q3) measure. The problem has been approached first in an *ab initio* manner involving the target protein sequence without any additional information [48, 150–154], e.g. on the relevant protein family or homologous sequences, and many prediction methods have been developed based on properties of amino acids. These methods yield 60–65% performance on independent benchmark sets. Major advances have been achieved by employing a homology-based approach to 2D prediction. PhD [72, 155] exploits knowledge on homologous sequences by using multiple sequence alignments and the associated profiles of secondary structure assignments for training a neural network. This procedure does not need knowledge of the tertiary structure of proteins that are related to the target sequence. Such approaches profit significantly from the increased number of available sequences and, thus, larger and better multiple alignments, but also from much improved search procedures to recruit homologous sequences for the training. The collection of homologous sequences is done via iterative profile-based search methods, such as PSI-BLAST (see Chapter 2). A secondary structure method based on this approach is PsiPRED [156] achieving around 80% of correct three-state predictions on benchmark sets and on selected blind-prediction examples [157].

The second improvement in this area is the use of more involved consensus prediction methods. With a range of different methods at hand, it is possible to derive a consensus prediction as has been done in SOPMA [158], JPRED [159, 160], or CODE [161] by systematically optimizing the performance on a training set by using decision-tree methods or machine-learning approaches such as support-vector machines [162].

Having an improved secondary structure prediction available with, say, more than 70% expected correctness can improve 3D prediction. There are several approaches: such as matching the predicted secondary structures with the template structures [163, 164], using secondary structure prediction as a scoring term to evaluate sequence and threading alignments [147] or incorporating a respective scoring term into the objective function of a threading algorithm. From an algorithmic point of view all alternatives are easy to realize with negligible computational overhead. The evaluation of alignments requires training a neural network using an additional scoring term. The optimization involves adding another profile term to the objective function. Nevertheless, the use of secondary structure predictions generally improves the performance of remote homology detection as well as alignment quality [147, 165].

On the other hand, for really challenging instances of fold recognition the approach cannot always help: if no homologous sequences can be found, the secondary structure predictions tend to be of poorer quality; and using wrong secondary structure assignment can disturb the scoring function and mislead the alignment process.

Assessment of secondary structure predictions: these methods involve statistical analyses of applying the methods to known structures and recomputing their secondary structure as well as blind predictions (see also Section 6.5.1).

A state-of-the-art assessment of predictions requires an up-to-date set of secondary structure assignments and a measure to compare predictions with the correct assignments. Evaluation sets of non-homologous domains with secondary structure assignments have been proposed by Barton et al. [166, 326]. This dataset is an extension of other proposed test sets [155, 167]. It contains 496 non-homologous (SD cutoff of 5, which is more stringent than mutually less than 25% sequence identity) domains. The database contains 82 847 residues, 28 678 alpha-helix (H), 17 741 beta-strand (E), and 36 428 coil (C) residues.

The standard measure for accuracy is Q3 defined as the number of residues correctly predicted divided by the number of residues. Q3 can also be computed for the residues of individual proteins Q3P or of individual secondary structures QH, QE, and QC. Other methods use certain types of correlation coefficients or take the type and position of secondary structure elements into account (proposed by Rost [168] and further developed for the CASP experiments [169]). One example is the SOV (segment overlap) score, i.e. the average over all secondary structure segments of the ratio of the overlap of predicted and true secondary segment and the overall length of the two segments.

More measures and programs to compute some of these measures for comparison and evaluation of new methods can be found on the CASP website [327] and a summary of measures can be found at the EVA site [328].

6.3.2
Knowledge-based 3D structure prediction

Knowledge or similarity-based methods exploit similarities between a given sequence and (sequences of) proteins of known 3D structure: the underlying assumption is that the more similar the two sequences, the more similar are the two corresponding 3D structures. The similarity is often generalized to additionally include quasi-physical properties, such as preferences of amino acids to occur in certain states, e.g. buried inside the protein or solvent-exposed at its surface. In general, similarity approaches try to align a given sequence to 3D structures from a library of structures (or folds) of known proteins, such that an alignment score quantifying the similarity is optimized. These search procedures find the fold from the library that is most compatible to the sequence in question.

Depending on the type of the scoring function, fast dynamic programming methods can be used to find optimal alignments or heuristic search procedures have to be employed to find approximate solutions. In practice, a two-step approach [170] can be applied to first identify a small set of candidate folds out of a large library using a simple scoring function and fast threading tool (e.g. CCP and ToPLign/123D) and then use a more detailed scoring function and a more involved alignment algorithm (pair-interaction potentials and ToPLign/RDP) to compute more accurate alignments and to distinguish among the remaining template candidates.

If there is significant structural similarity between the query sequence and proteins in the fold library used for threading, similarity searching followed by homology modeling is the method of choice, as it can yield plausible structures quickly. If the considered protein has a novel fold, i.e. a fold not yet represented in the fold library, the method will not be able to come up with the new fold as a structural model for the sequence, but at best can predict that none of the folds in the library applies.

The following section discusses methods for computing sequence-structure alignments for both types of scoring functions.

6.3.2.1 The algorithmic viewpoint on sequence-structure alignment and threading

In general any threading of a sequence of length n onto a structure with m residues can be represented by a simple alignment of the two sequences. As is well known the number $f(n, n)$ of alignments of sequences of length n is finite but huge [171]:

$$f(n, n) \sim \frac{(1 + \sqrt{2})^{2n+1}}{\sqrt{n}}, \quad \text{as } n \to \infty.$$

Therefore, exhaustive enumeration of possible alignments followed by an evaluation of individual alignments is prohibitive. Fortunately, for scoring functions that take the form of threading profiles the optimal alignment among the exponential number of possible alignments can be found efficiently with dynamic programming methods as in sequence alignment (see Chapter 2 of this volume). Depending on the gap cost function, threadings with profiles can be computed in quadratic (linear and affine gap penalty functions) with the Gotoh [6] or in cubic time (general gap costs) with the Needleman and Wunsch [7] algorithm.

The protein folding process and its determining factors are still poorly understood. But even if all factors could be modeled appropriately, folding simulations can span only a tiny fraction of the actual time required to fold a protein which is on the order of milliseconds to seconds. In addition, the contributing factors need to be modeled extremely accurately in order to avoid error propagation during the vast amount of computation needed for completely folding a protein. Also the structure prediction problem appears to be difficult [172–174].

It is generally assumed that contact interactions between amino acids or their groups are important stabilizing contributions to protein stability. Such interactions preclude the use of dynamic programming methods for the optimization (see Section 6.2). In fact, since the involved optimization problem is NP-hard, optimal solutions are generally not computable within reasonable computing time and resources. Nevertheless, there are some approaches trying to do so, despite the fact that optimal solutions are likely to turn out not to be of much practical value due to the deficiencies of the scoring function, e.g. [175].

In contrast, knowledge-based discrimination methods can be often successful. This section discusses the two major approaches to knowledge-based 3D structure prediction: purely sequence based methods, and methods exploiting structural knowledge. The latter approaches can be classified into sequence methods using secondary structure information and methods using 3D structural templates (threading). An *ab initio* approach somewhat related to threading is fragment assembly, where fragments are identified via local sequence and structure conservation and then these fragments are assembled into known or new folds thereby minimizing some energy function [5].

Other theoretical structure prediction methods, such as simulating folding pathways, lattice models, simulations and predictions, inverse folding (i.e. designing the "best" sequence for a fold/structure), physical methods (based on finding minimum energy conformations for "realistic" physico-chemical potentials), are not discussed here.

6.3.2.2 Sequence methods

Structure prediction can be achieved by considering sequence information alone: given a sequence a search is performed against a fold library by comparing the associated sequences of the folds with the query. If a significant hit is found, the fold is predicted as a model for the query sequence. The sequence alignment can be used to actually build a model from the identified template structure. Sequence alignment searches are certainly the methods most relying on homology information and profit a lot from the increase in sequence and family information. Therefore, sensitive sequence profile [176–181] search methods such as PSI-BLAST [9] and intermediate sequence search using, in addition, information from vast sequence databases are competitive protein structure methods. In order to predict a structure, PSI-BLAST is used to search all sequence databases to collect related sequences, to compute a multiple alignment, to derive a consensus profile from it, and re-search the databases using the profile instead of the query sequence until a protein of known structure is hit. An E-value is used to indicate the confidence in the hits found.

Other successful structure prediction methods based on sequence alone are FASTA [15, 18], hidden Markov models (HMMs) [25, 26, 28, 179], intermediate sequence search [182], and iterative profile search [29]. Sequence analysis methods are discussed in detail in Chapter 2. The results of such methods in the CASP experiment are described in [157, 183, 184].

6.3.2.3 Sequence and secondary structure

Experimentally determined or predicted secondary structure information can successfully be phased into sequence-based methods for structure prediction. GenThreader uses profile sequence alignment and various other criteria evaluated on the resulting alignments including compatibility of predicted and template secondary structure using a neural network to come up with a confidence measure for the alignment [147, 157]. Koretke et al. [184] used consensus secondary structure predictions to align predicted and template secondary structure elements using the program MAP [164].

6.3.2.4 Tertiary structure-based methods

Structure-based methods are reviewed in [185] and can be classified into two types: profile methods based on dynamic programming (see Chapter 2 of this volume) for computing optimal alignments of structural environments and threading approaches using full pair-interaction potentials [30, 125, 138, 186] (see Section 6.2). The latter methods use optimal exhaustive search procedures [175], branch and bound approximation algorithms [187, 188] or other heuristic approaches for the optimization [189, 190]. Reviews and comparison of threading methods can be found in [186, 191, 192].

Profile threading methods As described in Section 6.2, algorithms that optimize threading profiles are the same as for pairwise sequence alignment and are based on dynamic programming. These methods have been introduced by Eisenberg et al. [136, 193, 194].

Several variants using multiple sequence information and sequence profiles generated for the query sequence for the threading process have been described and demonstrate to improve fold recognition performance [109, 195–203] similarly as has been observed for secondary structure prediction and sequence methods.

Combining secondary structure prediction with profile threading [165, 204] has been demonstrated to improve threading performance as well.

As methods based on dynamic programming cannot account for pairwise-interaction potentials the so-called frozen approximation approach [189] has been proposed. This method performs several iterations of profile environments. In the first iteration, the chemical environment is defined via the contact partners of the template. In subsequent rounds the aligned residues from the previous iteration replace the residues of the template. The idea is that target and template structure are similar enough such that the iterative process converges towards the optimal assignment.

A method employing 3D structural profiles for detecting remote homologies has been proposed by Sternberg et al. [205]. This method, called 3D-PSSM (three dimensional position specific scoring matrix), uses sequence profiles to search databases. The profiles are computed from multiple sequence alignments using the standard method used for generating PSI-BLAST profiles [9, 206]. In order to use structural information, the following approach has been proposed (see Figure 6.4): for any protein f in the fold library, determine the set S of related structures in the superfamily. For any sequence x in S compute the PSI-BLAST profile via a sequence

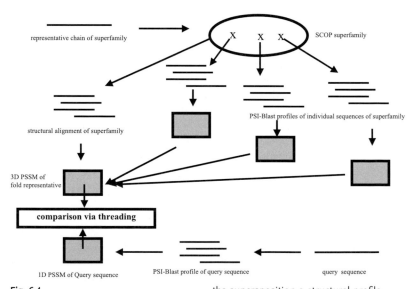

Fig. 6.4

Overview of the 3D-PSSM approach for protein structure prediction: For any protein (superfamily) PSI-BLAST searches and profiles are computed and using the structural superposition of the (structurally related) family members. On the basis of the superposition a structural profile representing the family is compiled. Given a query sequence, a corresponding sequence profile is computed, again using PSI-BLAST, and both profiles are aligned using a dynamic programming based profile threading method.

search of x against an appropriate database yielding a profile P_x. The profiles for the sequences in S are then combined using the structural alignments for the superfamily and the implied equivalences of residues. This is possible, despite the fact that the proteins in S do not necessarily show a reliable sequence similarity. The reason is that these proteins are structurally superposable and thus reliable residue equivalences can be determined. The idea behind the approach is to use all the information available for a group of structurally similar proteins in order to do the sequence structure (threading) search. The fold library and the associated profiles can be computed beforehand such that the search for a query protein can efficiently be done via computing a sequence profile and matching this profile against all 3D-PSSMs from the library using standard dynamic programming methods. During this search, in addition, solvent accessibility and secondary structure preferences are evaluated. The secondary structure prediction is performed with the program PsiPRED [156]. Thus, the approach successfully combines evolutionary and structural information available in the sequence and structure databases using the most advanced search (PsiPRED) and prediction methods (PsiPRED) as well as established profile building methods (PSI-BLAST) and structure classifications (SCOP). The major disadvantage of the method appears to be that there is probably not one single profile representing the properties of any individual member of a protein superfamily. Nevertheless, the performance of the method has been shown to improve on sequence search alone on a benchmark of 136 homologous protein pairs that cannot be detected with PSI-BLAST (see [205] for details of the benchmark construction and the results obtained). Here, 3D-PSSM was able to detect 18% of the pairs. The program is available as a web service at the Biomolecular Modeling Laboratory of the Imperial Cancer Research Fund [205]. In the latest CASP-4 and CAFASP-1 experiments both this approach and the associated automated prediction server were among the most successful groups.

Optimal threading with full pairwise/interaction potentials via exhaustive enumeration Usually, algorithms aiming at optimal solutions address the fragment threading problem: here, the structure is partitioned into a number k of secondary structure elements, which are considered fixed fragments of a definite length where no gaps are allowed in any threading alignment. The problem reduces to finding a mapping of sequence positions onto the first positions of all the fixed fragments of the structure. This mapping has to observe the linear ordering of the sequence residues, the structure fragments and the loop length constraints. The basic assumption behind this model for sequence–structure relation is that the structure is composed of important and irrelevant parts, the important parts are the fragments, which have to be aligned to constitute the essential parts of the fold and the irrelevant part are loop regions, which are usually not conserved even among similar folds and, thus, need not be aligned and the alignment score is not taken into account for the overall alignment score.

The fragment threading problem has a much smaller search space of size exponential in n and k instead of n and m for protein threading. In some cases, for small proteins or proteins with a small number of fragments an exhaustive enumeration of all mappings of sequence regions onto fragments is possible. Therefore, the optimal threading can be found for almost any conceivable scoring function, i.e. scoring functions that depend on the complete alignment. Nevertheless, to find meaningful threadings capturing shared biological and/or structural features in a true sequence-structure relationship between the two proteins, the scoring function has to model the respective features very accurately. As of yet, no scoring function appears to be accurate enough to discriminate between similar alignments and to select the 'correct' alignment from a large number of alternative alignments. For a broad class of empirical pair interaction potentials it has actually been demonstrated that almost always there exist alignments in the vicinity of the structurally closest alignments, that score better than the standard-of-truth according to the scoring function used. Such alignments are easily found via a greedy search of the alignment space or some simple perturbations of the correct alignment using a genetic algorithm [207].

Branch-and-bound algorithms for fragment threading An algorithm for finding the global optimum of gapped threadings between a query sequence and a structural core of consecutive fragments (fragment threading [208]) has been described by Lathrop and Smith [175]. The fragment threading formalization is contrasted to the full threading approach as addressed by recursive dynamic programming (RDP, see below). The method can deal with quite general scoring functions including full pair interaction potentials and imposes only some (length) restrictions on the loop regions. The method deals with sets of large numbers of threadings at a time using an efficient representation and estimates appropriate bounds for the maximum achievable score of all the alignments in the sets. Thus, large numbers of potential threadings can be eliminated by applying one of several bounding conditions. Alignments in the fragment threading formalization can be represented by defining sequence positions to be mapped onto the first structural position of any core element, i.e. a vector $\langle t_1, \ldots, t_n \rangle$ for n Core elements $C_1, \ldots, C_n$. The amino acid chain and loop restrictions define additional lower and upper bounds for the amino acid to be placed onto the start of a structural segment. A set of alignments is than represented as a vector of lower (b_i) and upper bounds (d_i) for the segments C_i and contains all threadings compatible with the constraint that the amino acid aligned to $C_i(1)$ is in the range (b_i, d_i). The algorithm enumerates all possible threadings by selecting such a set and a segment and a so called split point t in (b_i, t_i) via recursively producing three subsets ($b_i, t - 1$), (t), and ($t + 1, d_i$) and computing the associated bounds on the score of the subsets. If the set of alignments cannot improve the currently best alignment score known, the whole set can be discarded. The algorithm guarantees to come up with the optimal threading alignment [175], given enough time. If stopped early, the algorithm will present the best scoring alignment found so far. Even after an optimal alignment has been found, the algorithm can be continued in order to produce alternative optimal alignments solutions or present the proof that the alignment is optimal, which can only be stated for certain, once the whole search space has been covered. Typically, it takes about ten times as long to provide the proof as compared to computing the optimal alignment.

The performance of the algorithm heavily depends on the stringency of the bounds of eliminated large sets of poorly performing solutions before actually evaluating them individually.

Considering the potentially huge search space, Lathrop and Smith report implicit evaluation of up to $6.8 * 10^{28}$ alignments per second [175] (Table 6.1), which compares with about 140 explicit evaluations of the scoring function as observed by Bryant [209].

Despite the fact that, for a set of threading instances, the optimal alignment could be produced for a set of five commonly used scoring potentials, the relevance of globally optimal solution remains doubtful from a biological point of view, due to apparent deficiencies in the used scoring systems. Nevertheless, the approach demonstrated that optimal solutions can be produced for biologically relevant problem instances with, in some cases, structurally meaningful alignments that exhibit small deviations from the standard-of-truth alignment. Whether the algorithmic optimization or the tuning of the scoring functions is the most decisive factor for obtaining high quality alignments with a reasonable amount of computing resources remains to be seen.

Recently, Xu and Xu proposed a new branch-and-bound algorithm called PROSPECT [210–212] for the full threading problem which guarantees to find the globally optimal alignment for a scoring function incorporating pairwise contact terms (in addition to sequence terms, environment profile scores and gap penalties). The algorithm can avoid enumerating the complete search space by using appropriate bi-partitionings of the sequence and structure, such that topological complexity (TC) is minimized, i.e. the parts of the partitions exhibit a minimum number of inter-partition links (contacts). As all legal combinations of link assignments have to be considered the TC is an important factor determining the running time of the algorithm. In biological examples, the TC has been observed to be small (<8) in practice, and the algorithm has a time complexity of $O(n^{TC} \cdot n^{TC/2})$, where n is the target length and N the maximum allowed loop length (in the current version $N < 20$). Again, including predicted secondary structure information improves the recognition performance of the approach. The PROSPECT method allows exploiting constraints on the target protein such as known disulfide bonds, active site information and long-range (NOE/NMR) distance constraints. The method has been evaluated on benchmark sets [210]

Tab. 6.1
Protein structure resources and databases (coordinates, models, classifications, representative and benchmark sets) on the WWW
Major protein structure archives and structure coordinate resources:

Name	What	Where	Refs
Protein Data Bank (PDB)	The main database of protein structures	http://www.rcsb.org/pdb/ Rutgers University, NJ, UCSD, and NIST, Washington	[245, 284]
Molecular Modelling Database (MMDB)	Structure data (ASN.1, Entrez), Visualization (cn3D) and structure comparison tool (VAST), Structure-Taxonomoy DB	http://www.ncbi.nlm.nih.gov/Structure/ NCBI, NIH, Bethesda, S. Bryant et al.	[285, 286]
Molecules R US Utility	A full text search of the PDB database to identify structures (pdb ids) by name/keyword	http://molbio.info.nih.gov/cgi-bin/pdb	
PDB At A Glance	Hypertext-based classification of the PDB into biochemically meaningful contexts	http://cmm.info.nih.gov/modeling/pdb_at_a_glance.html	
PDBSUM	Summary and annotation of PDB entries	http://www.biochem.ucl.ac.uk/bsm/pdbsum/ University College, London, R. Laskowski	[287]
Swiss-Model	Wide range of protein analysis tools and protein models	http://www.expasy.ch/swissmod/SWISS-MODEL.html Swiss Institute of Bioinformatics (SIB), Geneva, CH, M. Peitsch et al.	[288]
ModBase	Automatically generated model structures (MODELLER)	http://guitar.rockefeller.edu/modeller/modeller.html Rockefeller University, NY, A. Sali et al.	[243]
Protein Resource Entailing Structural Information on Genomic Entities (PRESAGE)	Resource for information on structural genomics projects and for keeping track of the current status (experimental structure status, structural assignments, models, annotations) for proteins	http://presage.berkeley.edu UC Berkeley, S. Brenner et al.	[289]

6.3 Methods | 265

Name	What	Where	Refs
Structural Genomics Projects	Resources from various structural genomics projects, Target and model proteins	http://www.structuralgenomics.org http://www.x12c.nsls.bnl.gov/StrGen.htm	[246]
GeneCensus	Protein folds in (completely) sequenced genomes	http://bioinfo.mbb.yale.edu/genome/ Yale University, M. Gerstein et al.	
Macromolecular Movements	Database of Macromolecular Movements with Associated Tools for Geometric Analysis	http://bioinfo.mbb.yale.edu/MolMovDB/ Yale University, M. Gerstein et al.	[290]
BIND	Database of macromolecular interactions	http://www.bind.ca/ Samuel Lunenfeld Research Institute, Toronto, C. Hogue et al.	[291, 292]
DIP	Database of interacting proteins (Literature, yeast2hybrid)	http://dip.doe-mbi.ucla.edu UC Los Angeles, I. Xenarios et al.	[92, 93]

Selection of databases on protein structure and protein family classification:

Name	What	Where	Refs
Structural Classification			
Structural Classification of Proteins (SCOP), Cambridge, UK	Major structural classification of protein structures and domains (Murzin et al.)	http://scop.mrc-lmb.cam.ac.uk/scop/ Cambridge University, UK, A. Murzin et al.	[245, 284]
Structural Classification of Proteins (CATH), Cambridge, UK	Major structural classification of protein structures and domains (Thornton/Orengo et al.)	http://www.biochem.ucl.ac.uk/bsm/cath/ University College, London, J. Thornton et al.	[44, 45]
Dali structure comparison and FSSP structure classification	Protein classification via structural alignment, Dali Domain Dictionary (DDD)	http://www2.ebi.ac.uk/dali/ http://www2.ebi.ac.uk/dali/fssp/fssp.html EBI, Cambridge, L. Holm et al.	[102, 293, 294]
3Dee	Database of structural domains	http://jura.ebi.ac.uk:8080/3Dee/help/help_intro.html EBI, Cambridge, G. Barton et al.	
HOMSTRAD	Database of structure based alignments	http://www.cryst.bioc.cam.ac.uk/~homstrad/ K. Mizuguchi et al.	[112]

Tab. 6.1 (continued)

Name	What	Where	Refs
3d-ali	Database of structure based alignments	http://www.embl-heidelberg.de/argos/ali/ali.html S. Pascarella, P. Argos	[295, 296]
Database of 3-D Protein Structure Comparison and Alignment	3-D Protein Structure Comparison and Alignment via the Combinatorial Extension (CE) Method	http://cl.sdsc.edu/ce.html UC San Diego, CA, P. Bourne et al.	[297]
Representative Sets of Structures			
PDB40	40% sequence id SCOP and derived classifiaction information	http://astral.stanford.edu/ Stanford University/UC Berkeley, S. Brenner	[271]
pdb select	Non-redundant sets at various sequence id levels	ftp://ftp.embl-heidelberg.de/pub/databases/pdb_select/, http://www.cmbi.kun.nl/whatif/select/ EMBL, Hobohm et al.	[298]
culled pdb	Non-redundant sets at various sequence id levels	http://www.fccc.edu/research/labs/dunbrack/culledpdb.html Fox Chase Cancer Center, Philadelphia, R. Dunbrack	
FSSP/Dali	List of structurally related pairs according to Dali superposition	http://www.ebi.ac.uk/dali/fssp/TABLE2.html EBI, Cambridge, L. Holm	[299]
nrdb	Non redundant sequence DB	http://blast.wustl.edu/pub/nrdb/ Washington University, W. Gish	
HSSP	Database of multiple alignments of sequences to known structures	http://www.sander.ebi.ac.uk/hssp/ EMBL/EBI, R. Schneider, C. Sander	[74]
Protein Family			
PROSITE	Database of sequence patterns and profile indicative for protein function	http://www.expasy.org/prosite/ Swiss Institute of Bioinformatics (SIB), A. Bairoch	[86, 87]

Name	Description	URL / Source	Ref.
BLOCKS	Ungapped sequence alignment blocks	http://blocks.fhcrc.org/ J. Henikoff, S. Henikoff	[31, 141]
EMotif	Database of significant sequence patterns for biochemical properties and biological functions	http://motif.stanford.edu/emotif Stanford University, D. Brutlag et al.	[84]
PRINTS	Database of family fingerprints	http://www.bioinf.man.ac.uk/dbbrowser/PRINTS/ Manchester University, UK, T. Attwood et al.	[83]
PRODOM	Database of protein domains based on BLAST sequence alignment	http://www.toulouse.inra.fr/prodom.html INRA, FR, F. Corpet et al.	[75]
PFAM	Database or protein domains based on sequence alignment via HMM	http://www.sanger.ac.uk/Software/Pfam/ Sanger Centre, Cambridge, UK, A. Bateman et al.	[22, 24, 73]
Interpro	Combination of the major protein family databases (PRINTS, PROSITE, Pfam, ProDom)	http://www.ebi.ac.uk/interpro/ EBI, Cambridge, R. Apweiler et al.	[76, 300]
PROTOMAP	Clustering of the structure space	http://www.protomap.cs.huji.ac.il http://protomap.cornell.edu Cornell University, G. Yona et al.	[257, 301]
SMART	Database of multiple alignments of signalling domains	http://smart.embl-heidelberg.de EMBL, P. Bork et al.	[302]
SYSTERS	Clustering of Swissprot sequences via BLAST	http://www.dkfz-heidelberg.de/tbi/services/cluster/systersform DKFZ, Heidelberg, A. Krause, M. Vinbron	[303]
COGS	Clusters of orthologous groups (sequences occurring in several genomes)	http://ncbi.nlm.nih.gov/COG NCBI/NIH, E. Koonin et al.	[304]

and, for blind prediction, in the CASP experiment [213]. For achieving the results for the CASP targets, a neural network approach has been used to convert threading scores into a confidence measure based on a training set of structurally superposed proteins using SARF. In addition, the program has been run for a broad range of parameters using a supercomputer. The consensus predictions from these runs have been used to obtain higher confidence in the predictions [213].

Heuristic search procedures for full threading based on simulated annealing and genetic algorithms   Deficiencies of the scoring function obstruct the use of standard combinatorial search methods for finding the optimal threading, multiple alignment, and fragment assembly for protein structure prediction. Therefore, heuristic search procedures such as simulated annealing [36, 214–217], Monte Carlo methods [218–222], and genetic algorithms [173, 223–227] have been used. In general, these approaches are based on the Metropolis method [228]. Starting from an initial alignment, they generate alignment alternatives and accept the new alignments as replacement for the current alignment if they score better or otherwise only with a certain chance depending on the score and an annealing parameter (temperature). After a few steps of the procedure, the annealing parameter is changed according to a so-called cooling scheme and the procedure is continued until no further improvement on the current threading score can be achieved. Apart from the well known critical dependence on the particular cooling scheme to be employed, the approach is computer-intensive and will result in meaningful alignments only if the scoring function is very accurate.

The recent and very successful fragment assembly approach ROSETTA (sometimes called mini-threading method) [5, 37] has been proposed by Baker and co-workers [36]. Here, a simple simulated annealing procedure is used to assemble small segments with similar sequences from unrelated protein structures into possible protein-like structures, which are then subjected to an empirical Bayesian scoring function incorporating sequence, disulfide bonding, solvation, secondary structure packing, and residue pair interaction terms to select the most probable fold. Fragments are nine residues long and the search space is defined by using the conformations of nine residue fragments taken from an ensemble of 25 alternative structure fragments selected from a 25% non-redundant protein set via a sequence similarity criterion. Starting from the fully extended chain, a move substitutes the torsion angles from a randomly chosen neighbor fragment. Invalid (atom distances less than 2.5 Å) conformations are discarded, others are accepted according to a Metropolis criterion. A specific realization [37] of this approach employed for the CASP3 blind predictions constructs 1200 structures, each the result of 100 000 fragment angle substitutions in the simulated annealing procedure. This takes about 2 days of computing time on a fast workstation for a protein domain with 120 amino acids. The performance of the method is evaluated via extensive simulation runs on randomized sequences and fragments, with and without information from multiple alignments. Scoring filters have been performed and the number of conformations within a small rms of the native structure during 100 simulation runs has been recorded to calibrate the method. The current version of the approach has been used with great success for blind prediction in the CASP experiments on both targets with known similar template structures as well as for novel folds [37, 229].

Dynamic programming approaches, double dynamic programming As has been mentioned above, dynamic programming (DP) approaches cannot guarantee the optimal solution. Nevertheless, DP methods have been used in several ways. They optimize simplified scoring functions ignoring individual pair contacts, such as threading profiles. In iterative approaches they take fixed interaction partners defined through the alignment of the previous iteration (frozen approximation) [189, 230]. Hierarchical variants, such as double dynamic programming [231, 232] employ DP to determine locally optimal structural environment scores (the first level matrix) for any structural position i matched with some residue a (from the sequence) by optimally aligning sequence residues to the contact partners of position i. They exploit this information from the first level matrix in a subsequent DP (second level matrix) step to find an overall alignment (see also [101] for more detail). The major disadvantage of double dynamic programming is that it requires a dynamic programming optimization of time complexity $O(n * m)$ for any of the $n * m$ initial matched sequence structure position pair resulting in a $O(n^2 * m^2)$ overall time complexity. This is impractical already for medium sized proteins. Clearly, the method is too slow to be used in large scale threading experiments that align a sequence against a fold library of several thousand domains. Thus the method cannot be used to annotate large numbers of sequences. Moreover, the procedure is still a very rough approximation to the optimal solution, as dependencies of partial solutions cannot be handled fully.

Core fragment threading Core fragment threading or core-element alignment is a variant of the threading problem similar to the problem formulation used by Lathrop et al. for the optimal branch-and-bound algorithm [175, 208, 210–212, 233] (see Subsection 6.4.2.3.2 above). The idea is that the fold library is preprocessed into a library of structural cores. The query sequence is then aligned to the structural cores without allowing for gaps in the core elements and with some restrictions on the loop length connecting the core elements.

The core structures of template proteins are defined as consecutive segments of the structure that are conserved across multiple alignments of structural [195, 197]. Loop length restrictions are defined to account for the maximum length in observed for the structural neighbors and to avoid non-physical models.

The core-element alignment algorithm [234] starts with an initially random alignment of query sequence segments with the core elements. Given an alignment of the segments, alternative alignments are generated iteratively using preferences implied by the aligned fragments and an empirical pair potential [208] and other scoring terms via a Gibbs sampling strategy (see Appendix). Alternatives are produced by shifting or extending target sequence fragments. Gap penalties are not used aside from the requirement of satisfying the loop length constraints during the generation of alternatives. Similar to RDP (see below) this approach extends the "frozen approximation approach" [189] as the environment is dynamically updated during the Gibbs sampling optimization. The procedure is not guaranteed to converge to a unique and optimal solution. Therefore, a detailed estimation of the significance of a threading and its score is performed by a shuffling the query sequence followed by a re-alignment. Composition-corrected scores are computed as described in [208] and the probability is estimated that a random sequence would score higher than the query sequence assuming a normal distribution of scores and taking length dependencies into account [209].

The core fragment threading approach, using carefully defined structural core models, a calibrated scoring function, and a statistical estimation of *P*-values for the scores, appears to be one of the most successful methods in the CASP-3 experiments [196, 197].

Recursive dynamic programming (RDP) Another heuristic search procedure based on a branch-and-bound approach has been described in [188]. The approach tries to avoid the restrictions implied by a predefined structural core. Even if the structural core is correctly and completely defined, even the best alignment of the query sequence onto the structure core can be based on information of about half or less of the residues and structural positions of the template structure. Often, structures contain highly conserved loop regions, which if responsible for conserved functions (e.g. ATP binding loops [170]) are also conserved on the sequence level or contain detectable sequence motifs or functional patterns. Such loops can provide important initial hints on similarities and partial alignments exploitable for the assembly of overall alignments.

The recursive dynamic programming approach is based on this observation. It uses a number of so-called oracles to find initial hits of various origin, sequence conservation, conserved disulfide bonds, matching secondary structures, functional patterns, 123D hits, etc. Such an initial hit represents a partial alignment (see Figure 6.5a). From a partial alignment, the contacting regions in the structural template are determined. The already aligned parts then shed (see Figure 6.5b) some pseudo-energy preferences defined by the scoring system used onto the contacting positions. These preferences can be compiled into a structural preference profile for these

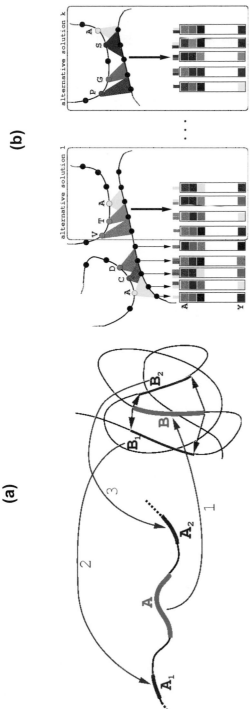

Fig. 6.5

This figure illustrates the RDP procedure: (a) illustrates the finding of initial hits and how to proceed with further searches based on the contact structure of the template and the remaining not yet aligned sequence parts. (b) illustrates the basic search step given a partial alignment: the partial alignment implies contact profiles for the remaining positions. Optimal threadings can be found using dynamic programming of sequence parts onto the profiles.

positions. The initial hit also partitions the query sequence into aligned and not aligned parts. Now, the procedure uses dynamic programming to find the best possible match from the not aligned parts of the query with the structural regions represented by the contact profile. This results in more partial alignments extending the partial alignment under consideration. The procedure is continued recursively until no further significant partial alignment can be found in the still unaligned parts. The alignment is then complete and the rest of the structure and the query is considered to be insertions and treated as gaps in the alignment.

This recursive process starts with a set of initial partial alignments each partitioning the problem into a set of subproblems, which need to be solved to obtain an overall solution. From all solutions for each member of the set, the best overall solution(s) are considered as solutions to the overall problem.

Especially in the early phases of this process, many incorrect or suboptimal partial alignments have to be expected due to the partial information used so far. Therefore many alternatives are considered using a variety of oracles.

Technically, the search space of RDP can be represented as an AND–OR tree with alternating levels of AND and OR nodes. An AND node contains a set of subproblems to be solved in order to solve the problem represented by the node. An OR node contains a set of problems, where the solution of any of them could be used as the solution of the overall problem represented by the OR node.

Usually, the search is organized using a priority queue, holding the nodes in the order of most promising subproblems first. The search selects the most promising subproblem and determines solutions for it in the context of the partial assignments already made.

A detailed description and discussion can be found in [188]. The approach has been shown to improve on profile threading methods on several comprehensive benchmark sets containing structural pairs of increasing difficulty.

The approach is fast enough to allow threading a query sequence against a small fold library of several hundred domains. For large scale threading applications using thousands of template structures or for predicting thousands of sequences, e.g. for whole genome analysis, the normal mode of operation is to apply fast threading methods such as ToPLign/123D* first to select a smaller set of candidates, against which the query is threaded with RDP to obtain additional evidence or more accurate alignments.

Additional functional information and constraints derived from experiments can be exploited in similarity methods to balance inaccurate scoring functions or excessive computing requirements in order to improve fold recognition and alignments of sequences to structures. Distance constraints provide orthogonal measures for scoring alignments and structural models and means for drastically reducing the search space, and thus computing and memory resources. There are essentially two ways to incorporate such distance constraints into an improved prediction method: the first uses the constraints in evaluating the predicted models either via discarding invalid (i.e. incompatible with the measured constraints) models or by re-sorting the ranked list of models via a score combining threading score and fulfilled constraints. The second approach uses the constraints in the prediction algorithm itself thereby excluding partial model structures in conflict with the experimental constraints. Here a XML-based specification language ProML is employed to represent and access the additional information on the target sequence and the fold template in the prediction algorithm [32, 33].

The RDP method lends itself naturally to such an approach. The search procedure can be easily tailored to special needs via using specialized generating functions to compute the sets of partial solutions and specialized evaluation functions to select the best assembled solutions. If a set of distance constraints is given, the generating functions can be restricted such that only solutions respecting the set of constraints are generated, thereby focusing on consistent alignments in order to guide the search process towards valid alignments and discarding huge search spaces not containing any valid alignment. On the other hand, distance constraints can also be checked between aligned residues after having assembled partial solutions into overall solutions. Therefore, the evaluation function is also restricted with respect to the constraints in order to filter out invalid alignments. Again, the idea is to eliminate invalid alignments as early as possible to accelerate the computation and to improve the alignment quality of solutions.

Experimentally measured distance constraints can be of valuable use in similarity methods for structure prediction. Even for moderate numbers (sequence length/10) of distance constraints, we could gain one-third more recognized folds by re-ranking 123D alignments and about two-thirds by employing better alignments on a representative benchmark set [32].

6.4
Results

Protein structure methods as discussed in the previous sections can be employed in a variety of applications from whole genome analysis to detailed rational drug design projects. In basic research, protein structure prediction and the resulting structural models can help to assign possible functions of proteins and to better understand possible functional mechanisms and specificities, possibly in order to plan expression, mutation and recombination experiments, and understand genetic variation. In pharmaceutical and molecular medicine, research applications range from finding functional proteins in complex networks (probably with the help of new high throughput screening data, such as DNA expression measurements with arrays and chips) through the identification of possible drug target molecules, based on the protein function in metabolic and regulatory net-

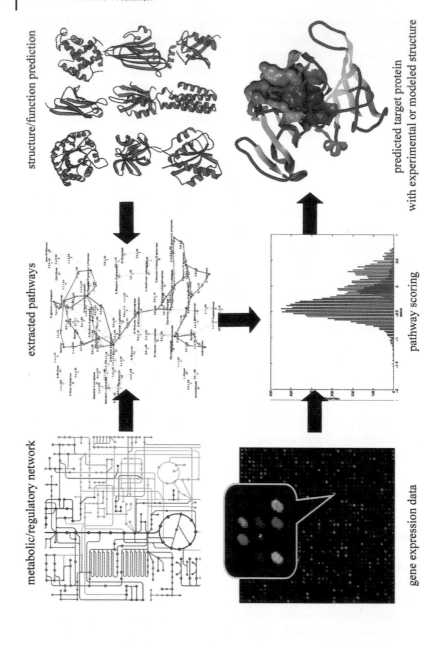

structure/function prediction

metabolic/regulatory network

extracted pathways

pathway scoring

gene expression data

predicted target protein
with experimental or modeled structure

work contexts, to modeling detailed structures for rationally guiding and designing new drug molecules. Figure 6.6 gives a schematic overview of structure-based methods in a larger functional genomics and target-finding setup.

The following sections briefly discuss some results and trends in using structure prediction for remote homology detection especially in the genomic context, to aid the structural genomics projects, to further whole genome annotation and to exploit the sequence-to-structure-to-function paradigm for functional predictions.

6.4.1
Remote homology detection

For the detection of remote homologies based on sequences alone two methods are state-of-the-art: iterative sequence searching with profiles PSI-BLAST, iterative sequence search and hidden Markov models. Haussler et al. have performed a range of experiments and evaluations, which clearly demonstrates the superiority of methods exploiting evolutionary information in the form of profiles [20, 235, 236]. Based on a standard benchmark derived form the PDB40 set they could show that these methods could identify more than twice as many related pairs of proteins (in the PDB40 set) according to the SCOP classification. For distant homologs (less than 30% sequence identity) the HMM based SAM-T98 [236] method outperforms pairwise methods such as FASTA and GAP-BLAST by a factor of three for related pair recognition. With respect to the recognition of all related pairs of proteins in a set of targets these sequence based method currently rival the best structure based methods. Also HMMs have been applied with great success in the CASP experiments [183, 229, 237].

State-of-the-art methods for protein structure prediction are continuously discussed and evaluated in the bi-annual CASP experiments (see the accompanying special issues of the Journal Proteins [329]), the LiveBench [330] and EVA [331] experiments. The major available methods, programs and servers are collected in Tables 6.2 and 6.3.

◀ **Fig. 6.6**
Protein structure prediction and analysis in a broader context of drug target finding and rational drug design. The figure illustrates one particular method of combining background information on biochemical networks, relevant metabolic or regulatory pathways, the mapping of available structural knowledge onto these pathways, and available expression measurements using a statistical scoring function. The method produces a candidate list of most interesting target proteins within the context and data in question. In an ideal situation, the most promising target proteins have an experimentally solved or a predicted model of high quality attached to them. Such candidates could then be used not only in functional assays and biochemical experiments but in addition to computer aided rational drug design and experiment planning.

Tab. 6.2
Selection of tools for structure analysis, superposition, classification, and visualization

Name	What	Where	Refs
Secondary Structure Computation and Prediction			
DSSP	3-state secondary structure annotation for protein structures	http://www.cmbi.kun.nl/swift/dssp/ W. Kabsch, C. Sander (distributed via CMBI, NL, G. Vriend)	[305]
STRIDE	Secondary structure, accessibility, and hydrogen bonding information	http://www.embl-heidelberg.de/stride/stride.html EMBL, D. Frishman, P. Argos	[306]
TOPS	Topology cartoons of secondary structure arrangements of proteins from DSSP input	http://www.sander.embl-ebi.ac.uk/tops EBI, Cambridge, UK, D. Gilbert et al.	[307]
Visualization and Structure Viewers			
Swiss-PDBViewer	Structure, superposition, alignment and modelling visualization	http://www.expasy.ch/spdbv/ SIB, CH, N. Guex et al.	[308]
Rasmol	Easy to use molecular structure viewer	http://www.bernstein-plus-sons.com R. Sayle	[309]
chime	Molecular structure viewer and plugin	http://www.mdlchime.com/chime/ MDL Information Systems Inc.	
Cn3d	Molecular structure viewer and plugin	http://www.ncbi.nlm.nih.gov/Structure/CN3D/ cn3d.shtml NCBI, Bryant et al.	
kinemage	Interactive molecule presentation and annotation system	http://kinemage.biochem.duke.edu Duke University, J&D. Richardson	

Name	Description	URL / Source	Ref
Molscript	Production of publication quality graphics for protein structures	http://www.avatar.se/molscript/ P. Kraulis	[310]
Ligplot	Publication quality graphics for protein ligand complexes	http://www.biochem.ucl.ac.uk/bsm/ligplot/ligplot.htm University College, London, J. Thornton et al.	[311]
Structural Alignment and Analysis			
ProFit	Superposition of two pdb structures or regions	http://www.bioinf.org.uk/software University of Reading, A. Martin	
Dali	Multiple structural alignment and protein clustering (DDD, FSSP)	http://www2.ebi.ac.uk/dali/ EBI, Cambridge, L. Holm	[294]
SSAP	Double Dynamic Programming based structure alignment	http://www.biochem.ucl.ac.uk/~orengo/ssap.html MRC and UCL, London, W. Taylor, C. Orengo	[101, 232, 312]
SARF2	Secondary structure element based superposition	http://www-lmmb.ncifcrf.gov/~nicka/sarf2.html NCI/NIH, N. Alexandrov	[103]
VAST	Alignment of vectors of secondary structures	http://www.ncbi.nlm.nih.gov/Structure/VAST/vast.shtml NCBI/NIH, Bryant et al.	
	Structural alignment of SCOP sequences	http://bioinfo.mbb.yale.edu/align/ Yale University, M.Gerstein Lab	
CE/CL	Combinatorial extension (CE) of optimal paths of local structural similarities/compound likeness (CL) searches	http://cl.sdsc.edu/ce.html UCSD, Bourne et al.	
Procheck	Checking protein quality and quality plots	http://www.biochem.ucl.ac.uk/~roman/procheck/procheck.html University College, London, J. Thornton et al.	
WHATIF/WHATCHECK	Structure quality check modules as part of the modelling package WhatIF	http://www.cmbi.kun.nl/whatif/ CMBI, Nijmwegen, NL, G. Vriend	[313]

Tab. 6.2 (continued)

Name	What	Where	Refs
Alpha shapes	General geometry package usable to analyse packing and accessibility	http://alpha.geomagic.com/alpha/ NCSA, H. Edelsbrunner	
Voronoi tesselation	General geometry package usable to analyse packing and accessibility	http://www.geom.umn.edu/locate/qhull Qhull program, University of Minnesota, Geometry Center	[314]
Scoring Potentials, PDF (probability density functions)			
ProStar	Archive and evaluation of scoring potentials	http://prostar.carb.nist.gov/ CARB, Rockville, MD, J. Moult et al.	

Tab. 6.3
Structure prediction servers and programs

Secondary Structure Prediction			
PHD	Multiple alignment trained neural network prediction	http://www.embl-heidelberg.de/predictprotein/pre-dictprotein.html EMBL/Columbia University, B. Rost	[72]
predator	multiple alignment based	http://www.embl-heidelberg.de/argos/predator/predator_info.html EMBL/MIPS/GSF, Munich, D. Frishman	
PsiPRED	PSI-BLAST generated sequence profiles and NN	http://insulin.brunel.ac.uk/psipred/ Brunel University, UK, D.Jones	[156]
JPred	Consensus of several predictions	http://barton.ebi.ac.uk/servers/jpred.html EBI, Cambridge, UK, G. Barton et al.	[159, 160]
CODE	Decision tree based consensus of several predictions	http://cartan.gmd.de/ToPLign/ToPLign.html GMD, Sankt Augustin, J. Selbig et al.	[161]
Threading programs			
PDB-BLAST	Produce sequence profile with PSI-BLAST and use it in PSI-BLAST run against PDB	http://bioinformatics.ljcrf.edu/pdb_blast/ A. Godzik, L. Rychlewski	[283]
TOPITS, Predict_Protein	alignment of secondary structure motifs	http://www.embl-heidelberg.de/predictprotein/ EMBL, B. Rost	[163]
threader, threader2	Threading with pair interaction potentials	http://insulin.brunel.ac.uk/threader/threader.html Brunel University, UK, D. Jones	[190]
Bryant Gibbs Threading	Gibbs sampling for sequence-structure alignment with contact potentials	http://www.ncbi.nlm.nih.gov/Structure/RESEARCH/threading.html NCBI, NIH, S. Bryant et al.	[234]

Tab. 6.3 (continued)

Name	Description	URL / Source	Reference
GenThreader (mGenThreader)	Sequence (PSI-BLAST profile) alignment with neural network based evaluation of alignments with structural scores	http://insulin.brunel.ac.uk/psipred/ Brunel University, D. Jones	[147, 157]
3D-PSSM	Sequence profiles are threaded against structure profiles using secondary structure preferences and solvation potentials	http://www.bmm.icnet.uk/~3dpssm ICRF, London, M. Sternberg et al.	[205, 283]
BioinBGU	Consensus of several profile alignment methods with different scoring matrices and profiles	http://www.cs.bgu.ac.il/~bioinbgu Ben-Gurion University, Israel, D. Fischer	
FFAS	Produce sequences with PSI-BLAST and search against PDB	http://bioinformatics.burnham-inst.org/FFAS/ A. Godzik, L. Rychlewski	[200, 283, 315]
SAM-T99	Build HMM for sequence; use HMM to predict secondary structure and to search the PDB. Conversely, HMMs for PDB structure are used to score the sequence	http://www.cse.ucsc.edu/research/compbio/HMM-apps/T99-query.html UC Santa Cruz, CA, K. Karplus	[20, 183, 237]
FUGUE	Library of profiles and substitution matrices from HOMSTRAD structural alignment DB. Sequence or profile is searched against library.	http://www-cryst.bioc.cam.ac.uk/~fugue/prfsearch.html University of Cambridge, UK, K. Mizuguchi, T. Blundell et al.	
P-Map	PSI-BLAST approach	http://www.dnamining.com/Pages/products1.html dnaMining informatics software LLC, Bruno	
ssPsi		http://www.sbc.su.se/~arne/sspsi Stockolm University, A. Elofsson	
FORREST	HMMs for structural families from CATH	http://abs.cit.nih.gov/forest2/ Analytical Biostatistics Section, NIH, V. Di Francesco et al.	

loopp	Learning, Observing and Outputting Protein Patterns	http://ser-loopp.tc.cornell.edu/loopp.html Cornell University, R. Elber et al.	[199]
frsvr	Threading with 3D profiles	http://www.doe-mbi.ucla.edu/people/frsvr/frsvr.html UCLA, D. Fischer, L. Eisenberg	
PROFIT	Dynamic programming method for contact potential threading	http://lore.came.sbg.ac.at/home.html University of Salzburg, M. Sippl	[137]
123D+	Profile threading with CCPs (see section 4) using secondary structure predictions	http://www-lmmb.ncifcrf.gov/~nicka/123D.html NIH/NCI, Frederick, MD, N. Alexandrov, R. Zimmer	
123D*	Profile threading with CCPs (see Section 6.4) using PSI-BLAST sequence/structure profiles and optimized scoring term weighting	http://cartan.gmd.de/ToPLign/123D.html GMD, Sankt Augustin, R. Zimmer et al.	[137, 138, 149, 273]
ToPLign	CCP (123D*) and pair contact potential threading with (recursive) dynamic programming (RDP, branch-and-bound algorithm), profiles, and p-values	http://cartan.gmd.de/ToPLign.html GMD, Sankt Augustin, R. Zimmer et al.	[138, 149, 188, 273]
ROSETTA	Combination of threading/sequence features with ab initio simulation: assembly of structure fragments with sequence similarity using MC simulations	http://depts.washington.edu/bakerpg (Information only) Washington University, D. Baker et al.	[5, 36, 37, 316]
Swiss-model	Variety of alignment, analysis and modeling tools, automated modeling server	http://www.expasy.ch Swiss Bioinformatics Institute, Geneva, CH, N. Guex; M. Peitsch	

6.4.2
Structural genomics

The goal of structural genomics projects is to solve experimental structures of all major classes of protein folds systematically independent of some functional interest in the proteins [238, 239]. The aim is to chart the protein structure space efficiently. Functional annotations and/or assignment are made afterwards.

Several estimates on the number of different folds have been presented. Some estimates have been based on the number of new and related folds and an analysis of structural classifications and their growth during the past years [240]. Other estimates can be derived from systematic predictions on complete genomes [241–243]. A recent analysis [244] of this type, classifying the new entries of the central structural depository (PDB [245, 332]), has revealed that the fraction of proteins establishing a new family, superfamily, or fold remains constant (at about 15%) over the years (1990–1997), as compared to new proteins belonging to a known family (15%), known proteins from different species (20%), and new (modified) structures from already resolved proteins (about 50%). In addition, new structural entries for sequences with no detectable sequence similarity to known folds (no significant hit via PSI-BLAST searches) yielded about 30% new folds, about 25% new superfamilies with an already known fold, and 45% new families homologous to a known superfamily.

Bioinformatics can support structural genomics by selecting targets for experimental high throughput structure determination, in order to make the best use of the solved structures; i.e. the chance of determining novel folds among the structures to be solved should be optimized. Several clusterings, based on a range of approaches, are used for the purpose of selecting and prioritizing appropriate targets for further X-ray or NMR investigation (see the protein structure initiative [333]) and for keeping track of structural knowledge on proteins, including models and predictions (Presage server [334]). One goal of the overall structural genomics endeavor is to have a solved structure within a certain structural distance to any possible target sequence, which would allow computing a reliable model for all target sequences. This requirement depends on the future enhancements of the modeling procedures on the one hand and on the metric of structural similarity underlying the structure space and the accompanying clusterings, on the other hand. Once a map of the structure space is available, this knowledge should provide additional insights on what the function of the protein is in the cell and with what other partners it might interact. Such information should add to information gained from high-throughput screening and biological assays.

So far, glimpses of what will be possible could be obtained by analyzing complete genomes or large sets of proteins from expression experiments with the structural knowledge available today, i.e. more or less complete representative sets and a quite coarse coverage of structure space [243, 246, 247].

Recently, several successful individual structure-based functional predictions within the structural genomics projects have been reported: an ATPase or ATP-mediated switch in one example [248], a new NTPase in *M. jannaschii* in a second example [249]; and other test cases are the HIT family [250], *E. coli* ycaC [251], HdeA [252], and yjgF gene products [253]. For a recent review on the current capabilities and prospects of function predictions from structure for bona fide hypothetical proteins on a genome-wide scale, see [254].

6.4.3
Selecting targets for structural genomics

The number of different folds is much smaller than the number of proteins. Structure is more conserved than sequence, as nature modularly re-uses successful architectural designs, which allow for some sequence variation. A (continuously increasing) fraction of the possible folds is known. Structural genomics projects aim at producing a representative set of folds. There are increasing chances of finding an appropriate fold for a sequence with unknown structure in the current database of representative folds. Determining a correct fold facilitates the application of homology-modeling techniques to produce a full-atom model structure.

In order to enable the structural genomics projects to focus on the most promising targets for structure determination, structure prediction methods have been applied to estimate the chances of particular proteins to adopt a novel fold, i.e. a fold not yet contained in the structure databases. This should be done for many sequences based on the sequence information alone. Two different approaches have been published. One is based on a clustering of proteins and the distribution of the clusters in sequence and structure space [255]. The other is based on an estimate of erroneous predictions as a function of the corresponding computed sequence-structure compatibility score resulting from a prediction experiment [256]. In this work, Eisenberg and coworkers use a threading program (sequence derived properties SDP [199]) to assign folds to all ORFs of the archaebacterium *P. aerophilum*. For that purpose they determine a confidence level for an alignment describing the likelihood that it is correct. This confidence is derived from the probability distribution of incorrect matches as judged by DALI/FSSP alignments and z-scores (see Chapter 2 of this volume for the definition of z-score) using a test set of 3285 domains and 10 784 655 pairs achieving a high enough SDP z-score. For the calibrated confidence levels more than 40% of the non-trans-membrane proteins could be assigned a fold with 90% and more than 10% with 99% confidence. In addition, about 11% have an as yet unobserved fold with 90% confidence and for 99% confidence still 14 proteins have been predicted to adopt a novel fold.

Portugaly and Linial used the exhaustive PROTOMAP clustering of sequences [257] and the structural classification SCOP [81] to determine a combined sequence–structure classification of proteins. Based on PROTOMAP 2.0 (72 623 SwissProt sequences clustered into 13 454 clusters, 1999) and SCOP version 1.37 (2294 domains, 834 families, 593 superfamilies, 427 folds, Oct. 1997) and a mapping of structures to sequences, the sequence clusters are classified into vacant and occupied, depending on whether the clusters contain known folds. The PROTOMAP clustering structures the sequences space with a graph (the PROTOMAP graph), the edges between the sequence clusters indicate their proximity. Using several measures derived from the cluster space (distances between the clusters, density of clusters, maximal vacant volume) statistical models are constructed to characterize the vacant clusters containing yet undetermined known folds and vacant clusters with probably novel folds. The procedure is illustrated in Figure 6.7. The clusters having the highest probability for representing an as yet unobserved fold constitute promising targets for experimental structure determination.

6.4.4
Genome annotation

Since only 20–40% of the protein sequences in a genome such as *M. genitalium*, *M. Janaschii*, *M. tuberculosis* have a sequence similarity that can exhibit their paralogy to proteins of known function [242, 258], we need to be able to make conclusions on proteins that exhibit much lower similarities to suitable model proteins. As the similarity between query sequence and model sequence decreases below a threshold of, say, 25% we cannot make safe conclusions on a common evolutionary origin of the query sequence and the model sequence any more. However, it turns out that, in many cases, we can still reliably predict the protein fold, and in several cases we can even generate detailed structural models of protein binding sites. Thus, especially in this similarity range, protein structure prediction can help to ascertain aspects of protein function [243, 246, 247].

6.4.5
Sequence-to-structure-to-function paradigm

Having 300 000 sequences but only a few thousand structures in the databases requires a method to establish relationships of sequences to homologous proteins with known function and/or structure.

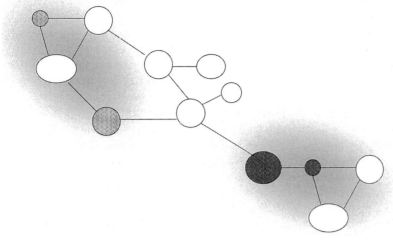

Fig. 6.7

Illustration of the PROTOMAP approach and an example of a PROTOMAP graph of protein structure space. Nodes represent sequence clusters; edges represent proximities between clusters in sequence. Lightly and darkly shaded nodes are clusters that contain sequences known to fold into structure A and B respectively. Using these clusters the remaining protein space can be heuristically partitioned into areas that are likely to fold into known structures and those that are likely to fold into novel structures.

The relationship between structure and function is a true many-to-many relation. Recent studies have shown that particular functions can be mounted onto several different protein folds [85] and, conversely, several protein fold classes can perform a wide range of functions [259]. This limits our potential of deducing function from structure. But it is still possible to use aforementioned knowledge on the range of folds supporting a particular function and the range of functions implemented by particular folds in order to make functional prediction from structure.

Homologous proteins have evolved from a common ancestor and have nearly always a similar 3D structure. This does not necessarily imply similar function: [129, 259, 260]. Probabilistic sequence searches have their limits despite the significant improvements due to PSSMs [9, 203, 205], secondary structure information [156–7, 165, 184] and HMMs [25, 179, 261], which have been systematically evaluated by several authors [20, 129, 147, 195, 205].

If a protein has a known fold, which in addition is annotated with a particular or a few specific functions, the model derived for the sequence in question based on the respective fold can be evaluated with respect to compatibility with some of the proposed functions.

In general, this sequence-to-structure-to-function paradigm involves predicting a structural model for a given sequence (either by sequence and/or structural means, e.g. threading) and, afterwards, identifying a 3D pattern, e.g. an active site, indicative for a particular function in the constructed model.

Especially for enzymes, common mechanisms and associated active sites are often observed. This allows searching for appropriate 3D patterns in a database of structures and for checking models to contain such a pattern (for a well known example, see SER-HIS-ASP serine protease catalytic triad [262]).

Sometimes ligand-binding pockets are identifiable on protein surfaces. With such a binding pocket hypothesis, docking tools can be applied in order to screen potential compounds to bind the protein site.

Protein–protein interactions can be predicted with various methods: analyzing known protein interfaces and inferring common motifs via surface and sequence conservation criteria [263–265], protein fusions [95, 266] or, less successful so far, correlated mutations [267].

A systematic assessment of the relationship between protein function and structure has been performed by Hegyi and Gerstein [259] via relating yeast enzymes classified by the Enzyme Commission (EC numbers) to structural SCOP domains. In this study it has been found that different structural folds have different "propensities" for various functions. Most versatile functions (hydrolases and O-glycosyl glucosidases) have been identified to be mounted onto seven different folds, whereas the most versatile folds (e.g. TIM-barrel and Rossmann folds) realize up to 16 different functions.

In addition to conventional sequence motifs (Prosite, BLOCKS, PRINTS, etc.), the compilation of structural motifs indicative of specific functions from known structures has been proposed [268]. This should improve even the results obtained with multiple (one-dimensional sequence) patterns exploited in the BLOCKS and PRINTS databases. Recently, the use of models to define approximate structural motifs (sometimes called fuzzy functional forms, FFFs [269]) has been put forward to construct a library of such motifs enhancing the range of applicability of motif searches at the price of reduced sensitivity and specificity. Such approaches are supported by the fact that, often, active sites of proteins necessary for specific functions are much more conserved than the overall protein structure (e.g. bacterial and eukaryotic serine proteases), such that an inexact model could have a partly accurately conserved part responsible for function. As the structural genomics projects produce a more and more comprehensive picture of the structure space with representatives for all major protein folds and with the improved homology search methods linking the related sequences and structures to such representatives, comprehensive libraries of highly discriminative structural motifs are envisionable.

Whether or not it is possible to infer protein function, however detailed, from a experimental structure, or from an approximate model, or even from only a partial fold model is a key question in the context of large-scale functional annotation. Currently, the situation is that sometimes such derivations are possible; in more cases additional evidence for functions predicted with sequence methods or from functional motifs can be established; in most cases detailed functional classification from inaccurate structures cannot be guessed.

Though it is not easy to derive functions from resolved protein structures, the availability of structural information improves the chances as compared to relying on sequence methods alone.

6.5
Validation of predictions

Theoretical predictions are risky. Therefore for almost all such prediction experimental validation is required. Nevertheless, often the models can indicate appropriate ways for validation or further experiments. These experiments can be expected to be time-consuming, and expensive. Furthermore, the protein actually needs to be available for the suggested experiments. All of this limits the applicability of experimental validation. Therefore, it is mandatory to reduce errors as much as possible and to indicate the expected error range via computer-based predictions. This is not a trivial problem for structure prediction, though. An estimation of the performance and accuracy of the respective methods can be obtained from large scale comparative benchmarking, from successful blind predictions and from a community wide assessment experiment (CASP [109, 229]/ CAFASP [283]). These are addressed in turn in the following:

6.5.1
Benchmark set tests

Structure prediction methods have to be validated using comprehensive and representative benchmark sets. A couple of benchmark sets have been proposed [188, 199, 270] consisting of a fold library and a list of structurally similar protein pairs without significant sequence similarity. Such a benchmark [335] involves using the method in question to thread one protein of each pair against the library and to count the number of successful recognitions of the structural match or the number of correct identifications of related protein pairs. A couple of methods have been evaluated this way.

More convincing are benchmarks representative for the currently known folds. Thus, given a structural classification of all the structures in the PDB (e.g. SCOP) and a list of proteins without high sequence similarities, say less than 40% pairwise identities (e.g. PDB40 [336], which can be obtained from, e.g. the ASTRAL server [271] or from a couple of other sources for such representative lists [337]), one can easily determine a complete list of structural similarities from the structural classification. I.e. a pair of proteins from PDB40 are considered structurally related if they belong to the same structural class (depending on the level, the same superfamily or the same fold). Other lists of structurally related pairs (via structural alignment of the 3D structures) are available (e.g. from the DALI/FSSP database [338]). A comprehensive benchmark then involves threading each sequence of the PDB40 against all other members [20, 272, 273]. Two benchmarking protocols can be defined: 1. (*Fold recognition*) for each threading run identify the best fitting fold (usually the one with the best score, sometimes after applying filtering or reranking of alignments or models in a post-processing step) and predict it as the closest structural homologue in the fold library and use it as structure template for model building. Of interest in this protocol is the number or percentage of correctly predicted structures. 2. (*Related pair recognition*) for the all-to-all threadings and the respective scores define a score threshold above which threadings are considered predictions of structural similarity. As only one score threshold is applied for the individual threading runs, the scores need to be normalized with respect to the length and other features of the query protein in order to render the scores comparable to each other. Computing p-values from the scores is a method to achieve comparability among experiments and to assign a significance measure to each prediction [273]. This leads to a list of predicted pairs with associated p-values, which can be compared to the list of standard-of-truth pairs. In the simplest case this leads to the number of correctly predicted pairs (TP), of incorrectly predicted pairs (FP), and the number of missed correct pairs (FN). Indeed, the ranked list of scored pairs contains much more information and allows yielding a complete overview of the performance of the method for all possible threshold values of the p-values. This is represented as a so-called sensitivity/specificity plot. Given a p-value threshold the following values of sensitivity *Sens* and specificity *Spec* can be easily read off the ranking list:

$$Sens = TP/(TP + FN) \quad \text{and} \quad Spec = TP/(TP + FP),$$

where TP = true positives, FN = false negatives, FP = false positives, the predicted pairs (positives) change according to the p-value threshold, the negatives remain constant. The sensitivity is the fraction of correct pairs above the threshold as compared to all correct pairs, and the specificity ($1 - $ error rate) is the fraction of correct pairs above the threshold as compared to all pairs above the threshold. Now the set of all possible (*Sens, Spec*) pairs for all possible thresholds can be simply obtained for the ranking list by iteratively decreasing the threshold from the maximum achieved p-value, to the next highest p-value, down to the lowest p-value observed. For each consecutive threshold value the (*Sens, Spec*) values are updated according to whether the new pair above the threshold is a true (TP) or false (FP) structural pair. Example plots for evaluating the fold recognition performance of 123D with five different significance criteria are shown in Figure 6.8. From such a plot, the expected performance of a certain method using certain parameter settings can be read off for any given error rate to be acceptable for the user.

6.5.2
Blind prediction experiments (CASP)

A blind prediction is a computer-based construction of 3D models of a protein sequence for which no structure is known at the time. CASP (comparative assessment of structure prediction methods [109, 274–276]) is a worldwide contest of protein structure prediction that takes place every two years. During the CASP experiments, a set of automated numerical evaluation tools have been implemented [110, 111, 277, 278] to cope with the large number of predictions in a way that is as objective as possible. In fact, the experiment is also devoted to the research and development of such unbiased methods. However, there is still quite some controversy on the criteria to judge protein structure predictions and the corresponding models [339].

For the CASP2 experiment, Marchler-Bauer and Bryant defined a large number of quantitative criteria [110] to be used by the assessor to judge and rank the models and prediction teams. These criteria include several measures for fold recognition specificity, for threading alignment specificity, threading contact specificity, and model accuracy.

For an evaluation of the CASP3 predictions, Sippl used the ProSup rigid sequence-independent 3D superposition software [111, 279]. The basic assumption is that the best prediction corresponds to the best structural match between the predicted model and the experimental structure, but that the best structural match cannot always be determined without ambiguity automatically. Therefore, all possible structural matches between model and structure are considered and ambiguities are resolved in favor of the predictor to determine the closest correspondence between the prediction and structural matches. Ranking a set of structures according to their similarity to a reference fold or model defines a totally ordered similarity measure. Unfortunately, structural superpositions are characterized by two contradicting quantities, the number of equivalenced residues and the best possible rms achieved by optimally superposing the equivalent residues, which can be individually optimized compromising the other. A solution proposed by Sippl for CASP3 follows a similar strategy as used in the protein superposition tool SARF [103, 280], i.e. considering only superpositions with a rms smaller than a constant value (say 3.0 Å) and maximizing the number of residues within that rms range. Usually, this leads to an easily interpreted measure of structural similarity and corresponding rankings of structures [111]. In order to evaluate a prediction, the maximum number of equivalent residues is determined for all possible structural matches between model and structure. The conclusion of the numerical evaluation is that it is difficult to clearly discriminate between appropriate and inappro-

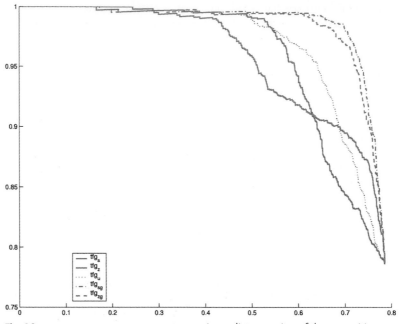

Fig. 6.8

Sensitivity-specificity plots. The figure shows example sensitivity/specificity plots for several fold recognition criteria (raw threading score, *z*-score, *p*-value, *gap*-score, *gap-z*-score) for a benchmark experiment of 123D* threading on a PDB40 domain library using sequence profiles generated for the query sequence via PSI-BLAST. Such plots allow an immediate overview of the recognition performance (sensitivity, *x*-axis) over the whole range of error rates (specificity, *y*-axis). In this case, one can achieve a maximum sensitivity of about 78% (at any error rate), more than 60% sensitivity for a specificity of more than 95% for all but one criterion, for which about 8% error (Spec = 0.92) has to be tolerated in order to recognize more than 60% of the folds.

priate structural matches as often the ranked list of PDB structures matching a target structure shows a continuum of matched residues. In addition, different criteria, such as equivalent residues according to the best structural match or according to the match producing the maximum number of equivalence, or minimum shift error of aligned residues, lead to a general trend of good and bad models, but in most cases also to debatable decisions on which models are best or better than others. The comparison of the results with structural matches obtained from other methods, e.g. the Dali structural alignment, where a *z*-score is computed from the two quantities and used for ranking, leads to quite different results [111].

Therefore, an alternative approach by Hubbard [277] avoids coming up with a single number qualifying a prediction, but rather a plot of rms vs coverage of a number of predictions. The plot is produced such that for any prediction there is a set of points (drawn as a line) representing for each coverage value the minimum rms out of a sampling of superpositions. Such plots allow for a comparative general overview of all predictions: one can easily identify outstanding predictions for a given structural similarity (i.e. rms) or coverage (i.e. a given number or percentage of aligned residues) and, in addition, assess predictions for the whole range of structural similarities or coverages. The plots for the CASP3 targets and the submitted predictions are available at the CASP Web site [340].

Another approach used for CASP3 is sequence-dependent superposition – in contrast to the two approaches mentioned above which are sequence independent relying on the best structural superposition of the 3D model without reference to the natural 1:1 alignment between model and template. In difficult cases, sequence-dependent superposition iteratively excludes residues being modeled less accurate from the computation [278]. This procedure does not necessarily lead to unambiguous solutions. The method allows defining and displaying another set of evaluation criteria for fold recognition and individual alignment accuracy such as longest continuous segments and largest superimposable sets of residues. These criteria can also be restricted to certain subsets of the structure, such as secondary structure elements, buried or exposed residues etc. Results of these measures for the CASP predictions are also available from the CASP Web site.

Due to the debatable outcome of the various methods for numerical evaluation of protein structure prediction the final assessment of models based on the quantitative measures is kind of a creative task similar to the inherent ambiguities encountered in structural classifications [196]. This is especially true for the range of sequence and structure similarities encountered for CASP fold recognition targets, where similarities and partial similarities are hard to identify even for the best human experts in the field. Indeed, a slightly more general definition of the SCOP fold similarity has been used for the CASP3 assessment [196]. The overall result of the structure prediction experiment relevant for this Chapter (fold recognition), which can be considered as an approximation to the overall state-of-the-art in the field, is as follows (according to Table I in [196]): 22 of the 26 target domains show substantial similarities to known structures, three proteins are classified as new fold, one domain has not been classified. Eleven domains are likely to be distantly homologous to known structures. Nine of the 11 homologous relationships have been detected, four of which with accompanying overall correct alignments and models. The other 11 domains with known folds are predicted as follows: five folds identified, one correct topology predicted, two partial structure correctly predicted, and only three folds have not been identified by any method. Two of the three new folds have been predicted correctly as new. Overall, 19 of the target domains have been solved by one or more of the prediction teams, only six could not be solved, and one domain remains unclassified.

In accordance with the focus of this chapter on current proteins structure prediction methods, and despite some considerable improvements of *ab initio* methods (e.g. the fragment assembly approach ROSETTA [37], sometimes called "mini-threading approach" and an approach based exclusively on a global optimization of a potential energy function [35]) as compared to CASP2, CASP3 has shown no emphasis on "physical methods", i.e. methods concentrating on the balance of forces determining structure or folding pathways leading to native structures [109]. It is also apparent that currently the homology-based methods have much more predictive power and it seems to be like that for some time to come.

6.6
Conclusion: strengths and limitations

6.6.1
Threading

Threading allows, given a query sequence, to search for sequences with a similar fold but without apparent sequence similarity. Therefore, threading has three major objectives: first, provide orthogonal evidence of possible homology for distantly related protein sequences, second, detect possible homology in cases where sequence methods fail, and third, to improve structural models for the query sequence via structurally more accurate alignments.

Towards all three goals continuous improvements have been achieved during the up-to-date four rounds of the CASP experiment. The CASP experiments are the main indicator of the progress in protein structure prediction. Whereas the evaluation of progress is not as easy as first expected, it turns out that, for fold recognition, detectable progress can be observed from CASP1 to CASP2, where there is no significant progress between CASP2 and CASP3, but the same performance was achieved on more difficult targets. In CASP4, again, manually applied knowledge-based approaches to distant fold recognition outperformed automated methods and servers. Improvements have been made especially on the prediction of new folds where both manually (A. Murzin) and the *ab inito*/fragment assembly approach [5] performed well on a couple of targets with novel folds. The latter approach performed well on average also for fold recognition and comparative modeling targets [281]. But there appears to be a certain limit of current fold recognition methods, which is still well below the limit of detectable structural similarity (via structural comparisons). In addition, in CASP3 several groups produced reasonable models of up to 60 residues for *ab initio* target fragments, in CASP4, Baker could demonstrate good average performance of the fragment assembly approach.

We provide a summary of an estimate of the current state-of-the-art, in CASP3. Out of 43 protein targets, 15 could be classified as comparative homology modeling targets, i.e. related folds and accompanying alignments could be derived beyond doubt. For more than half of the 21 more difficult cases, a reasonable model could be predicted by at least one of the participating prediction teams. In addition, the CAFASP subsection of the assessment has demonstrated that 10 out of 19 folds could be solved via completely automatic application of the best threading methods. The complete set of results obtained during the CASP experiments, including the recent CASP4 assessment (summer/winter 2000), can be viewed and analyzed through the CASP Web page at the prediction center [341].

Methods for refining rough structural models towards the true native structure of the query protein are not straightforward, but are an active area of research [282].

6.6.2
Strengths

The strengths of structural predictions is that structural and in some cases functional classifications and annotations can be made, where such an annotation is not possible at all with pure sequence-based methods. In general, if there is detectable sequence similarity the corresponding annotation should overrule contradicting – even high confidence – structural predictions. In cases with insignificant sequence similarity, a significant threading alignment can provide additional evidence for an annotation. In addition to extending the range of remote homology searches, a knowledge-based structure prediction always results in a structural model, which can be inspected and evaluated with other means. In particular, such models can be completed to full atom models accessible to energy minimization and theoretical model validation procedures. The model can also trigger the validation and cross-checking with additional experimental facts and in turn can indicate further experiments either to validate the model or to check for putative functions of interactions with other partners.

In addition, experimental methods become available for checking predicted models in a cheap, fast and efficient manner. One such method uses chemical cross-links of a certain length and specificity and allows for validating distance constraints between residues and groups implied by a model. The method uses a protocol involving chemical cross-linking, digestion by specific hydrolases, measurement of peptide masses via (MALDI-TOF) mass spectrometry, inferring distance constraints for the protein under investigation and checking the experimental distances against the proposed model [32]. This appears to be a very promising approach to boost the significance and thus the practical application of theoretical structure prediction approaches, especially for proteins, which are not easily amenable to experimental structure determination, i.e. proteins difficult to crystallize, membrane proteins, or highly flexible proteins.

6.6.3
Limitations

Still structure prediction is an unsolved problem and the methods available are rough and heuristic. The outcome of predictions is not easily interpreted by the non-specialist and can easily be over-interpreted by biologists if taken too seriously. The computation of suitable p-values to allow for a straightforward judgment of the respective prediction remains problematic for many methods and parameter settings. In addition, models derived from similarities on the fold level often simply are not as structurally similar as one could hope for, i.e. not very usable for detailed structural analysis. Variations and differences apart from conserved structural cores have to be expected frequently even for highly significant structural predictions.

This is one reason why in general reliable function predictions cannot be inferred from structural models and mere significance of structural similarity at the level of folds. The current state-of-the-art of understanding structure-function relationships requires additional evidence to make such predictions. This, however, could be available via additional experimental measurements (e.g. expression data) or knowledge on biochemical and protein interaction networks.

Protein interaction/fusion:

1. Recently, the *Rosetta stone* method has been introduced. This method uses over 20 completely sequenced genomes and analyzes evolutionary correlations of two domains being fused into one protein in one species and occurring in separate proteins in another species. From these classifications the method establishes pairwise links between functionally related proteins [94] and elicits putative protein–protein interactions [95].

2. The *phylogenetic profile* method analyzes the co-occurrence of genes in the genomes of different organisms, for the same purpose [96].

One major reason for the currently not satisfying performance of knowledge-based prediction methods is the obvious deficiency in the currently used empirical scoring functions. As has been demonstrated there are many methods to improve the capabilities of the potentials by calibrating parameters or weightings of the scoring system [149]. This hints to clear deficiencies of the available potentials. As more data become available the database-derived potential can be expected to become more accurate.

The other major disadvantage today is that prediction methods do not efficiently exploit the available heterogeneous information on function, motifs, family information, network topology, sequence and structure space topology during the prediction process. This is clearly demonstrated by the fact that the prediction performance and alignment accuracy of human experts armed with the above mentioned prediction methods has been much better than automated versions of the methods alone [109, 283]

As has been briefly mentioned above the sensitivity of current methods can be expected to be quite low for the detection of related structural pair recognition and can be estimated at a quite satisfying performance of about 50% for finding a suitable structural template. The rather high error rate of around 50% has to be tolerated to achieve these sensitivities. This seems to be acceptable if the correctly identified models are new targets and of sufficient interest for further investigation to compensate for the false hints. If appropriate experimental assays are available, probably even a hit rate of 1 in 10 could be acceptable in target finding applications. As mentioned above, together with additional theoretical evaluation and evidence, and appropriate experimental techniques and validation, the success rate can be expected to be much higher.

6.7
Accessibility

Many of the resources, programs, databases, and servers discussed in this Chapter and relevant for protein structure prediction are freely available over the internet. The field, and thus the available resources, change quickly according to the progress made on collected experimental data and new methods. The Tables in the Appendix provide a current overview.

Table 6.1 contains a selection of protein structure resources and databases (coordinates, models, classifications, representative and benchmark sets) available on the WWW with the major protein structure archives and structure coordinate resources.

Table 6.2 lists a selection of tools for structure analysis, superposition, classification, and visualization.

Table 6.3 contains a list of structure prediction servers and programs. See also the CAFASP home page for further information and for results achieved for CASP4 targets and structures newly submitted to the PDB (check LiveBench [342] and EVA [343] for current updates and evaluations).

Acknowledgments

This work has been supported by the BMBF project TargId under contract no 0311615 and DFG-funded projects ZI 616/1-1 and ZI 616/2-1.

References

1 BRANDEN, C. and J. TOOZE, *Introduction to Protein Structure.* **1991**: Garland Publishing Inc.

2 CREIGHTON, T. E., *Proteins: Structures and Molecular Properties.* **1983**: W. H. Freeman.

3 ANFINSEN, C. B., *Principles that govern the folding of protein chains.* Science, **1973**. 181(96): p. 223–30.

4 LEVINTHAL, C., Chim. Phys., **1968**. 65: p. 44–45.

5 SIMONS, K. T., C. STRAUSS, and D. BAKER, Prospects for ab initio protein structural genomics. *J Mol Biol,* **2001**. 306(5): p. 1191–9.

6 GOTOH, O., An improved algorithm for matching biological sequences. *J Mol Biol,* **1982**. 162(3): p. 705–8.

7 NEEDLEMAN, S. B. and C. D. WUNSCH, A general method applicable to the search for similarities in the amino acid sequence of two proteins. *J Mol Biol,* **1970**. 48(3): p. 443–53.

8 SMITH, T. F. and M. S. WATERMAN, Identification of common molecular subsequences. *J Mol Biol,* **1981**. 147(1): p. 195–7.

9 ALTSCHUL, S. F., et al., Gapped BLAST and PSI-BLAST: a new generation of protein database search programs. *Nucleic Acids Res,* **1997**. 25(17): p. 3389–402.

10 ALTSCHUL, S. F. and E. V. KOONIN, Iterated profile searches with PSI-BLAST – a tool for discovery in protein databases. *Trends Biochem Sci,* **1998**. 23(11): p. 444–7.

11 ALTSCHUL, S. F., et al., Basic local alignment search tool. *J Mol Biol,* **1990**. 215(3): p. 403–10.

12 SCHAFFER, A. A., et al., IMPALA: matching a protein sequence against a collection of PSI-BLAST-constructed position-specific score matrices. *Bioinformatics,* **1999**. 15(12): p. 1000–11.

13 ZHANG, Z., et al., Protein sequence similarity searches using patterns as seeds. *Nucleic Acids Res,* **1998**. 26(17): p. 3986–90.

14 PEARSON, W. R. and D. J. LIPMAN, Improved tools for biological sequence comparison. *Proc Natl Acad Sci U S A,* **1988**. 85(8): p. 2444–8.

15 PEARSON, W. R., Flexible sequence similarity searching with the FASTA3 program package. *Methods Mol Biol,* **2000**. 132: p. 185–219.

16 PEARSON, W. R., Empirical statistical estimates for sequence similarity searches. *J Mol Biol,* **1998**. 276(1): p. 71–84.

17 PEARSON, W. R., Searching protein sequence libraries: comparison of the sensitivity and selectivity of the Smith-Waterman and FASTA algorithms. *Genomics,* **1991**. 11(3): p. 635–50.

18 PEARSON, W. R., Rapid and sensitive sequence comparison with FASTP and FASTA. *Methods Enzymol,* **1990**. 183: p. 63–98.

19 BARRETT, C., H. R., KARPLUS, K., Scoring hidden Markov models. *Comput Appl Biosci.,* **1997**. 13(2): p. 191–9.

20 PARK, J., et al., Sequence comparisons using multiple sequences detect three times as many remote homologues as pairwise methods. *J Mol Biol,* **1998**. 284(4): p. 1201–10.

21 KARCHIN, R. and R. HUGHEY, Weighting hidden Markov models for maximum discrimination. *Bioinformatics*, **1998**. 14(9): p. 772–82.

22 BATEMAN, A., et al., Pfam 3.1: 1313 multiple alignments and profile HMMs match the majority of proteins. *Nucleic Acids Res*, **1999**. 27(1): p. 260–2.

23 McCLURE, M. A., C. SMITH, and P. ELTON, Parameterization studies for the SAM and HMMER methods of hidden Markov model generation. *Ismb*, **1996**. 4: p. 155–64.

24 SONNHAMMER, E. L., et al., Pfam: multiple sequence alignments and HMM-profiles of protein domains. *Nucleic Acids Res*, **1998**. 26(1): p. 320–2.

25 EDDY, S. R., Profile hidden Markov models. *Bioinformatics*, **1998**. 14(9): p. 755–63.

26 EDDY, S. R., Hidden Markov models. *Curr Opin Struct Biol*, **1996**. 6(3): p. 361–5.

27 EDDY, S. R., Multiple alignment using hidden Markov models. *Ismb*, **1995**. 3: p. 114–20.

28 EDDY, S. R., G. MITCHISON, and R. DURBIN, Maximum discrimination hidden Markov models of sequence consensus. *J Comput Biol*, **1995**. 2(1): p. 9–23.

29 BROCCHIERI, L. and S. KARLIN, A symmetric-iterated multiple alignment of protein sequences. *J Mol Biol*, **1998**. 276(1): p. 249–64.

30 SIPPL, M. J., Calculation of conformational ensembles from potentials of mean force. An approach to the knowledge-based prediction of local structures in globular proteins. *J Mol Biol*, **1990**. 213(4): p. 859–83.

31 HENIKOFF, S., J. G. HENIKOFF, and S. PIETROKOVSKI, Blocks+: a non-redundant database of protein alignment blocks derived from multiple compilations. *Bioinformatics*, **1999**. 15(6): p. 471–9.

32 HOFFMANN, D. S., V. WEFING, STEPHAN, ALBRECHT, M., HANISCH, D., Zimmer, R., *A new method for the fast solution of Protein-3D-Structures, combining experiments and bioinformatics.* in *Caesarium 2000.* **2001**. Bonn: Springer Verlag.

33 HANISCH, D., R. ZIMMER and T. LENGAUER, *ProML – the Protein Markup Language for specification of protein sequences, structures and families,* Proceedings GCB2001, to appear.

34 MURZIN, A. G. and A. BATEMAN, Distant homology recognition using structural classification of proteins. *Proteins*, **1997**. Suppl(1): p. 105–12.

35 LEE, J., et al., Calculation of protein conformation by global optimization of a potential energy function. *Proteins*, **1999**. Suppl(3): p. 204–8.

36 SIMONS, K. T., et al., Assembly of protein tertiary structures from fragments with similar local sequences using simulated annealing and Bayesian scoring functions. *J Mol Biol*, **1997**. 268(1): p. 209–25.

37 SIMONS, K. T., et al., Ab initio protein structure prediction of CASP III targets using ROSETTA. *Proteins*, **1999**. 37(S3): p. 171–176.

38 LATHROP, R. H., The protein threading problem with

sequence amino acid interaction preferences is NP-complete. *Protein Eng*, **1994**. 7(9): p. 1059–68.

39 CHOTHIA, C., Principles that determine the structure of proteins. *Annu Rev Biochem*, **1984**. 53: p. 537–72.

40 RICHARDSON, J. S., The anatomy and taxonomy of protein structure. *Adv Protein Chem*, **1981**. 34: p. 167–339.

41 PAULING, L. C. and R. B. COREY, Configurations of polypeptide chains with favored orientations around single bonds: two new pleated sheets. *PNAS*, **1951**. 37: p. 729–740.

42 PAULING, L. C., R. B. COREY and H. R. BRANSON, The structure of proteins: two hydrogen-bonded helical configurations of the polypeptide chain. *PNAS*, **1951**. 37: p. 205–211.

43 LO CONTE, L., et al., SCOP: a structural classification of proteins database. *Nucleic Acids Res*, **2000**. 28(1): p. 257–9.

44 BRAY, J. E., et al., The CATH Dictionary of Homologous Superfamilies (DHS): a consensus approach for identifying distant structural homologues. *Protein Eng*, **2000**. 13(3): p. 153–65.

45 ORENGO, C. A., et al., CATH – a hierarchic classification of protein domain structures. *Structure*, **1997**. 5(8): p. 1093–108.

46 CHOU, P. Y. and G. D. FASMAN, Conformational parameters for amino acids in helical, beta-sheet, and random coil regions calculated from proteins. *Biochemistry*, **1974**. 13(2): p. 211–22.

47 LIM, V. I., Structural principles of the globular organization of protein chains. A stereochemical theory of globular protein secondary structure. *J Mol Biol*, **1974**. 88(4): p. 857–72.

48 GARNIER, J., D. J. OSGUTHORPE, and B. ROBSON, Analysis of the accuracy and implications of simple methods for predicting the secondary structure of globular proteins. *J Mol Biol*, **1978**. 120(1): p. 97–120.

49 SHOEMAKER, B. A., J. J. PORTMAN, and P. G. WOLYNES, Speeding molecular recognition by using the folding funnel: the fly-casting mechanism. *Proc Natl Acad Sci U S A*, **2000**. 97(16): p. 8868–73.

50 HARDIN, C., et al., Associative memory hamiltonians for structure prediction without homology: alpha-helical proteins. *Proc Natl Acad Sci U S A*, **2000**. 97(26): p. 14235–40.

51 SOCCI, N. D., J. N. ONUCHIC, and P. G. WOLYNES, Protein folding mechanisms and the multidimensional folding funnel. *Proteins*, **1998**. 32(2): p. 136–58.

52 ONUCHIC, J. N., Z. LUTHEY-SCHULTEN, and P. G. WOLYNES, Theory of protein folding: the energy landscape perspective. *Annu Rev Phys Chem*, **1997**. 48: p. 545–600.

53 WOLYNES, P. G., Folding funnels and energy landscapes of larger proteins within the capillarity approximation. *Proc Natl Acad Sci U S A*, **1997**. 94(12): p. 6170–5.

54 WOLYNES, P., Z. LUTHEY-SCHULTEN, and J. ONUCHIC, Fast-folding experiments and the topography of protein folding energy landscapes. *Chem Biol*, **1996**. 3(6): p. 425–32.

55 ONUCHIC, J. N., et al., Protein folding funnels: the nature of the transition state ensemble. *Fold Des*, **1996**. 1(6): p. 441–50.

56 ONUCHIC, J. N., et al., Toward an outline of the topography of a realistic protein-folding funnel. *Proc Natl Acad Sci U S A*, **1995**. 92(8): p. 3626–30.

57 DINNER, A. R., et al., Understanding protein folding via free-energy surfaces from theory and experiment. *Trends Biochem Sci*, **2000**. 25(7): p. 331–9.

58 LAZARIDIS, T. and M. KARPLUS, Effective energy functions for protein structure prediction. *Curr Opin Struct Biol*, **2000**. 10(2): p. 139–45.

59 DINNER, A. R., et al., Factors that affect the folding ability of proteins. *Proteins*, **1999**. 35(1): p. 34–40.

60 CHAN, H. S. and K. A. DILL, The protein folding problem. *Physics Today*, **1993**. 46(2): p. 24–32.

61 DILL, K. A. and H. S. CHAN, From Levinthal to pathways to funnels. *Nat Struct Biol*, **1997**. 4(1): p. 10–9.

62 CHAN, H. S. and K. A. DILL, Protein folding in the landscape perspective: chevron plots and non-Arrhenius kinetics. *Proteins*, **1998**. 30(1): p. 2–33.

63 DILL, K. A., et al., Principles of protein folding – a perspective from simple exact models. *Protein Sci*, **1995**. 4(4): p. 561–602.

64 YUE, K., et al., A test of lattice protein folding algorithms. *Proc Natl Acad Sci U S A*, **1995**. 92(1): p. 325–9.

65 RECKTENWALD, A., D. SCHOMBURG, and R. D. SCHMID, Protein engineering and design. Method and the industrial relevance. *J Biotechnol*, **1993**. 28(1): p. 1–23.

66 AEHLE, W., et al., Rational protein engineering and industrial application: structure prediction by homology and rational design of protein-variants with improved 'washing performance' – the alkaline protease from Bacillus alcalophilus. *J Biotechnol*, **1993**. 28(1): p. 31–40.

67 BORNSCHEUER, U. T. and M. POHL, Improved biocatalysts by directed evolution and rational protein design. *Curr Opin Chem Biol*, **2001**. 5(2): p. 137–43.

68 ROOT, M. J., M. S. KAY, and P. S. KIM, Protein Design of an HIV-1 Entry Inhibitor. *Science*, **2001**. 11: p. 11.

69 GUEROIS, R. and L. SERRANO, Protein design based on folding models. *Curr Opin Struct Biol*, **2001**. 11(1): p. 101–6.

70 KARLIN, S. and S. F. ALTSCHUL, Methods for assessing the statistical significance of molecular sequence features by using general scoring schemes. *Proc Natl Acad Sci U S A*, **1990**. 87(6): p. 2264–8.

71 KARLIN, S. and S. F. ALTSCHUL, Applications and statistics for multiple high-scoring segments in molecular sequences. *Proc Natl Acad Sci U S A*, **1993**. 90(12): p. 5873–7.

72 ROST, B., C. SANDER, and R. SCHNEIDER, PHD – an automatic mail server for protein secondary structure prediction. *Comput Appl Biosci*, **1994**. 10(1): p. 53–60.

73 BATEMAN, A., et al., The Pfam protein families database. *Nucleic Acids Res*, **2000**. 28(1): p. 263–6.

74 SANDER, C. and R. SCHNEIDER, Database of homology-derived protein structures and the structural meaning of sequence alignment. *Proteins*, **1991**. 9(1): p. 56–68.

75 CORPET, F., et al., ProDom and ProDom-CG: tools for protein domain analysis and whole genome comparisons. *Nucleic Acids Res*, **2000**. 28(1): p. 267–9.

76 APWEILER, R., et al., InterPro – an integrated documentation resource for protein families, domains and functional sites. *Bioinformatics*, **2000**. 16(12): p. 1145–50.

77 HOLM, L. and C. SANDER, Touring protein fold space with Dali/FSSP. *Nucleic Acids Res*, **1998**. 26(1): p. 316–9.

78 DIETMANN, S., et al., A fully automatic evolutionary classification of protein folds: Dali Domain Dictionary version 3. *Nucleic Acids Res*, **2001**. 29(1): p. 55–7.

79 SIDDIQUI, A. S., U. DENGLER, and G. J. BARTON, 3Dee: a database of protein structural domains. *Bioinformatics*, **2001**. 17(2): p. 200–1.

80 DENGLER, U., A. S. SIDDIQUI, and G. J. BARTON, Protein structural domains: analysis of the 3Dee domains database. *Proteins*, **2001**. 42(3): p. 332–44.

81 MURZIN, A. G., et al., SCOP: a structural classification of proteins database for the investigation of sequences and structures. *J Mol Biol*, **1995**. 247(4): p. 536–40.

82 HUBBARD, T. J., et al., SCOP, Structural Classification of Proteins database: applications to evaluation of the effectiveness of sequence alignment methods and statistics of protein structural data. *Acta Crystallogr D Biol Crystallogr*, **1998**. 54(1 (Pt 6)): p. 1147–54.

83 ATTWOOD, T. K., et al., PRINTS-S: the database formerly known as PRINTS. *Nucleic Acids Res*, **2000**. 28(1): p. 225–7.

84 HUANG, J. Y. and D. L. BRUTLAG, The EMOTIF database. *Nucleic Acids Res*, **2001**. 29(1): p. 202–4.

85 KASUYA, A. and J. M. THORNTON, Three-dimensional structure analysis of PROSITE patterns. *J Mol Biol*, **1999**. 286(5): p. 1673–91.

86 BAIROCH, A., PROSITE: a dictionary of sites and patterns in proteins. *Nucleic Acids Res*, **1991**. 19 Suppl: p. 2241–5.

87 HOFMANN, K., et al., The PROSITE database, its status in 1999. *Nucleic Acids Res*, **1999**. 27(1): p. 215–9.

88 ITO, T., et al., A comprehensive two-hybrid analysis to explore the yeast protein interactome. *Proc Natl Acad Sci U S A*, **2001**. 98(8): p. 4569–74.

89 UETZ, P., et al., A comprehensive analysis of protein-protein interactions in Saccharomyces cerevisiae. *Nature*, **2000**. 403(6770): p. 623–7.

90 UETZ, P. and R. E. HUGHES, Systematic and large-scale two-hybrid screens. *Curr Opin Microbiol*, **2000**. 3(3): p. 303–8.

91 CAGNEY, G., P. UETZ, and S. FIELDS, High-throughput screening for protein-protein interactions using two-hybrid assay. *Methods Enzymol*, **2000**. 328: p. 3–14.

92 XENARIOS, I., et al., DIP: the database of interacting proteins. *Nucleic Acids Res*, **2000**. 28(1): p. 289–91.

93 XENARIOS, I., et al., DIP: The Database of Interacting Proteins: 2001 update. *Nucleic Acids Res*, **2001**. 29(1): p. 239–41.

94 MARCOTTE, E. M., et al., A combined algorithm for genome-wide prediction of protein function. *Nature*, **1999**. 402(6757): p. 83–6.

95 MARCOTTE, E. M., et al., Detecting protein function and protein-protein interactions from genome sequences. *Science*, **1999**. 285(5428): p. 751 3.

96 PELLEGRINI, M., et al., Assigning protein functions by comparative genome analysis: protein phylogenetic profiles. *Proc Natl Acad Sci U S A*, **1999**. 96(8): p. 4285–8.

97 DIAMOND, R., On the multiple simultaneous superposition of molecular structures by rigid body transformations. *Protein Sci*, **1992**. 1(10): p. 1279–87.

98 SIPPL, M. J., On the problem of comparing protein structures. Development and applications of a new method for the assessment of structural similarities of polypeptide conformations. *J Mol Biol*, **1982**. 156(2): p. 359–88.

99 KABSCH, W., A solution for the best rotation to relate two sets of vectors. *Acta Crystallogr*, **1976**. A32: p. 922–923.

100 ORENGO, C. A. and W. R. TAYLOR, SSAP: sequential structure alignment program for protein structure comparison. *Methods Enzymol*, **1996**. 266: p. 617–35.

101 TAYLOR, W. R., T. P. FLORES, and C. A. ORENGO, Multiple protein structure alignment. *Protein Sci*, **1994**. 3(10): p. 1858–70.

102 HOLM, L. and C. SANDER, Dictionary of recurrent domains in protein structures. *Proteins*, **1998**. 33(1): p. 88–96.

103 ALEXANDROV, N. N., SARFing the PDB. *Protein Eng*, **1996**. 9(9): p. 727–32.

104 ALEXANDROV, N. N. and D. FISCHER, Analysis of topological and nontopological structural similarities in the PDB: new examples with old structures. *Proteins*, **1996**. 25(3): p. 354–65.

105 FISCHER, D., et al., An efficient automated computer vision based technique for detection of three dimensional structural motifs in proteins. *J Biomol Struct Dyn*, **1992**. 9(4): p. 769–89.

106 SHATSKY, M., et al., Alignment of flexible protein structures. *Proc Int Conf Intell Syst Mol Biol*, **2000**. 8: p. 329–43.

107 BACHAR, O., et al., A computer vision based technique for 3-D sequence-independent structural comparison of proteins. *Protein Eng*, **1993**. 6(3): p. 279–88.

108 FALICOV, A. and F. E. COHEN, A surface of minimum area metric for the structural comparison of proteins. *J Mol Biol*, **1996**. 258(5): p. 871–92.

109 MOULT, J., et al., Critical assessment of methods of protein structure prediction (CASP): round III. *Proteins*, **1999**. Suppl(3): p. 2–6.

110 MARCHLER-BAUER, A. and S. H. BRYANT, Measures of threading specificity and accuracy. *Proteins*, **1997**. Suppl(1): p. 74–82.

111 LACKNER, P., et al., Automated large scale evaluation of protein structure predictions. *Proteins*, **1999**. 37(S3): p. 7–14.

112 MIZUGUCHI, K., et al., HOMSTRAD: a database of protein structure alignments for homologous families. *Protein Sci*, **1998**. 7(11): p. 2469–71.

113 MIZUGUCHI, K., et al., JOY: protein sequence-structure representation and analysis. *Bioinformatics*, **1998**. 14(7): p. 617–23.

114 SOWDHAMINI, R., et al., Protein three-dimensional structural databases: domains, structurally aligned homologues and superfamilies. *Acta Crystallogr D Biol Crystallogr*, **1998**. 54(1 (Pt 6)): p. 1168–77.

115 SCHMIDT, R., M. GERSTEIN, and R. B. ALTMAN, LPFC: an Internet library of protein family core structures. *Protein Sci*, **1997**. 6(1): p. 246–8.

116 HADLEY, C. and D. T. JONES, A systematic comparison of protein structure classifications: SCOP, CATH and FSSP. *Structure Fold Des*, **1999**. 7(9): p. 1099–112.

117 WODAK, S. J. and J. JANIN, Location of structural domains in protein. *Biochemistry*, **1981**. 20(23): p. 6544–52.

118 HOLM, L. and C. SANDER, Parser for protein folding units. *Proteins*, **1994**. 19(3): p. 256–68.

119 SWINDELLS, M. B., A procedure for detecting structural domains in proteins. *Protein Sci*, **1995**. 4(1): p. 103–12.

120 ROSE, G. D., Automatic recognition of domains in globular proteins. *Methods Enzymol*, **1985**. 115: p. 430–40.

121 JONES, S., et al., Domain assignment for protein structures using a consensus approach: characterization and analysis. *Protein Sci*, **1998**. 7(2): p. 233–42.

122 TAYLOR, W. R., Protein structural domain identification. *Protein Eng*, **1999**. 12(3): p. 203–16.

123 MURVAI, J., K. VLAHOVICEK, and S. PONGOR, A simple probabilistic scoring method for protein domain identification. *Bioinformatics*, **2000**. 16(12): p. 1155–6.

124 SIPPL, M. J., et al., Helmholtz free energies of atom pair interactions in proteins. *Fold Des*, **1996**. 1(4): p. 289–98.

125 HENDLICH, M., et al., Identification of native protein folds amongst a large number of incorrect models. The calculation of low energy conformations from potentials of mean force. *J Mol Biol*, **1990**. 216(1): p. 167–80.

126 BAUER, A. and A. BEYER, An improved pair potential to recognize native protein folds. *Proteins*, **1994**. 18(3): p. 254–61.

127 THOMAS, P. D. and K. A. DILL, Statistical potentials extracted from protein structures: how accurate are they? *J Mol Biol*, **1996**. 257(2): p. 457–69.

128 THOMAS, P. D. and K. A. DILL, An iterative method for extracting energy-like quantities from protein structures. *Proc Natl Acad Sci U S A*, **1996**. 93(21): p. 11628–33.

129 RUSSELL, R. B., et al., Recognition of analogous and homologous protein folds – assessment of prediction success and associated alignment accuracy using empirical substitution matrices. *Protein Eng*, **1998**. 11(1): p. 1–9.

130 RICE, D. W. and D. EISENBERG, A 3D-1D substitution matrix for protein fold recognition that includes predicted secondary structure of the sequence. *J Mol Biol*, **1997**. 267(4): p. 1026–38.

131 REVA, B. A., et al., Residue-residue mean-force potentials for protein structure recognition. *Protein Eng*, **1997**. 10(8): p. 865–76.

132 MELO, F. and E. FEYTMANS, Novel knowledge-based mean force potential at atomic level. *J Mol Biol*, **1997**. 267(1): p. 207–22.

133 JONES, D. T. and J. M. THORNTON, Potential energy functions for threading. *Curr Opin Struct Biol*, **1996**. 6(2): p. 210–6.

134 VAJDA, S., M. SIPPL, and J. NOVOTNY, Empirical potentials and functions for protein folding and binding. *Curr Opin Struct Biol*, **1997**. 7(2): p. 222–8.

135 MOULT, J., Comparison of database potentials and molecular mechanics force fields. *Curr Opin Struct Biol*, **1997**. 7(2): p. 194–9.

136 BOWIE, J. U., R. LUTHY, and D. EISENBERG, A method to identify protein sequences that fold into a known three-dimensional structure. *Science*, **1991**. 253(5016): p. 164–70.

137 ALEXANDROV, N. N., R. NUSSINOV, and R. M. ZIMMER, Fast protein fold recognition via sequence to structure alignment and contact capacity potentials. *Pac Symp Biocomput*, **1996**: p. 53–72.

138 ZIMMER, R., M. WOHLER, and R. THIELE, New scoring schemes for protein fold recognition based on Voronoi contacts. *Bioinformatics*, **1998**. 14(3): p. 295–308.

139 DAYHOFF, M., R. M. SCHWARTZ, and B. C. ORCUTT, *A model of evolutionary change in proteins,* in *Atlas of protein sequence and structure*. **1978**. p. 345–352.

140 GONNET, G. H., M. A. COHEN, and S. A. BENNER, Exhaustive matching of the entire protein sequence database. *Science*, **1992**. 256(5062): p. 1443–5.

141 HENIKOFF, J. G. and S. HENIKOFF, Blocks database and its applications. *Methods Enzymol*, **1996**. 266: p. 88–105.

142 HENIKOFF, S. and J. G. HENIKOFF, Performance evaluation of amino acid substitution matrices. *Proteins*, **1993**. 17(1): p. 49–61.

143 ALTSCHUL, S. F., A protein alignment scoring system sensitive at all evolutionary distances. *J Mol Evol*, **1993**. 36(3): p. 290–300.

144 ALTSCHUL, S. F., Amino acid substitution matrices from an information theoretic perspective. *J Mol Biol*, **1991**. 219(3): p. 555–65.

145 KANN, M. and R. A. GOLDSTEIN, *Optimizing for success: A new score function for distantly related protein sequence comparison.* in *RECOMB2000*. **2000**. Tokyo: ACM Press.

146 MAIOROV, V. N. and G. M. CRIPPEN, Contact potential that recognizes the correct folding of globular proteins. *J Mol Biol*, **1992**. 227(3): p. 876–88.

147 JONES, D. T., GenTHREADER: an efficient and reliable protein fold recognition method for genomic sequences. *J Mol Biol*, **1999**. 287(4): p. 797–815.

148 LEMMEN, C., et al., Application of parameter optimization to molecular comparison problems. *Pac Symp Biocomput*, **1999**: p. 482–93.

149 ZIEN, A., R. ZIMMER, and T. LENGAUER, A simple iterative approach to parameter optimization. *J Comput Biol*, **2000**. 7(3–4): p. 483–501.

150 CHOU, P. Y. and G. D. FASMAN, Prediction of the secondary structure of proteins from their amino acid sequence. *Adv Enzymol Relat Areas Mol Biol*, **1978**. 47: p. 45–148.

151 PTITSYN, O. B., et al., Prediction of the secondary structure of the L7, L12 proteins of the E. coli ribosome. *FEBS Lett*, **1973**. 34(1): p. 55–7.

152 LEVIN, J. M., B. ROBSON, and J. GARNIER, An algorithm for secondary structure determination in proteins based on sequence similarity. *FEBS Lett*, **1986**. 205(2): p. 303–8.

153 GARNIER, J., J. F. GIBRAT, and B. ROBSON, GOR method for predicting protein secondary structure from amino acid sequence. *Methods Enzymol*, **1996**. 266: p. 540–53.

154 BIOU, V., et al., Secondary structure prediction: combination of three different methods. *Protein Eng*, **1988**. 2(3): p. 185–91.

155 ROST, B. and C. SANDER, Prediction of protein secondary structure at better than 70% accuracy. *J Mol Biol*, **1993**. 232(2): p. 584–99.

156 JONES, D. T., Protein secondary structure prediction based on position-specific scoring matrices. *J Mol Biol*, **1999**. 292(2): p. 195–202.

157 JONES, D. T., et al., Successful recognition of protein folds using threading methods biased by sequence similarity and predicted secondary structure. *Proteins*, **1999**. 37(S3): p. 104–111.

158 GEOURJON, C. and G. DELEAGE, SOPMA: significant improvements in protein secondary structure prediction by consensus prediction from multiple alignments. *Comput Appl Biosci*, **1995**. 11(6): p. 681–4.

159 CUFF, J. A. and G. J. BARTON, Application of multiple sequence alignment profiles to improve protein secondary structure prediction. *Proteins*, **2000**. 40(3): p. 502–11.

160 CUFF, J. A., et al., JPred: a consensus secondary structure prediction server. *Bioinformatics*, **1998**. 14(10): p. 892–3.

161 SELBIG, J., T. MEVISSEN, and T. LENGAUER, Decision tree-based formation of consensus protein secondary structure prediction. *Bioinformatics*, **1999**. 15(12): p. 1039–46.

162 CHRISTIANINI, N., SHAWE-TAYLOR, J., *An introduction to support vector machines*. **2000**: Cambridge University Press.

163 ROST, B., TOPITS: threading one-dimensional predictions into three-dimensional structures. *Proc Int Conf Intell Syst Mol Biol*, **1995**. 3: p. 314–21.

164 RUSSELL, R. B., R. R. COPLEY, and G. J. BARTON, Protein fold recognition by mapping predicted secondary structures. *J Mol Biol*, **1996**. 259(3): p. 349–65.

165 ALEXANDROV, N. N. and V. V. SOLOVYEV, Statistical significance of ungapped sequence alignments. *Pac Symp Biocomput*, **1998**: p. 463–72.

166 CUFF, J. A. and G. J. BARTON, Evaluation and improvement of multiple sequence methods for protein secondary structure prediction. *Proteins*, **1999**. 34(4): p. 508–19.

167 KING, R. D. and M. J. STERNBERG, Identification and application of the concepts important for accurate and reliable protein secondary structure prediction. *Protein Sci*, **1996**. 5(11): p. 2298–310.

168 ROST, B., C. SANDER, and R. SCHNEIDER, Redefining the goals of protein secondary structure prediction. *J Mol Biol*, **1994**. 235(1): p. 13–26.

169 ZEMLA, A., et al., A modified definition of Sov, a segment-based measure for protein secondary structure prediction assessment. *Proteins*, **1999**. 34(2): p. 220–3.

170 ZIMMER, R. and R. THIELE, *Fast Fold Recognition and accurate sequence-structure alignment*. in *GCB96*. **1997**: Springer Verlag.

171 WATERMAN, M. S., *Introduction to Computational Biology*. **1995**: Chapman & Hall.

172 UNGER, R. and J. MOULT, Finding the lowest free energy conformation of a protein is an NP-hard problem: proof and implications. *Bull Math Biol*, **1993**. 55(6): p. 1183–98.

173 UNGER, R. and J. MOULT, Genetic algorithms for protein folding simulations. *J Mol Biol*, **1993**. 231(1): p. 75–81.

174 FRAENKEL, A. S., Complexity of protein folding. *Bull Math Biol*, **1993**. 55(6): p. 1199–210.

175 LATHROP, R. H. and T. F. SMITH, Global optimum protein threading with gapped alignment and empirical pair score functions. *J Mol Biol*, **1996**. 255(4): p. 641–65.

176 GRIBSKOV, M., A. D. McLACHLAN, and D. EISENBERG, Profile analysis: detection of distantly related proteins. *Proc Natl Acad Sci U S A*, **1987**. 84(13): p. 4355–8.

177 HENIKOFF, J. G. and S. HENIKOFF, Using substitution probabilities to improve position-specific scoring matrices. *Comput Appl Biosci*, **1996**. 12(2): p. 135–43.

178 NEUWALD, A. F., et al., Extracting protein alignment models from the sequence database. *Nucleic Acids Res*, **1997**. 25(9): p. 1665–77.

179 KROGH, A., et al., Hidden Markov models in computational biology. Applications to protein modeling. *J Mol Biol*, **1994**. 235(5): p. 1501–31.

180 TATUSOV, R. L. K., E. V., LIPMAN, D. J., A genomic perspective on protein families. *Science*, **1997**. 278(5338): p. 631–7.

181 ARAVIND, L. and E. V. KOONIN, Gleaning non-trivial structural, functional and evolutionary information about proteins by iterative database searches. *J Mol Biol*, **1999**. 287(5): p. 1023–40.

182 ABAGYAN, R. A. and S. BATALOV, Do aligned sequences share the same fold? *J Mol Biol*, **1997**. 273(1): p. 355–68.

183 KARPLUS, K., et al., Predicting protein structure using only sequence information. *Proteins*, **1999**. 37(S3): p. 121–125.

184 KORETKE, K. K., et al., Fold recognition using sequence and secondary structure information. *Proteins*, **1999**. 37(S3): p. 141–148.

185 JONES, D. T., Progress in protein structure prediction. *Curr Opin Struct Biol*, **1997**. 7(3): p. 377–87.

186 SIPPL, M. J. and H. FLOCKNER, Threading thrills and threats. *Structure*, **1996**. 4(1): p. 15–9.

187 THIELE, R., R. ZIMMER, and T. LENGAUER, Recursive dynamic programming for adaptive sequence and structure alignment. *Proc Int Conf Intell Syst Mol Biol*, **1995**. 3: p. 384–92.

188 THIELE, R., R. ZIMMER, and T. LENGAUER, Protein threading by recursive dynamic programming. *J Mol Biol*, **1999**. 290(3): p. 757–79.

189 GODZIK, A., A. KOLINSKI, and J. SKOLNICK, Topology fingerprint approach to the inverse protein folding problem. *J Mol Biol*, **1992**. 227(1): p. 227–38.

190 JONES, D. T., W. R. TAYLOR, and J. M. THORNTON, A new approach to protein fold recognition. *Nature*, **1992**. 358(6381): p. 86–9.

191 WESTHEAD, D. R., et al., Protein fold recognition by threading: comparison of algorithms and analysis of results. *Protein Eng*, **1995**. 8(12): p. 1197–1204.

192 FISCHER, D., et al., Assigning amino acid sequences to 3-dimensional protein folds. *Faseb J*, **1996**. 10(1): p. 126–36.

193 LUTHY, R., J. U. BOWIE, and D. EISENBERG, Assessment of protein models with three-dimensional profiles. *Nature*, **1992**. 356(6364): p. 83–5.

194 ZHANG, K. Y. and D. EISENBERG, The three-dimensional profile method using residue preference as a continuous function of residue environment. *Protein Sci*, **1994**. 3(4): p. 687–95.

195 PANCHENKO, A. R., A. MARCHLER-BAUER, and S. H. BRYANT, Combination of threading potentials and sequence profiles improves fold recognition. *J Mol Biol*, **2000**. 296(5): p. 1319–31.

196 MURZIN, A. G., Structure classification-based assessment of CASP3 predictions for the fold recognition targets. *Proteins*, **1999**. 37(S3): p. 88–103.

197 PANCHENKO, A., A. MARCHLER-BAUER, and S. H. BRYANT, Threading with explicit models for evolutionary conservation of structure and sequence. *Proteins*, **1999**. Suppl(3): p. 133–40.

198 ELOFSSON, A., et al., A study of combined structure/sequence profiles. *Fold Des*, **1996**. 1(6): p. 451–61.

199 FISCHER, D. and D. EISENBERG, Protein fold recognition using sequence-derived predictions. *Protein Sci*, **1996**. 5(5): p. 947–55.

200 JAROSZEWSKI, L., et al., Fold prediction by a hierarchy of sequence, threading, and modeling methods. *Protein Sci*, **1998**. 7(6): p. 1431–40.

201 ROST, B., R. SCHNEIDER, and C. SANDER, Protein fold recognition by prediction-based threading. *J Mol Biol*, **1997**. 270(3): p. 471–80.

202 YI, T. M. and E. S. LANDER, Recognition of related proteins by iterative template refinement (ITR). *Protein Sci*, **1994**. 3(8): p. 1315–28.

203 DEFAY, T. R. and F. E. COHEN, Multiple sequence information for threading algorithms. *J Mol Biol*, **1996**. 262(2): p. 314–23.

204 VON ÖHSEN, N. and R. ZIMMER, Improving profile-profile alignment via log average scoring. Algorithms in Bioinformatics (WABI 2001), Springer LNCS 2149, p. 11–26.

205 KELLEY, L. A., R. M. MACCALLUM, and M. J. STERNBERG, Enhanced genome annotation using structural profiles in the program 3D-PSSM. *J Mol Biol*, **2000**. 299(2): p. 499–520.

206 HENIKOFF, S. and J. G. HENIKOFF, Position-based sequence weights. *J Mol Biol*, **1994**. 243(4): p. 574–8.

207 ZIMMER, R. and J. HALFMANN, *Optimizing Homology Models of protein structures with potentials of mean force,* in *Monte Carlo Approach to Biopolymers and protein Folding*, P. B. Grassberger, G., Nadler, W., Editor. **1998**, World Scientific. p. 45–68.

208 Bryant, S. H. and C. E. Lawrence, An empirical energy function for threading protein sequence through the folding motif. *Proteins*, **1993**. 16(1): p. 92–112.

209 Bryant, S. H. and S. F. Altschul, Statistics of sequence-structure threading. *Curr Opin Struct Biol*, **1995**. 5(2): p. 236–44.

210 Xu, Y. and D. Xu, Protein threading using PROSPECT: design and evaluation. *Proteins*, **2000**. 40(3): p. 343–54.

211 Xu, Y. X., D., Uberbacher, E. C., An efficient computational method for globally optimal threading. *Journal of Computational Biology*, **1998**. 5: p. 597–614.

212 Xu, Y. X., D., Crawford, O. H., Einstein, J. R., A computational method for NMR-constrained protein threading. *Journal of Computational Biology*, **2000**. 7: p. 449–467.

213 Xu, D., O. H. Crawford, P. F. LoCascio, and Y. Xu, Application of PROSPECT in CASP4: Characterizing protein strctures with new folds. *Proteins*, **2001**. to appear (Special issue).

214 Kirkpatrick, S., C. D. Gelatt, and M. P. Vecchi, Optimization by simulated annealing. *Science*, **1983**. 220(4598): p. 671–680.

215 Otten, R. H. J. M. and L. P. P. P. van Ginneken, *The annealing algorithm*. **1989**: Kluwer Academic.

216 Sali, A. and T. L. Blundell, Definition of general topological equivalence in protein structures. A procedure involving comparison of properties and relationships through simulated annealing and dynamic programming. *J Mol Biol*, **1990**. 212(2): p. 403–28.

217 Ishikawa, M., et al., Multiple sequence alignment by parallel simulated annealing. *Comput Appl Biosci*, **1993**. 9(3): p. 267–73.

218 Kalos, M. H. and P. A. Whitlock, *Monte Carlo Methods*. **1986**: John Wiley & Sons, Inc.

219 Grassberger, P. B., G., Nadler, W., *Monte Carlo Approach to Biopolymers and Protein Folding*. **1998**: World Scientific Publ. Co.

220 Guda, C., et al., A new algorithm for the alignment of multiple protein structures using Monte Carlo optimization. *Pac Symp Biocomput*, **2001**: p. 275–86.

221 Ortiz, A. R., A. Kolinski, and J. Skolnick, Tertiary structure prediction of the KIX domain of CBP using Monte Carlo simulations driven by restraints derived from multiple sequence alignments. *Proteins*, **1998**. 30(3): p. 287–94.

222 Ortiz, A. R., A. Kolinski, and J. Skolnick, Nativelike topology assembly of small proteins using predicted restraints in Monte Carlo folding simulations. *Proc Natl Acad Sci U S A*, **1998**. 95(3): p. 1020–5.

223 Goldberg, D. E., *Genetic Algorithms in Search, Optimization, and Machine Learning*. **1988**: Addison Wesley.

224 Pedersen, J. T. and J. Moult, Genetic algorithms for protein structure prediction. *Curr Opin Struct Biol*, **1996**. 6(2): p. 227–31.

225 May, A. C. and M. S. Johnson, Improved genetic algorithm-based protein structure comparisons: pairwise and multiple superpositions. *Protein Eng*, **1995**. 8(9): p. 873–82.

226 ZHANG, C. and A. K. WONG, A genetic algorithm for multiple molecular sequence alignment. *Comput Appl Biosci*, **1997**. 13(6): p. 565–81.

227 SZUSTAKOWSKI, J. D. and Z. WENG, Protein structure alignment using a genetic algorithm. *Proteins*, **2000**. 38(4): p. 428–40.

228 METROPOLIS, N. R., A. W., ROSENBLUTH, M. N., TELLER, A. H., TELLER, E., Equation of state calculations by fast computing machines. *J. Chem. Phys.*, **1953**. 21(6): p. 1087–1093.

229 SIPPL, M., et al., *Assessment of the CASP4 fold recognition category*, in *to appear in Proteins*. **2001**.

230 GODZIK, A. and J. SKOLNICK, Sequence-structure matching in globular proteins: application to supersecondary and tertiary structure determination. *Proc Natl Acad Sci U S A*, **1992**. 89(24): p. 12098–102.

231 TAYLOR, W. R. and C. A. ORENGO, A holistic approach to protein structure alignment. *Protein Eng*, **1989**. 2(7): p. 505–19.

232 TAYLOR, W. R. and C. A. ORENGO, Protein structure alignment. *J Mol Biol*, **1989**. 208(1): p. 1–22.

233 GREER, J., Comparative model-buildung of the mammalian serine proteases. *JMB*, **1981**. 153: p. 1027–1042.

234 BRYANT, S. H., Evaluation of threading specificity and accuracy. *Proteins*, **1996**. 26(2): p. 172–85.

235 BARRETT, C., R. HUGHEY, and K. KARPLUS, Scoring hidden Markov models. *Comput Appl Biosci*, **1997**. 13(2): p. 191–9.

236 KARPLUS, K., C. BARRETT, and R. HUGHEY, Hidden Markov models for detecting remote protein homologies. *Bioinformatics*, **1998**. 14(10): p. 846–56.

237 KARPLUS, K., et al., Predicting protein structure using hidden Markov models. *Proteins*, **1997**. Suppl(1): p. 134–9.

238 KIM, S. H., Shining a light on structural genomics. *Nat Struct Biol*, **1998**. 5 Suppl: p. 643–5.

239 MONTELIONE, G. T. and S. ANDERSON, Structural genomics: keystone for a Human Proteome Project [news]. *Nat Struct Biol*, **1999**. 6(1): p. 11–2.

240 ORENGO, C. A., D. T. JONES, and J. M. THORNTON, Protein superfamilies and domain superfolds. *Nature*, **1994**. 372(6507): p. 631–4.

241 FISCHER, D. and D. EISENBERG, Assigning folds to the proteins encoded by the genome of Mycoplasma genitalium. *Proc Natl Acad Sci U S A*, **1997**. 94(22): p. 11929–34.

242 HUYNEN, M., et al., Homology-based fold predictions for Mycoplasma genitalium proteins. *J Mol Biol*, **1998**. 280(3): p. 323–6.

243 SANCHEZ, R. and A. SALI, Large-scale protein structure modeling of the Saccharomyces cerevisiae genome. *Proc Natl Acad Sci U S A*, **1998**. 95(23): p. 13597–602.

244 BRENNER, S. E. and M. LEVITT, Expectations from structural genomics. *Protein Sci*, **2000**. 9(1): p. 197–200.

245 BERNSTEIN, F. C., et al., The Protein Data Bank: a computer-based archival file for macromolecular structures. *J Mol Biol*, **1977**. 112(3): p. 535–42.

246 SALI, A., 100 000 protein structures for the biologist [see comments]. *Nat Struct Biol*, **1998**. 5(12): p. 1029–32.

247 SKOLNICK, J., J. S. FETROW, and A. KOLINSKI, Structural genomics and its importance for gene function analysis. *Nat Biotechnol*, **2000**. 18(3): p. 283–7.

248 ZAREMBINSKI, T. I., et al., Structure-based assignment of the biochemical function of a hypothetical protein: a test case of structural genomics. *Proc Natl Acad Sci U S A*, **1998**. 95(26): p. 15189–93.

249 HWANG, K. Y., et al., Structure-based identification of a novel NTPase from Methanococcus jannaschii. *Nat Struct Biol*, **1999**. 6(7): p. 691–6.

250 LIMA, C. D., M. G. KLEIN, and W. A. HENDRICKSON, Structure-based analysis of catalysis and substrate definition in the HIT protein family. *Science*, **1997**. 278(5336): p. 286–90.

251 COLOVOS, C., D. CASCIO, and T. O. YEATES, The 1.8 A crystal structure of the ycaC gene product from Escherichia coli reveals an octameric hydrolase of unknown specificity. *Structure*, **1998**. 6(10): p. 1329–37.

252 YANG, F., et al., Crystal structure of Escherichia coli HdeA. *Nat Struct Biol*, **1998**. 5(9): p. 763–4.

253 VOLZ, K., A test case for structure-based functional assignment: the 1.2 A crystal structure of the yjgF gene product from Escherichia coli. *Protein Sci*, **1999**. 8(11): p. 2428–37.

254 EISENSTEIN, E., et al., Biological function made crystal clear – annotation of hypothetical proteins via structural genomics. *Curr Opin Biotechnol*, **2000**. 11(1): p. 25–30.

255 PORTUGALY, E. and M. LINIAL, Estimating the probability for a protein to have a new fold: A statistical computational model. *Proc Natl Acad Sci U S A*, **2000**. 97(10): p. 5161–6.

256 MALLICK, P., et al., Selecting protein targets for structural genomics of Pyrobaculum aerophilum: validating automated fold assignment methods by using binary hypothesis testing. *Proc Natl Acad Sci U S A*, **2000**. 97(6): p. 2450–5.

257 YONA, G., et al., A map of the protein space – an automatic hierarchical classification of all protein sequences. *Ismb*, **1998**. 6: p. 212–21.

258 FISCHER, D. and D. EISENBERG, Predicting structures for genome proteins. *Curr Opin Struct Biol*, **1999**. 9(2): p. 208–11.

259 HEGYI, H. and M. GERSTEIN, The relationship between protein structure and function: a comprehensive survey with application to the yeast genome. *J Mol Biol*, **1999**. 288(1): p. 147–64.

260 RUSSELL, R. B., P. D. SASIENI, and M. J. E. STERNBERG, Supersites within superfolds. Binding site similarity in the absence of homology. *J Mol Biol*, **1998**. 282(4): p. 903–18.

261 DURBIN, R., S. R. EDDY, A. KROGH, and G. MITCHISON, *Biological Sequence Analysis*. **1998**: Cambridge University Press.

262 WALLACE, A. C., R. A. LASKOWSKI, and J. M. THORNTON, Derivation of 3D coordinate templates for searching structural databases: application to Ser-His-Asp catalytic triads in the serine proteinases and lipases. *Protein Sci*, **1996**. 5(6): p. 1001–13.

263 JONES, S. and J. M. THORNTON, Protein-protein interactions: a review of protein dimer structures. *Prog Biophys Mol Biol*, **1995**. 63(1): p. 31–65.

264 TSAI, C. J., et al., Protein-protein interfaces: architectures and interactions in protein-protein interfaces and in protein cores. Their similarities and differences. *Crit Rev Biochem Mol Biol*, **1996**. 31(2): p. 127–52.

265 LICHTARGE, O., H. R. BOURNE, and F. E. COHEN, An evolutionary trace method defines binding surfaces common to protein families. *J Mol Biol*, **1996**. 257(2): p. 342–58.

266 MARCOTTE, E. M., et al., A combined algorithm for genome-wide prediction of protein function [see comments]. *Nature*, **1999**. 402(6757): p. 83–6.

267 PAZOS, F., et al., Correlated mutations contain information about protein-protein interaction. *J Mol Biol*, **1997**. 271(4): p. 511–23.

268 LASKOWSKI, R. A., et al., Protein clefts in molecular recognition and function. *Protein Sci*, **1996**. 5(12): p. 2438–52.

269 FETROW, J. S. and J. SKOLNICK, Method for prediction of protein function from sequence using the sequence-to-structure-to-function paradigm with application to glutaredoxins/thioredoxins and T1 ribonucleases. *J Mol Biol*, **1998**. 281(5): p. 949–68.

270 FISCHER, D., et al., Assessing the performance of fold recognition methods by means of a comprehensive benchmark. *Pac Symp Biocomput*, **1996**: p. 300–18.

271 BRENNER, S. E., P. KOEHL, and M. LEVITT, The ASTRAL compendium for protein structure and sequence analysis. *Nucleic Acids Res*, **2000**. 28(1): p. 254–6.

272 BRENNER, S. E., C. CHOTHIA, and T. J. HUBBARD, Assessing sequence comparison methods with reliable structurally identified distant evolutionary relationships. *Proc Natl Acad Sci U S A*, **1998**. 95(11): p. 6073–8.

273 ZIEN, A., R. ZIMMER, and T. LENGAUER, *Empirical p-values for Threading scores*. in *ISMB1999*. **1999**. Heidelberg.

274 MARCHLER-BAUER, A. and S. H. BRYANT, A measure of progress in fold recognition? *Proteins*, **1999**. 37(S3): p. 218–225.

275 SIPPL, M., P. LACKNER, F. S. DOMINGUES, and W. A. KOPPENSTEINER, An attempt to analyse progress in fold recognition from CASP1 to CASP3. *Proteins*, **1999**. 37(S3): p. 226–230.

276 VENCLOVAS, C., et al., Some measures of comparative performance in the three CASPs. *Proteins*, **1999**. Suppl(3): p. 231–7.

277 HUBBARD, T. J., RMS/Coverage graphs: A qualitative method for comparing three-dimensional protein structure predictions. *Proteins*, **1999**. 37(S3): p. 15–21.

278 ZEMLA, A., et al., Processing and analysis of CASP3 protein structure predictions. *Proteins*, **1999**. Suppl(3): p. 22–9.

279 FENG, Z. K. and M. J. SIPPL, Optimum superimposition of protein structures: ambiguities and implications. *Fold Des*, **1996**. 1(2): p. 123–32.

280 ALEXANDROV, N. N. and N. GO, Biological meaning, statistical significance, and classification of local spatial similarities in nonhomologous proteins. *Protein Sci*, **1994**. 3(6): p. 866–75.

281 MURZIN, A. G., Progress in protein structure prediction. *Nat Struct Biol*, 2001. 8(2): p. 110–2.

282 KOLINSKI, A., et al., A method for the improvement of threading-based protein models. *Proteins*, 1999. 37(4): p. 592–610.

283 FISCHER, D., et al., CAFASP-1: critical assessment of fully automated structure prediction methods. *Proteins*, 1999. Suppl(3): p. 209–17.

284 BERMAN, H. M., et al., The Protein Data Bank. *Nucleic Acids Res*, 2000. 28(1): p. 235–42.

285 WANG, Y., et al., MMDB: 3D structure data in Entrez. *Nucleic Acids Res*, 2000. 28(1): p. 243–5.

286 MARCHLER-BAUER, A., et al., MMDB: Entrez's 3D structure database. *Nucleic Acids Res*, 1999. 27(1): p. 240–3.

287 LASKOWSKI, R. A., et al., PDBsum: a Web-based database of summaries and analyses of all PDB structures. *Trends Biochem Sci*, 1997. 22(12): p. 488–90.

288 PEITSCH, M. C., ProMod and Swiss-Model: Internet-based tools for automated comparative protein modelling. *Biochem Soc Trans*, 1996. 24(1): p. 274–9.

289 BRENNER, S. E., D. BARKEN, and M. LEVITT, The PRESAGE database for structural genomics. *Nucleic Acids Res*, 1999. 27(1): p. 251–3.

290 GERSTEIN, M. and W. KREBS, A database of macromolecular motions. *Nucleic Acids Res*, 1998. 26(18): p. 4280–90.

291 BADER, G. D., et al., BIND – The Biomolecular Interaction Network Database. *Nucleic Acids Res*, 2001. 29(1): p. 242–5.

292 BADER, G. D. and C. W. HOGUE, BIND – a data specification for storing and describing biomolecular interactions, molecular complexes and pathways. *Bioinformatics*, 2000. 16(5): p. 465–77.

293 HOLM, L. and C. SANDER, Dali/FSSP classification of three-dimensional protein folds. *Nucleic Acids Res*, 1997. 25(1): p. 231–4.

294 HOLM, L. and C. SANDER, Dali: a network tool for protein structure comparison. *Trends Biochem Sci*, 1995. 20(11): p. 478–80.

295 PASCARELLA, S., F. MILPETZ, and P. ARGOS, A databank (3D-ali) collecting related protein sequences and structures. *Protein Eng*, 1996. 9(3): p. 249–51.

296 PASCARELLA, S. and P. ARGOS, A data bank merging related protein structures and sequences. *Protein Eng*, 1992. 5(2): p. 121–37.

297 SHINDYALOV, I. N. and P. E. BOURNE, Protein structure alignment by incremental combinatorial extension (CE) of the optimal path. *Protein Eng*, 1998. 11(9): p. 739–47.

298 HOBOHM, U. and C. SANDER, Enlarged representative set of protein structures. *Protein Sci*, 1994. 3(3): p. 522–4.

299 HOLM, L. and C. SANDER, Mapping the protein universe. *Science*, 1996. 273(5275): p. 595–603.

300 APWEILER, R., et al., The InterPro database, an integrated documentation resource for protein families, domains and functional sites. *Nucleic Acids Res*, 2001. 29(1): p. 37–40.

301 YONA, G., N. LINIAL, and M. LINIAL, ProtoMap: automatic classification of protein sequences and hierarchy of protein families. *Nucleic Acids Res*, **2000**. 28(1): p. 49–55.

302 SCHULTZ, J., et al., SMART, a simple modular architecture research tool: identification of signaling domains. *Proc Natl Acad Sci U S A*, **1998**. 95(11): p. 5857–64.

303 KRAUSE, A., et al., WWW access to the SYSTERS protein sequence cluster set. *Bioinformatics*, **1999**. 15(3): p. 262–3.

304 TATUSOV, R. L., et al., The COG database: a tool for genome-scale analysis of protein functions and evolution. *Nucleic Acids Res*, **2000**. 28(1): p. 33–6.

305 KABSCH, W. and C. SANDER, Dictionary of protein secondary structure: pattern recognition of hydrogen-bonded and geometrical features. *Biopolymers*, **1983**. 22(12): p. 2577–637.

306 FRISHMAN, D. and P. ARGOS, Knowledge-based protein secondary structure assignment. *Proteins*, **1995**. 23(4): p. 566–79.

307 GILBERT, D., et al., Motif-based searching in TOPS protein topology databases. *Bioinformatics*, **1999**. 15(4): p. 317–26.

308 GUEX, N. and M. C. PEITSCH, SWISS-MODEL and the Swiss-PdbViewer: an environment for comparative protein modeling. *Electrophoresis*, **1997**. 18(15): p. 2714–23.

309 SAYLE, R. A. and E. J. MILNER-WHITE, RASMOL: biomolecular graphics for all. *Trends Biochem Sci*, **1995**. 20(9): p. 374.

310 KRAULIS, P., MOLSCRIPT: A Program to Produce Both Detailed and Schematic Plots of Protein Structures. *Journal of Applied Crystallography*, **1991**. 24: p. 946–950.

311 WALLACE, A. C., R. A. LASKOWSKI, and J. M. THORNTON, LIGPLOT: a program to generate schematic diagrams of protein-ligand interactions. *Protein Eng*, **1995**. 8(2): p. 127–34.

312 TAYLOR W. R., Protein structure comparison using iterated double dynamic programming. *Protein Sci*, **1999**. 8(3): p. 654–65.

313 RODRIGUEZ, R., et al., Homology modeling, model and software evaluation: three related resources. *Bioinformatics*, **1998**. 14(6): p. 523–8.

314 BARBER, C. B., D. P. DOBKIN, and H. HUHDANPAA, *The Quickhull Algorithm for Convex Hulls*. ACM Transactions on Mathematical Software, **1995**.

315 RYCHLEWSKI, L., et al., Comparison of sequence profiles. Strategies for structural predictions using sequence information. *Protein Sci*, **2000**. 9(2): p. 232–41.

316 SIMONS, K. T., et al., Improved recognition of native-like protein structures using a combination of sequence-dependent and sequence-independent features of proteins. *Proteins*, **1999**. 34(1): p. 82–95.

317 Sternberg Group, ICRF: http://www.bmm.icnet.uk/~3dpssm/, http://www.bmm.icnet.uk/people/rob/CCP11BBS/, EBI Genequiz: http://jura.ebi.ac.uk:8765/ext-genequiz// genequiz.html Lion Bioscience, BioScout: http:// www.lionbioscience.com

318 http://scop.mrc-lmb.cam.ac.uk/scop/

319 http://www.biochem.ucl.ac.uk/bsm/dhs

320 http://www.tigr.org/tdb/mdb/mdbcomplete.html

321 http://www.fp.mcs.anl.gov/~gaasterland/genomes.html,
http://www.sanger.ac.uk/Projects/, http://www.nhgri.nih.gov/
genome_hub.html

322 http://www.structuralgenomics.org/, http://
www.nigms.nih.gov/funding/psi.html, http://asdp.bnl.gov/
asda/Struc_Genomics/

323 http://userpage.chemie.fu-berlin.de/~psf/

324 http://www.rsgi.riken.go.jp/

325 http://www.structuralgenomics.org

326 http://barton.ebi.ac.uk

327 http://PredictionCenter.llnl.gov/local/ss_eval/sspred_
evaluation.html

328 http://maple.bioc.columbia.edu/eva/doc/measure_sec.html

329 CASP1: Proteins: Vol 23, No 3, 1995; CASP 2: Proteins:
Supplement 1, 1997; CASP 3: Proteins: Supplement 3, 1999,
CASP 4: Proteins: forthcoming, 2001

330 http://bioinfo.pl/LiveBench/

331 http://maple.bioc.columbia.edu/eva/

332 http://www.pdb.org

333 http://www.structuralgenomics.org

334 http://presage.berkeley.edu

335 http://www.doe-mbi.ucla.edu/~fischer/BENCH/bench-
mark1.html

336 http://astral.stanford.edu/

337 ftp://ftp.embl-heidelberg.de/pub/databases/pdb_select/, http://
www.cmbi.kun.nl/whatif/select/, http://www.fccc.edu/
research/labs/dunbrack/culledpdb.html

338 http://wwww2.ebi.ac.uk/dali/fssp/TABLE2.html

339 FEBS Advanced Course, Frontiers of Protein Structure
Prediction, IRBM, Pomezia, 1997, http://predict.sanger.ac.uk/
irbm-course97/

340 http://PredictionCenter.llnl.gov/casp3/results/th

341 http://PredictionCenter.llnl.gov/CASP*

342 http://bioinfo.pl/LiveBench/

343 http://maple.bioc.columbia.edu/eva/

7

Protein–Ligand Docking in Drug Design

Matthias Rarey

7.1
Introduction

Since more and more three-dimensional protein structures become available due to X-ray crystallography, NMR spectroscopy, and homology modeling, software tools using this information in drug design projects become more important [1, 2]. The key problem these software tools have to tackle is the docking problem, namely predicting energetically favorable complexes between a protein and a putative drug molecule, also called ligand in this context.

Figure 7.1 (s. p. 317) shows the complex between the protein thrombin and a well-known inhibitor called NAPAP. Thrombin plays a central role in the blood coagulation cascade and has long been an important target in thrombosis research. Like many other enzymes, thrombin catalyzes a specific reaction by forming a complex with a substrate molecule, a peptide in this case. The inhibitor NAPAP is designed to bind to the same region on the protein surface, the active site, than the substrate but with a higher binding affinity. Therefore, in the presence of NAPAP, the natural reaction on the substrate molecule cannot be catalyzed by thrombin, the enzyme is blocked in its activity.

In the first steps of a drug design project the goal is to find a so-called lead structure, a small molecule which binds to a given target protein and can be further developed to a drug. If the three-dimensional structure of the target protein is known, docking algorithms can be applied to virtually search for potential leads. The two main questions arising are what the complex between a protein and a potential lead looks like and how strong the binding affinity of the lead is with respect to other candidates.

Several aspects make the docking problem hard to solve. First of all, one has to deal with the scoring problem [3–5], namely the calculation of binding affinity given a protein-ligand complex. Today, no general applicable scoring function is available giving an accurate prediction of the binding affinity. Secondly, a large number of degrees of freedom has to be considered during a docking calculation. The most important degrees of freedom are the relative orientation of the two molecules and the conformation of the ligand molecule. Beside these, the protein conformation may also vary, water molecules can be located between the molecules and the protonation states of the molecules can change. Finally, the various variables are subject to a complicated network of constraints. A small change in one variable may decide whether an overlap between the two molecules occur or a hydrogen bond between two groups can be formed or not. This implies that the scoring functions for protein-ligand docking typically contain lots of local minima and are difficult to optimize.

Although no general solution of the docking problem is available today, several algorithms for various kinds of docking problems have been developed and successfully applied. This chapter will give an overview of these algorithms, additional review articles about software for structure-based drug design can be found in the literature [1, 6–13].

Before we start our methods overview, we will summarize the types of docking problems and describe some typical application scenarios.

7.1.1
A taxonomy of docking problems

Docking problems can best be classified by the type of input molecules. In macromolecular docking, two macromolecules like proteins or DNA are docked. The major characteristics of these problems are that the complex typically has a large contact area and that the molecules – although still flexible – have a fixed overall shape. These features imply that methods based on geometric properties like shape complementarity alone can already be used efficiently for creating energetically favorable complexes. A survey of methods for protein-protein docking is given in Chapter 8 of this book and elsewhere [14].

The second class contains docking problems in which a small molecule is docked to a macromolecule. The macromolecule can be a DNA fragment or a protein, in the latter case the problem is called protein–ligand docking. In general, the small molecule is an organic molecule, for example a natural substrate or an inhibitor like NAPAP mentioned in the example above.

Small-molecule docking differs substantially from macromolecular docking in the fact that the ligand is typically not fixed in its overall shape. The conformational flexibility of the ligand molecule is of importance and geometric properties alone are often not sufficient to determine low-energy complexes. Even in cases where the molecule is more or less fixed in conformational space, the shape of the molecule is not characteristic enough to find low-energy complexes based on shape alone. Therefore, different algorithms are developed for small-molecule docking compared to macromolecular docking.

Fig. 7.1
Complex between the
protein thrombin and one
of its inhibitors NAPAP.
Only the active site of
thrombin is shown (atoms
drawn with sticks colored
by atom, molecular surface
is shown in green, NAPAP
is shown in red).

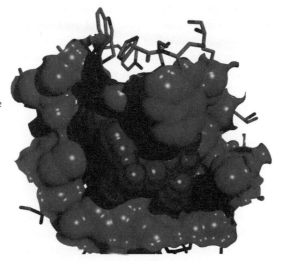

For various reasons pertaining to important drug properties like their bioavailability, most drugs are small molecules. Therefore, small-molecule docking is of great interest in pharmaceutical research. Because pharmaceutical targets are often proteins, most docking algorithms are developed for protein-ligand docking.

In principle, docking of small molecules to DNA can be handled by the same algorithms. Differences occur in the handling of DNA specific binding mechanisms like covalent binding or binding of so-called intercalators [15]. An intercalator is a small molecule that binds to the DNA between two subsequent bases. Since a structural change in the DNA is necessary for binding, conformational flexibility of the DNA must be taken into account. This can be done by docking into a distorted DNA structure [16]. A better way is to deal with the DNA flexibility directly. For example, Zacharias et al. [17] developed an algorithm based on normal mode analysis for handling structural changes of the DNA during docking calculations.

Protein-ligand docking problems can be further distinguished by the size of the ligand molecule. The typical ligand molecule occurring in drug design docking problems has about 5–12 rotatable bonds, thus, often the question of placing only a part of the ligand, a fragment, in the active site of the protein emerges. In contrast to a typical drug molecule, the conformational space of a fragment is quite small and must not necessarily be taken explicitly into account. Therefore, algorithms handling only the relative orientation of two molecules (rigid-body docking algorithms) can be applied. Docking fragments to proteins is a subproblem handled in several more complex docking algorithms. Some examples are the protein-ligand docking algorithms in which flexibility is handled by dividing the ligand into smaller fragments, *de-novo* design algorithms in which new molecules are created from a fragment database, or combinatorial docking algorithms in which combinatorial libraries of molecules are docked by combining placements for individual building blocks of the library.

Orthogonal to this classification by the type of input molecules, the second important parameter for categorizing docking algorithms is the time spent per prediction. The number of molecules which have to be processed covers a range from single molecules to several ten thousands and depends on the concrete application scenario.

7.1.2
Application scenarios in structure-based drug design

Docking problems arise in several different scenarios in structure-based drug design. The overall goal is always to plan the next experiment in order to find or create a lead structure or optimize specific properties of it.

If compounds binding to the target protein are known, the three-dimensional structures of their complexes with the target protein are of interest. They may give hints as to the catalytic mechanism of the enzyme or opportunities for optimizing the binding affinity of a compound. Because only a small set of compounds can be inspected in detail, computing time spent for predicting a protein–ligand complex plays a minor role. Typically, creating a set of starting geometries with a protein-ligand docking algorithm is the first step succeeded by molecular dynamic simulations and detailed energy calculations.

In some cases, a medium sized set of compounds (a hundred up to several thousand) are considered as candidates for lead structures. This set can emerge for example from a screening experiment, from filtering databases with other software tools, or from varying a known inhibitor or lead structure in several ways. Here, the required output is a ranked list of the compounds giving the medicinal chemist a hint which compounds should preferably be investigated further.

Virtual screening is the *in silico* variant of high-throughput screening (HTS) used to find initial lead structures from large compound collections. A virtual screening experiment can be performed on the basis of a docking algorithm. Although the accuracy of computer-predicted binding affinities is much lower than of experimental determined affinities, virtual screening has some advantages. There is no uncertainty about the molecular structure tested, the calculation is cheaper than the real experiment, and, most importantly, the calculation can be performed on compounds that are not yet purchased or synthesized. Obviously, for virtual screening a fast docking algorithm is needed. Compound collections with up to half a million molecules are typical for screening exercises and it can be assumed that this number will increase in the next years.

If no lead structures are available yet, a *de novo* ligand design can be performed as an alternative to a screening experiment. The immanent problem with this approach is to create molecules which can by synthesized easily. In order to achieve this goal, fragment-based approaches are preferably used. The central step in these approaches is to dock a library of small fragments into the active site of the target protein. Because these fragment libraries can be quite large, time-efficient docking algorithms are necessary for *de novo* ligand design.

A few years ago, combinatorial libraries of small organic compounds were introduced to drug design [18, 19]. Combinatorial chemistry offers a way to create large collections of diverse compounds which can then be used in screening experiments. Although the chemistry behind combinatorial libraries can be quite complicated, the overall methodology is quite intuitive and simple. The starting point of a combinatorial library is a specific reaction or a small set of reactions and a set of molecules (reagents). The reactions connect two or more of the reagents in a well-defined way resulting a single molecule (product) of the library. By combinatorially combining all reagents, an exponential set of products is synthesized. Usually the reactions result in a molecular fragment which is in common between all molecules of the library. This fragment is then called the core of the library, the fragments attached to the core are called R-groups.

Besides HTS libraries in which diversity and drug-likeness of the individual molecules are the main design criteria, targeted libraries play a role in drug design. A targeted library is designed such that the expected number of compounds binding to a specific target protein is increased. Docking algorithms can be used for solving various kinds of library design problems: Evaluation of a library (docking and scoring of all library molecules), preselection of R-group molecules to reduce the size of the library, or the *de novo* design of a library based on reactions and a database of available compounds. As in *de novo* ligand design, docking molecular fragments is a subproblem in several combinatorial library design problems. Due to the internal structure of a combinatorial library (the same set of fragments are connected in a well defined way in different combinations), drastic speedups can be achieved compared to docking a classical compound collection.

7.2
Methods for protein–ligand docking

The following section contains a survey of algorithms applied to the various types of docking problems as well as a short summary on scoring functions. The methods are typically related to specific software tools also mentioned here.

7.2.1
Rigid-body docking algorithms

Rigid-body docking algorithms have historically been the first approaches for screening sets of ligands by their fit to a given target protein. The protein as well as the ligand are held fixed in conformational space which reduces the problem to the search for the relative orientation of the two molecules with lowest energy.

7.2.1.1 Clique-search based approaches

The docking problem can be understood as a problem of matching characteristic features of the two molecules in three-dimensional space [20]. A match is an assignment of a ligand feature to a protein feature. Such a feature can either be a piece of volume of the active site of the protein or the ligand or an interaction between the molecules. The search procedure maximizes the number of matches under the constraint that they are compatible in three-dimensional space, i.e. that they can be realized simultaneously. In other words, compatibility means that a transformation can be found which simultaneously superimposes all ligand features to the matched protein features. In order to search for compatible matches, the following graph G is used: The vertices of G are all possible matches between the protein and the ligand, the edges connect pairs of vertices representing compatible matches. Mostly, compatibility is implemented as distance compatibility within a fixed tolerance ε: The matches $(p_1, l_1), (p_2, l_2)$ are distance-compatible if and only if $|d(p_1, p_2) - d(l_1, l_2)| < \varepsilon$. A necessary condition for a set of matches to be simultaneously realizable is that all pairs of matches are distance compatible. Therefore, an algorithm for enumerating cliques (fully connected subgraphs) can be applied to G. By superimposing the matched features of a clique one obtains an initial orientation of the ligand molecule in the active site of the protein.

One of the historically first and today probably most widely used software tool for molecular docking, the DOCK program [21], is based on the idea of searching distance-compatible matches. Starting with the molecular surface of the protein [22, 23, 24], a set of spheres is created inside the active site as shown in Figure 7.2. The spheres represent the volume which could be occupied by the ligand molecule. The ligand is either represented by spheres inside the ligand or directly by its atoms. In DOCK, an enumeration algorithm searches for sets with up to four distance-compatible matches. Each set is used for an initial fit of the ligand into the active site. Then the set is augmented by further compatible matches and the position of the ligand is optimized and scored. Since its first introduction, the DOCK software has been extended in several directions. The matching spheres can be labeled with chemical properties [25] and distance bins are used to speed up

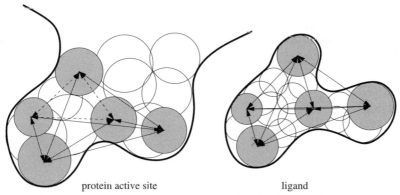

protein active site   ligand

Fig. 7.2
On the left, a protein active site is shown   covered with spheres is shown. An
filled with spheres as they are used in the   example for a distance compatible
DOCK algorithm. On the right, a ligand   matching is highlighted in grey. All
                                                 distances compared are shown by arrows.

the search process [26, 27]. Recently, the search algorithm for distance-compatible matches changed to the clique-detection algorithm introduced by Kuhl [20, 28].

An interesting algorithmic extension was introduced by Knegtel [29]. Instead of using a single protein, an ensemble of protein structures is used for the docking calculation. With this approach limited protein flexibility can be taken into account. Two of the flexible ligand docking approaches based on DOCK will be introduced below. Furthermore, several scoring functions are applied in combination with the DOCK algorithm [30–33].

While the initial DOCK algorithm uses volume as the feature to be matched, other approaches use chemical interactions. Mizutani et al. [34] presented the program ADAM in which hydrogen bonding is the feature used for matching. Possible matchings are enumerated and filtered based on distance compatibility.

Further examplex for distance-compatibility as an initial screen for fragment orientations are the well known *de novo* ligand design program LUDI [35, 36] and the rigid body docking program CLIX [37]. While LUDI's placement algorithm is based on matching hydrogen bond vectors and hydrophobic points, CLIX takes energetically favorable regions for functional groups of the ligand as the underlying model. These regions are precalculated with the computer program GRID [38] which will be introduced in Section 7.2.5. CLIX uses only two points for an initial matching. After fitting the two matched sites, the rotation about the common axis between the matched sites remains as an open degree of freedom. This rotation is then sampled in regular intervals.

7.2.1.2 Geometric hashing

Geometric hashing [39] originated from the area of computer vision and was first applied to molecular docking problems by Fischer et al. [40, 41]. In computer vision, the geometric hashing scheme was developed for the problem of recognizing (partially occluded) objects in camera scenes. For simplicity, we explain the geometric hashing algorithm for the two-dimensional case first and describe its application to the three-dimensional molecular docking afterwards.

Given a picture of a scene and a set of objects which can occur therein (called models in this context) both represented by points in 2D space, the goal is to recognize some of the models in the scene. In a preprocessing phase, a hash table is created from the set of models: For each model, each pair of points define a so-called basis. Then, for each basis, every third point belonging to the model is expressed in coordinates relative to the basis. A tuple (model, basis) is stored in a hash table addressed by the relative coordinates of the third point. The reason for having several bases for a model instead of a single one is that it is unknown in advance, whether a part of the model is occluded in the scene.

In the recognition phase the scene is analyzed as follows: Every pair of points is considered as a basis. Once the basis is defined, all other points can be expressed by relative coordinates with respect to the basis resulting in a query for the hash table created before. The query votes for all matching tuples (model, basis) stored in the hash table. Finally, models with many votes are extracted, a transformation is calculated from the matching points, and the match is verified.

Two aspects make geometric hashing attractive for molecular docking problems: it is time-efficient and it deals with partial matchings standing for partially occluded objects in the terms of pattern recognition. The latter is extremely important because in most docking applications, not all of the ligand features are matched with the protein since parts of the ligand surface are in contact with bulk water.

In order to apply geometric hashing to molecular docking, Fischer et al. [40, 41] used the sphere representation of DOCK as the underlying model. Because docking is performed in 3D space, three points (here spheres or atoms) are necessary to define a basis. As a consequence, the number of hash table entries increases with the fourth power of the number of ligand atoms resulting in unacceptably large hash tables. Therefore, the basis is described by only two points leaving one degree of freedom open (rotation around the axis defined by the two points). With this model in mind, the geometric hashing approach can be directly applied to the molecular docking problem.

7.2.1.3 Pose clustering

Pose Clustering [42] is a different approach from pattern recognition applied to the molecular docking problem [43]. Pose clustering was originally developed for detecting objects in pictures with an unknown camera location. The algorithm matches each triangle of object points with each triangle of points from the picture. From a match, a camera location can be computed such that the triangles superimpose. The camera locations are stored and clustered. If a large cluster is found, the object is identified and the orientation of the camera with respect to the object is determined.

For applying pose clustering to molecular docking, the LUDI model of molecular interactions [35, 36] is used as the underlying representation. For each interacting group, an interaction center and an interaction surface is defined (see Figure 7.3). The interaction surfaces of the protein are approximated by discrete points, which then form the scene in the pose clustering algorithm. The centers of the ligand interactions are the object points which have to be matched to the scene.

While in the pattern recognition problem each triangle of object points can be matched to each triangle of picture points, in the docking application, the matches are limited in various ways. First of all, the interaction types must be compatible and secondly, the triangle edges must be approximately of the same length. A hashing scheme is necessary to efficiently access matching protein interaction point triangles (picture points) for a given ligand interaction center triangle (object points). The hashing scheme stores edges between two points and addresses them by the two interaction types and the edge length. A list merging algorithm then creates all triangles based on lists of fitting triangle edges for two of the three query triangle edges. For a match, additional directionality constraints for the three interactions are checked. Then, a transformation is calculated that superimposes the two triangles. In the original application of pose clustering, the transformation is used to calculate the camera location. In molecular docking, the transformation directly describes the location of the fragment in the active site.

Finally, the transformations must be clustered. We use complete-linkage hierarchical clustering for this task. As a distance measure, the root-mean-square deviation of the atoms after applying the transformations is used. After a linear-time preprocessing phase, this quantity can be calculated from the transformations in constant time [43]. For each of the clusters generated, the typical post-processing steps are performed, like searching for additional interactions, checking for protein–ligand overlap, and scoring.

7.2.1.4 Superposition of point sets

In each of the introduced algorithms for rigid-body docking, the superposition of point sets is a fundamental subproblem that has to be solved efficiently and therefore is discussed here shortly:

The superposition problem can be described as follows: Given two sets X, Y with n vectors each, find a transformation $T = (\Omega, t)$ minimizing the root-mean-square deviation between X and the transformed vector Y:

$$RMSD_{X,Y}(T) = \sqrt{(1/n) \sum_i (x_i - \Omega y_i - t)^2}.$$

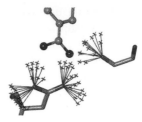

Fig. 7.3
Interaction surfaces of three hydrogen bond acceptors in thrombin (shown in red). Protein atoms are drawn with sticks, ligand atoms with balls & sticks. The left picture shows the interaction surfaces itself, the right picture shows the approximation by interaction points used in the pose clustering algorithm.

Let Ω describe a rotation around the centroid of Y, then $t' = (1/n) \cdot (\sum_i x_i - \sum_i y_i)$ minimizes $RMSD_{X,Y}(T)$ for all rotations Ω.

Optimizing Ω is a more difficult task. Ferro & Hermans [44] and later Sippl and Stegebuchner [45] proposed iterative algorithms rotating subsequently around the x-, y-, and z-axis. If the axis is fixed, the optimal rotation angle can be determined analytically. Kabsch [46, 47] formulated the problem as a constrained optimization problem. Using Lagrange multipliers and the calculation of eigenvectors, Kabsch was able to solve the problem directly. Kabsch's algorithm is very time-efficient and contained in several codes for molecular docking today.

7.2.1.5 Concluding remarks

Reviewing the development of more recent approaches to fragment docking, two enhancements were made. First, more elaborate algorithms for searching in three-dimensional space, ofton adapted from other disciplines, were applied. These methods allow a reasonable coverage of search space in short computing times. Second, more information about the physicochemical properties of binding are included directly into the search process. Therefore, energetically unfavorable fragment placements are directly avoided, even if they make sense from a steric point of view. Both enhancements are essential for achieving a good performance for large variety of molecular fragments.

7.2.2
Flexible ligand docking algorithms

The major limitation of the rigid-body docking algorithms is that the conformational flexibility of the ligand molecule is not considered. Often small molecules have large conformational spaces with several low energy states. Even for small molecules with only a few rotatable bonds significant differences between the bound conformation and the calculated low energy conformation can occur [48].

7.2.2.1 Conformation ensembles

In principle, ligand conformational flexibility can be incorporated by applying rigid-body docking algorithms to ensembles of rigid structures, each representing a different conformation of the same ligand. The size of the ensemble is critical, since the computing time increases linearly with the number of conformations and the quality of the result drops with increasing difference between the most similar conformation of the ensemble and the complex conformation.

Miller and Kearsley developed the Flexibase/FLOG docking algorithm based on conformation ensembles. Flexibases [49] store a small set of diverse conformations for each molecule from a given database. The conformations are created with distance geometry methods [50, 51, 52] which will be introduced later in the context of docking. Then a set of up to 25 conformations per molecule are selected by RMSD dissimilarity criteria. Each conformation of a molecule is then docked using the rigid-body docking tool FLOG [53] which is similar to the DOCK algorithm discussed above.

A different approach based on conformation ensembles was presented by Lorber and Shoichet [54]. Here, about 300 conformations per molecule are created on the average for a database of molecules. For each molecule, a rigid part, for example an aromatic ring system, is defined. The conformation ensemble is created such that the atoms of this rigid part are superimposed. Then, the DOCK algorithm is applied to the rigid part and all conformations are subsequently tested for overlap and finally scored. With this method, a significant speedup can be achieved compared to an independent docking of the conformations.

An important point in this scheme is the dependence between the conformation generation algorithm and the docking algorithm: The fewer conformations are created, the higher must be the tolerance in the matching phase of the docking algorithm. Therefore, the algorithms applied to the two subproblems are related making it necessary to tune parameters describing the coverage of conformational space in accord with the tolerance in protein-ligand overlap.

7.2.2.2 Flexible docking based on fragmentation

A possible way to handle conformational flexibility directly in a docking algorithm is by fragmentation. Here the ligand is divided into several fragments. Each fragment is either rigid or has only a small number of conformations which can be handled by a conformation ensemble. Obviously, the fragment-based docking approaches and *de novo* ligand design are closely related. The major difference is that while in the docking algorithm the fragments stem from a single molecule, the *de novo* ligand design algorithm picks a fragment out of a database.

7.2.2.2.1 Place & join algorithms

Looking at the way in which the fragments are reconnected during the docking calculation, we can distinguish between two classes of algorithms: In a place & join algorithm, multiple fragments are all docked independently. Then, placements for adjacent fragments in which the connecting atoms overlap are identified and reconnected.

The first algorithm of this kind was developed by DesJarlais et al. [55]. The ligand is manually divided into two fragments having one atom in common. Then, placement lists are created for each fragment using the docking algorithm DOCK. The algorithm searches through these lists for placement pairs in which the common atom is located approximately at the same point. Finally, the fragments are reconnected, energy minimized, and scored.

Sandak et al. [56–59] applied the geometric hashing paradigm to develop a place & join algorithm. As before, the ligand is divided into fragments with one overlapping atom, called the *hinge*. For each ligand atom triplet of a fragment, a hash table entry is created and addressed with the pairwise distances between the atoms. The entry contains a fragment identification as well as the location of the hinge. In the matching phase, protein sphere triplets are used to extract ligand atom triplets with similar distances. As a result a vote is counted for a hinge location for each match. Hinge locations with many votes are then selected, the fragments are reconnected accordingly and finally scored.

Place & join algorithms are advantageous in cases in which the molecule consists of a small set of medium-sized rigid fragments. If the fragments are too small, it is difficult to place them independently. Another difficulty is to generate correct bond lengths and angles at the connecting atom without destroying the previously found interactions of the fragments to the protein.

7.2.2.2.2 Incremental construction algorithms

The second kind of fragment-based docking algorithms, and the one which is used most frequently today, follows the incremental construction method. Instead of placing all fragments of the ligand independently, the incremental construction algorithm starts with placing one fragment (called base or anchor fragment) into the active site of the protein. Then, the algorithm adds the remaining parts of the ligand to the already placed fragment iteratively. Thus, an incremental construction algorithm has three phases: The selection of base fragments, the placement of base fragments, and the incremental construction phase. An incremental construction algorithm can start with several base fragments, but in contrast to the place & join algorithms the placements are not combined but taken as anchoring orientations to which the remaining parts of the ligand can be added.

Incremental construction originated from the area of *de novo* ligand design. Moon and Howe [60] presented the peptide design tool GROW based on this strategy. The first docking algorithm based on incremental construction was developed by Leach & Kuntz based on the DOCK program [61]. First, a single anchor fragment is selected manually and docked into the active site using a variant of the DOCK algorithm which handles hydrogen bonding features in the matching phase. A subset of placements is selected for which the incremental construction phase is started. For this phase, a backtracking algorithm is used enumerating the space of non-overlapping placements of the whole ligand in the active site. After adding a fragment to the current placement, a refinement routine is used to eliminate steric strain and to improve hydrogen bond geometries. The final placements are then filtered, refined and scored with a force-field based approach. Although there are several manual steps in this procedure, the work demonstrated that the incremental construction idea can be applied to the docking problem.

Leach published a second docking algorithm [62] similar to incremental construction with respect to the way that the degrees of freedom are fixed sequentially. Degrees of freedom considered are the ligand orientation and conformation described by a discrete set as well as a set of rotamers for selected protein side-chains. Leach used a variant of a branch & bound scheme, called *A* algorithm with dead-end elimination* (see Appendix), to search efficiently through the space of possible configurations.

The docking algorithm contained in the FLEXX package [63–65] is also based on incremental construction. FLEXX is a fully-automated approach to molecular docking developed for virtual screening purposes. In the first phase, a small number of base fragments are selected. An efficiently computable scoring function is used to select fragments which are suitable for placement They should contain a reasonably high number of interacting groups and at the same time a relatively low number of low-energy conformations. A necessary condition for a successful calculation is that the selected base fragment binds to the protein and is not mostly exposed to water in the final protein-ligand complex. In order to ensure this, a small set of base fragments distributed over the ligand is selected.

For placing the base fragments, the pose clustering algorithm is applied. Base fragment conformations are enumerated within the placement algorithm. The advantage of the pose clustering algorithm is that it is based on the molecular interactions instead of the shape of the fragment. This allows the handling of much smaller fragments than in shape-based algorithms down to the size of a single functional group. All calculated placements up to given number form the input to the incremental construction phase. In contrast to Leach's algorithm, a greedy strategy is applied selecting always the k placements with the highest estimated score (k is about 800). Each

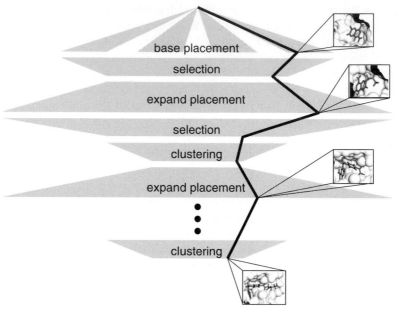

Fig. 7.4

Flexible ligand docking by incremental construction. The search tree resulting after the steps of the algorithm as described in the text are shown in grey. The black line illustrates the construction process for one placement of the complete ligand.

iteration of the incremental construction algorithm contains the following steps: adding the next fragment in all possible conformations to all placements from the previous iteration (or the base placement phase), searching for new protein-ligand interactions, optimizing the ligand position to improve the interaction geometries and reduce steric strain, selecting a subset of placements with high score, and clustering the placements to achieve a reasonable degree of diversity in the solution set. The overall search strategy is shown in Figure 7.4.

Ligand conformations within FlexX are based on the MIMUMBA model [66]: To each rotatable bond, a set of low energy torsion angles is assigned, previously derived from a statistical analysis of the Cambridge Structure Database (CSD) [67]. Ring system conformations are precomputed using the 3D structure generator Corina [68, 69]. For scoring protein-ligand complexes, a variant of Böhms empirical scoring function [70] is used.

Several extensions of the FlexX approach have been developed recently. The interaction model has been extended such that hydrophobic fragments can be handled with the pose clustering algorithm [71]. In order to place critical water molecules and metal ions located in the protein-ligand interface, the "particle concept" has been invented and integrated into the FlexX software. The particle concept allows for automatic placement and energetic consideration of approximately spherical objects during the docking calculation [72]. FlexE is an extension of FlexX regarding the handling of protein flexibility [73, 74]. Similar to Knegtels approach mentioned above, FlexE takes an ensemble of slightly differing protein structures as input. However the structures are not numerically merged to a single description. FlexE uses various graph algorithms to explicitly consider the different alternative conformations for protein parts like side-chains or small loop fragments and combine them to a single protein conformation best suited for the protein-ligand complex created during the docking calculation.

Two other approaches based on incremental construction have been published. The program Hammerhead [75] differs from FlexX in the construction strategy. Instead of adding small fragments (cut between each rotatable bond), the ligand is divided into a small set of large fragments. During the construction phase, the next fragment is added such that the connecting atom (or bond) overlaps and interactions to the protein can be formed. Therefore, there is no discrete sampling at the torsion angle of the added fragment. However, since the bond angle at the overlapping atom may vary, high-energy conformations will also be generated and the situation in which a fragment does not interact directly with the protein is more difficult to handle.

Makino and Kuntz [76] presented a fully automated incremental construction docking algorithm based on backtracking. A single anchor fragment is selected maximizing hydrogen bonding features. During the incremental construction phase, the number of conformations for each fragment is limited to reduce the size of the search space. This method is called a limited backtrack search. For scoring, the AMBER force field [77, 78] is used with a modification allowing the handling of multiple protonation states.

Incremental construction algorithms are the basis for widely used docking tools like DOCK and FlexX. The quality of the predicted structures strongly depends on the number of different placements considered in each iteration of the incremental construction process. Therefore, although the overall concept of incremental construction is simple, much effort must be put in the time-efficient analysis of partially placed ligands, for example the protein–ligand overlap test, the evaluation of the scoring function, and the postoptimization of the ligand placement.

In practice, incremental construction has been proven to be a reasonable compromise between accuracy and computing speed. This holds especially for virtual screening applications, in which computing speed is of central importance.

7.2.2.3 Genetic algorithms and evolutionary programming

Since the mid-1990s, genetic algorithms have been applied to the molecular docking problem in several approaches [79–83]. A genetic algorithm [84] is a general purpose optimization scheme which mimics the process of evolution. The individuals are configurations in the search space. A so-called fitness function is used to decide which individuals survive and produce offsprings.

Several elements must be modeled in order to use the idea of genetic algorithms for an application like molecular docking. First of all, a linear description of a configuration (the chromosome) is needed describing all degrees of freedom of the problem. Finding the chromosome description is the most difficult modeling part. A suitable description is free of redundancy and models constraints of the configuration space directly such that configurations violating constraints are never generated during the optimization.

Second, a fitness function has to be developed. The fitness function is closely related to scoring functions for molecular docking with one extension. Scoring functions normally work on 3D coordinates, therefore the chromosome of an individual has to be interpreted in order to apply the scoring function. This step is called the *genotype-to-phenotype conversion*. Since most of the computing time is spent for evaluating the fitness function, the conversion and the evaluation has to be done efficiently.

The optimization scheme itself is more or less independent from the application. Typically, several parameters have to be chosen like the population size, the number of generations, crossover and mutation rates, etc.. Here, it is important to achieve a reasonable trade-off between optimizing the fitness function and keeping the diversity in the population.

The genetic algorithm which is today probably most widely applied for molecular docking was developed by Jones et al. [79, 85] and is implemented in the software tool GOLD. A configuration in GOLD is represented by two strings. The first string stores the conformation of the ligand and selected protein side chains by defining the torsion angle of each rotatable bond. The second one stores a mapping between hydrogen bond partners in the protein and the ligand. For fitness evaluation, a 3D structure is created from the chromosome representation by first generating the ligand conformation. According to the mapping stored in the second string, hydrogen bond atoms are superimposed to hydrogen bond site points in the active site. Finally, a scoring function evaluating hydrogen bonds, the ligand internal energy as well as the protein–ligand van der Waals energy is applied as the fitness function.

Oshiro et al. [80] developed two variants of a docking method, both based on genetic algorithms and the DOCK approach. The variants differ in the way the relative orientation of the ligand to the protein is described. The first variant is similar to the GOLD algorithm and encodes the matching of ligand atoms to protein spheres in the chromosome. A superposition is used to generate the 3D orientation of the ligand. The second variant stores the relative orientation directly by a translation vector and three Euler angles. For scoring, a simplified version of the AMBER force field was used.

Gehlhaar et al. [82, 86] developed a docking algorithm based on evolutionary programming called EPDOCK. In contrast to a genetic algorithm, offsprings are created from one parent by mutation only. Each member of a population is competing for survival in a so-called tournament. EPDOCK contains a self-developed scoring function based on atomic pairwise linear potentials for steric interactions and hydrogen bonding.

7.2.2.4 Distance geometry

Distance geometry [51, 50] is a well-known technique from the area of structure determination via NMR technology. Instead of describing a molecule by coordinates in Euclidean space, it is described by a so-called distance matrix containing all interatomic distances. Based on distance matrices, a set of conformations can be described in a comprehensive form by calculating a distance interval for each atom pair.

The distance geometry methodology can be directly used in the docking algorithm based on clique search and distance compatibility (see Section 7.2.1.1). Two matches between protein site points and ligand atoms are compatible, if the site point distance lies within the distance interval of the ligand atom pair. The drawback of this approach is that the distance matrix is overdetermined. Fixing the atom-atom distance between a given atom pair to a single value causes other distance intervals to shrink. Since the exact new interval boundaries are difficult to calculate, the triangle and tetrangle inequality are used to approximate them [87, 88]. In other words, only a very limited number of distance matrices can be converted back to 3D space.

Ghose and Crippen [89] first worked on this approach on a more theoretical basis. Later, Smellie et al. [90] applied this methodology to real test cases. Billeter et al. [91] combined the description of molecules by distance constraints with an efficient algorithm for constrained optimization.

The screening software Specitope developed by Schnecke et al. [92] uses distance matrix comparisons as a first filter step. However, flexibility of molecules is not modeled by distance intervals. Instead, a weighting scheme is defined to scale down the contributions of more flexible atom pairs in the overall score.

7.2.2.5 Random search

Once a scoring function for evaluating protein-ligand complexes is available, random search algorithms can be applied to the docking problem. Random placements can be either created directly by randomly fixing all degrees of freedom or they can be derived from a (random) starting structure by random moves. In most cases the structure generation is combined with a numerical optimization driving the placements to the closest local minima. Most approaches of this kind are Monte Carlo algorithms discussed separately in Section 7.2.3.3.

Sobolev et al. [93] presented a random search algorithm in their docking program LIGIN. A large set of starting structures are created randomly and then optimized in two steps. The surface complementarity is optimized first, the hydrogen bond geometry in a second step. So far, LIGIN does not include ligand flexibility although the method is in principle able to handle it.

In order to avoid a repeated evaluation of very similar structures, Baxter et al. [94] use a method called *tabu search* in their docking software called

PRO_LEADS. Starting with an initial random structure, new structures are created by random moves. In tabu search optimization, a list (the tabu list) is maintained containing the best and most recently visited configurations. Moves which results in configurations close to one in the tabu list are rejected except if they are better than the best scoring one. The tabu-list technique improves the sampling properties of the random search algorithm by avoiding to revisit configurations. (see also Appendix, "Tabu search")

7.2.3
Docking by simulation

While all previous methods for the docking problem are based on some kind of combinatorial optimization algorithm, there are several approaches tackling the problem by simulation techniques. Instead of trying to enumerate a discrete low-energy subspace of the problem, these approaches begin their calculation with a starting configuration and locally move to configurations with lower energy.

7.2.3.1 Simulated annealing

Simulated annealing [95] is a well-known simulation technique which is also frequently used for solving complex optimization problems without any physical interpretation of the simulation itself. The overall simulation routine iterates the following steps. Starting with a configuration A with an energy or score value $E(A)$, a random local move to a new configuration B with energy $E(B)$ is calculated. The acceptance of the new configuration is based on the Metropolis criterion, which means that it is accepted if $E(B) \leq E(A)$ or with probability $P = e^{-(E(B)-E(A))/(k_B T)}$ otherwise where k_B is the Boltzmann constant. Over simulation time, the temperature T is reduced based on a so-called cooling schedule such that accepting configurations with increased energy becomes less likely. (see also Appendix, "Simulated annealing")

The AUTODOCK program for protein-ligand docking developed by Goodsell et al. [96, 97, 98] is based on this strategy. For energy calculation, molecular affinity potentials [38] are precalculated on a grid. Yue developed a program for optimizing distance constraints for rigid-body docking based on simulated annealing [99].

7.2.3.2 Molecular dynamics

In principle, molecular docking problems can be solved with molecular dynamics (MD) simulations. In fact, the earliest approaches for predicting protein-ligand interactions with the computer were based on MD calculations [100, 101].

In MD, a force field is used to calculate the forces on each atom of the simulated system. Then, following Newtonian mechanics, velocities and accelerations are calculated and the atoms are moved slightly with respect to a given time step. Introducing molecular dynamics and force fields is clearly beyond the scope of this article. However, some aspects of docking by MD simulations should be mentioned briefly. The simulation becomes more exact, the smaller the time step and the more atoms of the system are taken into account. Thus, MD simulations can become very time consuming and are therefore not appropriate for inspecting large sets of molecules. (see also Appendix, "Molecular dynamics")

In order to avoid very long computing times, methods performing greater moves of the ligand in a single step have been developed. This decreases the dependence of the outcome of the calculation from the starting structure and allows a better sampling of the search space. DiNola et al. [102, 103] developed a technique for this purpose called 'helicopter view'. For a limited time, the temperature of the system is increased for selected degrees of freedom (protein-ligand relative orientation) and the repulsive terms of the energy function are decreased. This enables the algorithm to escape from local minima in the energy function.

A similar method is used in a four-phase docking protocol developed by Given and Gilson [104]. First, a set of low-energy ligand conformations is created using MD with alternated heating (in order to perturb the structure) and cooling (in order to minimize the structure). Then the ligand is placed randomly into the active site and several times minimized. In the final phase, the most stable configurations are investigated further using MD with alternate heating and cooling. The goal of the last phase is to explore the search space around the stable conformations in more detail.

A frequently used technique to speed up MD simulations is the precalculation of force-field contributions from protein atoms on grids. The force on a ligand atom can then be efficiently calculated by a simple table-lookup instead of summing over all protein atom contributions. The potentials, however, can only be precalculated for atoms which do not change their orientation in space. In order to avoid completely neglecting protein flexibility, Luty et al. [105, 106] divided the protein into a rigid and a flexible part. Every atom sufficiently far away from the active site is considered as fixed in space, all force-field contributions of these atoms can be precalculated on a grid. During the MD simulation, only the active site atoms of the protein and the ligand atoms are allowed to move and have to be considered explicitly in the force-field calculation.

7.2.3.3 Monte-Carlo algorithms

In an MD simulation, the local movements of the atoms are performed due to the occurring forces. In contrast, in a Monte-Carlo (MC) simulation, the local moves of the atoms are performed randomly. The simulated annealing

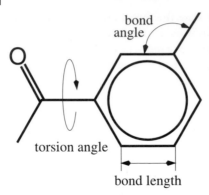

Fig. 7.5
Internal coordinate representation of a molecule. The three-dimensional structure can be constructed from bond lengths, bond angles, and torsion angles.

bond angle

torsion angle

bond length

algorithm discussed above is one special variant of an MC simulation. Two components are of major importance in the development of an MC algorithm: The description of the degrees of freedom and the energy evaluation.

Concerning the degrees of freedom, the aim is to describe them such that high-energy states are avoided. A good example of how to realize this concept is the description of the conformational space of a molecule by internal coordinates (bond lengths, bond angles and torsion angles) shown in Figure 7.5 instead of by Cartesian coordinates for all atoms. With internal coordinates, each type of variable is related to a different energy scale: Changing a bond length is energetically more expensive than changing a bond angle, which in turn is energetically more expensive than changing a torsion angle. Using internal coordinates, it is possible to navigate through the low-energy conformational space by defining the amounts by which the variables of each type are changed. Thus, internal coordinates are the more appropriate description form for MC algorithms. (see also Appendix, "Monte Carlo" Methods)

Most of the computing time during an MC simulation is spent in the calculation of the energy (or score) of a state. Therefore, this step has to be as time-efficient as possible. Often, energy potentials are precalculated on a grid to speed up this step.

There are several examples for MC-based docking algorithms. Hart and Read [107] developed an MC scheme combined with simulated annealing. In the MC run, random orientations are created and moved such that protein-ligand overlap is reduced. The results are then optimized in a simulated annealing optimization scheme. Protein–ligand overlap and scores are calculated based on precalculated grids. McMartins and Bohaceks QXP program [108] implements a similar approach which can be applied to molecular docking and structural superposition. Wallqvist and Covell [109] also use MC for optimizing the final ligand orientation. Instead of a random set of starting structures, a surface matching algorithm is used.

Abagyan et al. [110] developed the software package ICM combining the MC algorithm with an internal coordinate description explained above. In contrast to other approaches making purely random moves, ICM is able to make moves based on probability distributions for variable sets. For example, a probability distribution for a torsion angle can be defined such that low-energy torsion angles are more likely to occur than high-energy torsion angles. This biases the MC calculation towards low-energy states.

MC algorithms can be used to overcome the limitation of MD simulations to get stuck in local minima. In order to get low energy structures, MC can then be combined with energy minimization as described by Apostolakis et al. [111]: After a random creation of starting structures, a minimization with a modified van der Waals potential is performed. The energy function for the evaluation of the conformations is a sum of three terms: a force field energy, a hydrophobic solvation term which is proportional to the solvent-accessible surface of the complex and an electrostatic solvation term obtained from the solution of the linearized Poisson-Boltzmann equation. The best scoring configurations are further analyzed with a Monte-Carlo simulation method called MCM. MCM performs Monte-Carlo interleaved with minimization steps and uses the energy function mentioned above.

PRODOCK is an MC-based docking algorithm developed by Trosset and Scheraga [112, 113, 114]. The software uses either the AMBER IV or ECEPP/3 force field with a grid-based energy evaluation. For calculating energy values within the grid, Bezier splines were used allowing for a more accurate estimation of the energy value as well as information about the derivatives at this point. Like ICM, an internal description of the degrees of freedom is used. The MC simulation is interleaved with energy minimization steps like in Apostolakis' approach.

7.2.3.4 Hybrid methods

Due to the complexity of the docking problem, all methods have their pros and cons. Fragment-based approaches and genetic algorithms achieve a wide coverage of the configuration space, however simulation-based methods outperform others in finding exact local minima, i.e. predicting an exact placement. Combining different methods is therefore a reasonable approach which can result in methods containing the best of each.

Two approaches have been published recently which combine rapid fragment-based searching techniques with sophisticated MD or MC simulations.

Wang et al. [115] developed a multi-step approach based on rigid-body docking and MD. First a set of low-energy conformers is created. Each conformer is docked rigid into the active site using DOCK. The high scoring placements are then optimized using an MD-based simulated annealing optimization in combination with the AMBER force field.

Hoffmann et al. [116] combined the incremental construction algorithm in FLEXX with an MD-based procedure for post-optimization. First, FLEXX is used to create a sample of a few hundred ligand placements. The goal of

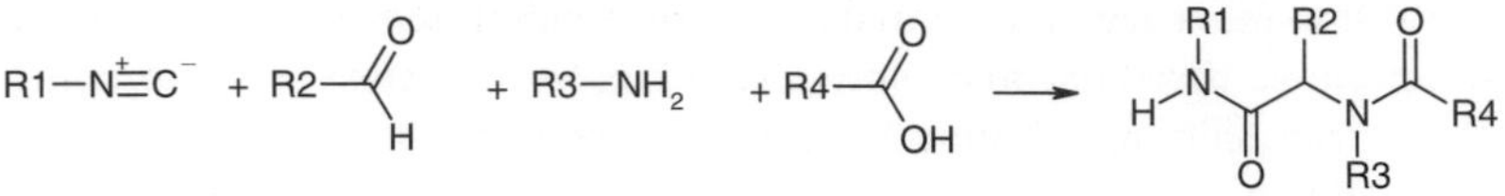

Fig. 7.6
Based on the Ugi reaction, combinatorial libraries with four different R-groups can be created. After the reaction, all library molecules have the core shown on the right in common but differ in the four R-groups attached to it.

the second phase is to improve the overall ranking of the solutions and to identify the correct placement. The placements are first energy-minimized using the CHARMm force field. For re-ranking, the software package CAMLab [117] is used. The final score consists of three contributions, a force field energy, an electrostatic part, and a nonpolar part of the solvation energy. The electrostatic part of the solvation energy is calculated by solving the Poisson equation using fast multigrid methods, while the nonpolar part is approximated by the total solvent-accessible surface.

7.2.4
Docking of combinatorial libraries

The development of combinatorial chemistry and its application to drug design [18, 19] has led to new search problems in the context of molecular docking. An example of a combinatorial library is given in Figure 7.6. The number of molecules which can be synthesized on the combinatorial chemistry bases has increased dramatically compared to classical methods. Therefore, any screening methodology has to face many more molecules.

Probably more important for the development of docking methods is the introduction of formal structure into this increased search space. If an un-structured compound collection is given, each molecule has to be analyzed independently in a screening experiment. Combinatorial libraries, however, follow a systematic build-up law for synthesizing molecules from a highly limited set of building blocks. This structure can be exploited to drastically reduce the runtime of virtual screening calculations.

In the context of combinatorial libraries, one can distinguish between three kinds of docking problems:

Combinatorial Docking Problem Given a library, calculate the docking score (and the geometry of the complex) for each molecule of the library.
R-Group Selection Problem Given a library, select molecules for the individual R-groups in order to form a smaller sublibrary with an enriched number of hits for the target protein.
***De novo* Library Design Problem** Given a catalog of molecules, design a library (including the rules of synthesis) optimizing the number of hits for the target protein.

Methods for these problems have emerged from the area of molecular docking and *de-novo* ligand design (see [118] for an overview on combinatorial docking methods). In the first case, the docking algorithms are applied to individual molecule fragments like R-groups or the core of the library and the resulting information is then combined yielding placements for individual library molecules. The methods for combining the placements differ. Like in the case of docking algorithms, they can also be classified as either a place & join or an incremental construction method. In the second case, the *de-novo* ligand design is constrained by predefined rules of synthesis.

Early algorithms for the combinatorial docking problem analyzed the similarity in given ligand datasets in order to speed up the search process. The focus in these papers is to structurally relate ligands within the dataset. One approach to do so is to generate a minimal tree structure representing the whole ligand dataset [119]. Another approach is to speed up conformational searching based on clustering similar molecules [120]. In both cases, the derived hierarchy of molecules can then be used in an incremental construction docking method.

The combinatorial docking tools PRO_SELECT [121] and CombiDOCK [122] are based on the incremental construction method. In both approaches, a library is formed by a template (or core) molecule with a set of attachment points to which one out of a predefined set of substituents can be connected. The template is then positioned in the active site without considering the substituents. Starting from a few orientations of the template, the substituents are placed into the active site of the protein independently. In case of PRO_SELECT, substituents are then selected based on score and additional criteria like 2D similarity and feasibility of synthesis. CombiDOCK calculates a final score for whole library molecules by combining fragment scores.

A new algorithm for the combinatorial docking problem based on the FlexX incremental construction algorithm was recently presented [123]. The algorithm is part of a new combinatorial docking extension of FlexX called FlexX^c. FlexX^c handles the library as a rooted tree in its closed form. Molecules from the library required during the docking calculations are created on the fly. With this method libraries with a few hundred thousand molecules can be handled in main memory during a docking calculation. The tree representation allows for handling complex libraries consisting of several R-groups which can be arbitrarily linked with the one limitation that no ring closures over different R-groups are allowed. The recursive combinatorial library algorithm is a natural extension of the incremental construction method to multiple molecules: In each incremental construction step, all possible R-group molecules are added sequentially.

Several approaches based on ligand *de novo* design software have been published for R-group selection problems. Kick et al. [124] applied a variant of the BUILDER program [125] to the preselection of substituents for a library targeted to Cathepsin D. Böhm [126] applied the LUDI program to the docking of two groups of fragments which can be connected pairwise in a single-step reaction to search for new thrombin inhibitors. In principle, all programs for fragment-based *de novo* ligand design can be applied in a similar way to the R-group selection problem.

Finally, we mention two methods for *de novo* library design. Caflisch [127] applied the MCSS technique generating fragment placements which are subsequently connected. The DREAM++ software [128] combines tools for fragment placement and selection. The selection process is done such that only a small set of well characterized organic reactions are needed to create the library.

7.2.5
Scoring protein–ligand complexes

A fundamental element of each software for molecular docking is the scoring function. The general question to answer is, given a protein and a ligand in aqueous solution, what is the ligand's free energy of binding to the protein? The scoring function topic is very complex, we will therefore give only a short overview about the main characteristics of the problem and the methods used. More comprehensive reviews can be found in [3, 4, 5, 129–132].

Computer methods are currently far from accurately calculating the free energy of binding. Typically, assumptions fixing some of the various degrees of freedom are made. Furthermore, in most approaches, it is assumed that the free energy of binding is dominated by a single, low-energy protein-ligand binding mode. Therefore, the energy is associated with a protein-ligand complex. Other aspects difficult to quantify like solvation effects, conformational changes in the protein, or loss of entropy during complex formation are either omitted or approximated only roughly.

With respect to the docking application, three kinds of problems can be defined which have to be solved by scoring functions with increasing complexity. The simplest problem is the ranking problem during a docking calculation. Here, different orientations of the same ligand in the same active site have to be evaluated and the scoring function has to distinguish energetically favorable orientations from less favorable ones. The most common problem in structure-based drug design is the ranking of a set of ligands with respect to the binding energy to a given protein. In contrast to the first problem, the scoring function has to deal not only with different orientations but also with different ligand structures. However, the protein is always the same and, at least in most cases, the ligands bind at approximately the same location, namely the active site. Also the necessary conformational changes of the protein during the binding of the ligand are likely to be similar simplifying the scoring problem. The most general case, namely predicting the free energy of binding of individual protein-ligand complexes with different ligands and proteins, is of course the most complicated one.

In order to dock a single molecule, the scoring function has usually to be evaluated several thousand times. Therefore, only functions with short computing times can be applied. Currently, there are three types of scoring functions frequently used in molecular docking calculations: force-field scoring, empirical scoring, and knowledge-based (sometimes also called PMF for potentials of mean force) scoring.

Force fields (see [133] for an overview) can be applied directly to create a score for a protein-ligand complex. This scoring method is normally used in docking algorithms based on simulation (see above and [134]). Normally, some modifications are performed: First a grid is used to speed up the calculation. Second the Lennard-Jones potential is exchanged by a function that is not as steep, in order to account for small structural changes in the protein. While force field scoring performs quite well on selecting docking modes, it is difficult to compare different molecules on this basis. The main reason is that entropic and solvation effects are not considered. A combination of a force field score with an approximation of the solvation energy can improve the results but is computationally expensive [116].

A frequently applied program for force field scoring is GRID [38]. Instead of atom-based potentials, the software calculates interaction potentials on a grid for frequently occurring molecular groups based on a force field energy function. These maps can be either used for scoring or for detection and visualization of important interaction centers in the active site of the protein.

Empirical scoring functions are based on a correlation of geometric parameters of the protein-ligand complex to measured free binding energies. The overall score is a sum of terms representing different physico-chemical effects contributing to the binding energy like hydrogen bonding, hydrophobic contacts, etc.. For each term a functional form and a set of geometric parameters derivable from the protein–ligand complex are chosen. Usually, weights for the individual terms are determined from a set of protein–ligand complexes with known structures and binding energies using multiple linear regression or partial least squares.

The first function of this type was developed by Böhm [70] for the *de novo* design program LUDI [35, 36]:

$$\Delta G = \Delta G_0 + \Delta G_{rot} \times N_{rot} + \Delta G_{hb} \sum_{neutral\ H-bonds} f(\Delta R, \Delta\alpha)$$

$$+ \Delta G_{io} \sum_{ionic\ int.} f(\Delta R, \Delta\alpha) + \Delta G_{lipo}|A_{lipo}|$$

Here, $f(\Delta R, \Delta\alpha)$ is a scaling function penalizing deviations from the ideal geometry of hydrogen bonds and ionic interactions, N_{rot} is the number of free rotatable bonds of the ligand, and A_{lipo} is the lipophilic contact area between the protein and the ligand. The terms ΔG_0, ΔG_{rot}, ΔG_{hb}, ΔG_{io}, and ΔG_{lipo} are adjustable parameters.

The Böhm function was further improved and customized for screening applications [70, 135, 136]. Since Böhm published his function, several other empirical scoring functions were developed [137–142]. The functions differ in the list of physico-chemical effects considered and the functional form for the individual terms.

Recently, three groups independently developed knowledge-based scoring functions for molecular docking [143–146]. This scoring method was previously used by Sippl [147, 148] for protein structure prediction and is successfully applied in this field. The idea underlying knowledge-based functions is that situations frequently seen in structures are energetically favorable. All three functions are based on inter-atom distances. For each atom type pair, the number of occurrences are counted depending on the distance. The resulting histograms are converted to an energy function. The score for a given complex is then calculated by adding the energy function values for all inter-atom pairs of the complex. Gohlke [146] combined the atom-pair score with a solvent-accessible surface term in order to take solvation into account. Mügge [145] added a volume correction factor to account for the ligand volume around each ligand atom which avoids interactions with the protein.

7.3
Validation studies and applications

All docking algorithms presented are heuristic algorithms giving no guarantee of the quality of the result. In addition, the underlying physico-chemical models are very rough. Therefore, an intensive validation of the docking algorithm is an essential requirement.

7.3.1
Reproducing X-ray structures

The first test usually performed with a docking tool is the reproduction of complexes with known structures. The results achieved with docking algorithms are quite difficult to compare for several reasons. First, there is no unique benchmark set and every method is published with a small set of test results only. Second, different levels of *a priori* knowledge are used in the preparation of the input data and in the calculation. Examples for *a priori* knowledge are the definition of the active site, the placement of polar hydrogens, the limitation of conformational or translational degrees of freedom, the removal of ligand parts located outside the active site, or simply the selection of considered test cases. Finally, it is a non-trivial task to reasonably quantify the quality of a docking calculation. For a single result, the RMSD (root mean square deviation) can be used. However, this value can be misleading if the molecule has nearly symmetric groups. In addition, the bound for what is called a 'good result' depends on the test case, important criteria are the size of the ligand and the overall hydrophobicity.

Two evaluation studies on large test sets can be found in the literature [85, 149]. Jones et al. applied the genetic algorithm GOLD to a set of 100 protein–ligand complexes taken from the PDB. Kramer et al. evaluated the FLEXX program with 200 protein–ligand complexes from the PDB. In general it can be said that about 65–70% of the test cases can be predicted in agreement (≤ 2.0 Å RMSD) with the crystal structure. This number varies with the flexibility of the ligand molecule: the larger and more flexible the ligand, the smaller is the chance to reproduce the crystallographic complex. This is not surprising, since the search space that has to be explored grows exponentially with the number of rotatable bonds in the ligand.

In both studies, the protein structure was taken in the complex conformation. One has to keep in mind that, in a real-case scenario, the protein can undergo conformational changes due to an induced fit during ligand binding. Obviously, the fraction of correct predictions will decrease with the amount of flexibility in the active site of the protein. In both papers, this problem is addressed in some additional calculations employing either protein structures crystallized without a ligand or structures of the same protein complexed with different ligands. A more detailed study on the influence of protein flexibility using the docking software PRO_LEADS can be found in [150].

Finally, we mention two papers which directly compare different algorithms for protein ligand docking. Vieth et al. [151] compared genetic algorithms (GA) with Monte-Carlo (MC) and Molecular Dynamics calculations (MD). They conclude that GA works best within a given number of energy evaluations while MD gives more accurate predictions. Westhead et al. [152] compared four heuristic search strategies: a genetic algorithm, an evolutionary program, simulated annealing and tabu search. Their analysis showed that the genetic algorithm finds the placements with lowest energy but the tabu search created placements closer to the crystallographically observed complexes.

The results achieved with a specific method depend strongly on the modeling of the problem into the algorithmic framework. The individual approaches based on genetic algorithms presented above illustrate how much freedom for modeling the problem there still is although the overall optimization technique is fixed. Therefore, the comparisons cannot be considered as a general statement for or against an optimization method.

7.3.2
Validated blind predictions

Most protein–ligand docking programs are designed empirically. First, the scoring function is derived from experimental data. Second, the algorithms are designed with background knowledge about protein–ligand complexes. Therefore, either testing the program on a large set of examples or performing blind predictions is essential for evaluating the capabilities of the software. Testing a large set of examples is similar to a blind prediction in the sense that it is unlikely that information about most of the complexes has flown into the software development.

Performing a validated blind prediction is a difficult task and requires the availability of the software, a test scenario, and the experimental facilities for validating the study. For protein–protein docking, a blind prediction of the complex between TEM-1 β-lactamase and the inhibitor BLIP was performed by 6 independent groups. All groups were able to identify the correct complex within 2 Å RMSD [153].

In CASP II (Critical Assessment of Methods of Protein Structure Prediction: Round II), a new prediction section on protein–ligand docking was introduced [154]. Several groups participated and submitted up to three models for some of the seven protein–ligand complexes (see [155–158] for reports). For each target, the 3D structure of the protein and the 2D structure of the ligand was given to the participants. All complexes were unpublished before the submission deadline.

The results show that accurate molecular docking is still a challenging problem. Although predictions within 3 Å were made for nearly all targets and within 2 Å for more than half of the cases, there is no method currently available that reliably and consistently predicts correct protein–ligand complex geometries. Dixon [154] concludes that, since the right solutions were often found but not ranked highest, there is a need to develop more accurate and reliable scoring functions.

7.3.3
Screening small molecule databases

The main application of molecular docking software is the virtual screening of compound databases for potential lead molecules. Testing docking software within this scenario is quite difficult and only a few papers discuss screening results with molecular docking software. Since software is often developed by groups not directly involved in drug design projects, experimental validation of the achieved screening results is not possible. Pharmaceutical companies that use docking software for preselecting compounds typically test the selected molecules such that a hit rate (the number of active molecules divided by the number of selected molecules) can be calculated. However, the number of active molecules within the whole data set before the preselection is still unknown after this experiment such that selectivity (the number of actives in the selected subset) can be measured but sensitivity (the number of selected actives with respect to the number of active in the database) cannot.

A useful measure for evaluating methods for preselecting compounds is the enrichment factor which is the hit rate of the method divided by the random hit rate (which is the total number of active molecules divided by the size of the data set). The enrichment factor shows how much the method improves the screening result compared to a random selection. Note that for calculating the enrichment factor, one has to know the total number of active molecules which can only be guessed in test scenarios, i.e., the whole database except the known actives is assumed to be inactive. In the following, a number of papers containing different types of evaluations studies are summarized.

Burkhard et al. [159] developed a software tool for rigid-body molecular docking called SANDOCK. The method is based on matching surfaces with mapped properties and is not described in detail here. The authors applied SANDOCK to searching for new thrombin inhibitors. One of the hits was *p*-amino-benzamidine, a small molecule with ten non-hydrogen atoms, was experimentally validated. The structure predicted for the complex was in agreement with the crystallographic structure determined afterwards.

Shoichet et al. [33] presented an extension of the DOCK scoring function taking solvation effects into account. In order to test the function, a screening run of the ACD with about 150 000 molecules was performed for three different targets using the rigid-body variant of DOCK. Kramer et al. [149] applied FlexX for screening a small set of about 550 molecules against 11 targets with known nanomolar inhibitors. The screening data set was created by mixing ligands of 200 experimentally solved protein–ligand complexes with drug-related molecules selected from the CSD (Cambridge Structural Database) [67]. Knegtel and Wagener [160] presented a screening application of DOCK for thrombin inhibitors and progesterone agonists. The compound collection used consists of about 1000 molecules. Since activity information was available for all molecules, enrichment factors could be calculated. They conclude that docking is effective in preselecting data sets, however the methods were not able to distinguish between true actives and similar inactives.

A comprehensive study on different scoring functions in combination with docking was recently presented by Charifson et al. [161]. In the study 13 different scoring functions in combination with two docking algorithms were applied to database searching. The database was formed from known inhibitors for three different targets mixed with 10 000 random molecules passing a drug-likeness filter. For each molecule, a set of conformers was generated and docked with DOCK and an in-house developed genetic algorithm called GAMBLER. The authors demonstrate that the number of hits found depend strongly on the scoring function used and that there is not a single function with best performance. As a consequence, a consensus scoring method was developed showing better performance than the individual functions.

7.3.4
Docking of combinatorial libraries

As in the case of database screening, the experimental validation of combinatorial docking methods is difficult to perform and therefore rarely found. The main reason is the lack of publicly available data on binding affinities for combinatorial libraries to proteins with known 3D structure.

Kick et al. [124] applied the *de novo* design program BUILDER [125] in a combinatorial fashion to the design of a focussed library for Cathepsin D. BUILDER is a fragment-based *de novo* design program related to the docking software DOCK discussed above. The program was used to evaluate putative R-groups for a given scaffold oriented in the active site of Cathepsin D. The created focussed library was synthesized and tested. Containing 30 compounds inhibiting Cathepsin D in the nanomolar range, the method improved the hit rate by a factor of 7.5 compared to a diverse library. The same library was used later to evaluate CombiDOCK [122].

A frequently used target protein for evaluating docking algorithms is thrombin already mentioned in the introduction. Böhm et al. [126] experimentally evaluated ligands proposed from a virtual combinatorial library by LUDI. They showed that five of ten proposed compounds exhibit binding affinities in the nanomolar range. Caflisch [127] and Murray [121] used human α-thrombin for demonstrating their combinatorial docking and design tools MCSS-CCLD and PRO_SELECT. They showed that ligands with well known binding patterns are created.

In order to evaluate FlexX^c [123], we used three different data sets for thrombin and DHFR for which experimental binding data was available (thrombin: [162, 163], DHFR: [164]). Molecules in the DHFR data set were very closely related such that no correlation between predicted and experimental binding affinity was found. For the first thrombin data set, a correlation of 0.6 was achieved. For the second thrombin data set, a molecule with high binding affinity was known which could be identified at a high rank.

7.4
Molecular docking in practice

In the previous sections, the algorithms for molecular docking and their performance on example data were discussed. In practical applications however, several additional issues arise. Pre- and post-processing steps are necessary on the input and output data and a docking algorithm most appropriate for the application scenario has to be chosen.

7.4.1
Preparing input data

Typically, the input data needed to perform a docking calculation is not in the appropriate format such that preprocessing steps are necessary. The situation is substantially different for the protein and the ligand molecule. All relevant application scenarios deal with only a single protein structure, the target protein, while typically large numbers of ligand molecules are involved. Therefore, manual intervention during protein preprocessing can be afforded while automatic tools are needed for the ligand molecules.

The most common starting point for the protein is the 3D structure in PDB [165] format containing coordinates for all heavy atoms. Additional information necessary for most docking programs are the location of the active site and the protonation states of polar atoms, and in some cases the location of the hydrogen atoms. In addition, ambiguities like alternate locations or conformations of amino acid side chains have to be resolved. Although automatic tools are available for most of these steps, manually driven approaches should be preferred since the exact definition of the protein structure has a large effect on the docking results. Often, many properties of the target protein are known or can be derived from already known ligands resulting in an improved model of the protein active site.

The ligand molecules are taken from a compound databases typically containing only structural formulae of the individual molecules. In order to use them in a docking calculation, 3D structures of the molecules have be generated. For this step, very powerful tools like Concord [166] or Corina [68, 69] exist. In addition, atom hybridization and protonation as well as bond types have to be determined. Typically, this task is performed by a series of subsequently executed rule-based scripts. It should be noted that due to the complexity of this task, the resulting structures are not free of errors in the assignment of atom and bond types. The docking software should therefore rely as little as possible on this information.

7.4.2
Analyzing docking results

The output of a docking program is a list of complex structures and an estimate for the binding affinity, the score, for each ligand. If the structures should be further examined, an energy minimization can be used for an initial refinement. Most docking programs are based on a discrete conformational model and therefore have to be tolerant against small overlaps between protein and ligand atoms. The minimization reduces these artefacts of combinatorial docking algorithms and moves the complex into the closest local minimum.

As already discussed in Section 7.2.5, estimating the binding affinity is a very complex task. The scores calculated by docking programs should therefore be seen only as a very crude measure. Depending on the number of ligand molecules, different techniques can be used for rescoring. If only few structures are of interest, time consuming methods like free energy perturbation can be applied to get a more detailed view on differences in binding energy between alternative ligands [167]. In most applications however, the number of ligands to be processed is too large for these techniques. One approach to overcome the limitations of single scoring functions is consensus scoring [161]. Depending on the screening performance achieved for a known set of ligands, a scoring function or a combination of functions can be chosen for calculating a final ranking for molecules with unknown binding affinity.

7.4.3
Choosing the right docking tool

In Section 7.1.2, the most frequent docking scenarios occurring in drug design projects have been introduced. Since a variety of docking algorithms have been developed, the question arises which algorithm or software is most useful for the individual scenarios.

If only a small set of ligands should be docked, one can choose from the full spectrum of methods since then the computing time is only a minor issue. The most promising approaches are probably the hybrid methods combining a very efficient search technique in a first phase with a very exact scoring and optimization technique in the second phase. The first phase is of importance since it guarantees a broad sampling within the solution space. The second phase then tries to find the local minima and scores them as reliably as possible.

If the set of ligands increases, a tradeoff between accuracy and computing time has to be made. Rigid-body docking tools are most efficient, but they can only be used if the ligands of interest have only a small number of low-energy conformations. If this is not the case, the chance of missing hits due to missing conformations is too high. In the frequent case that the ligands are flexible (about more than five rotatable bonds), a tool handling conformational flexibility during docking must be used. Using a rigid-body tool as a prefilter is not helpful here. A good prefilter has a low false-negative rate resulting in a low number of hits removed. The false-positive rate is of less importance, since all molecules passing the filter will be analyzed in more detail in the subsequent screening step. A rigid-body docking tool however has the opposite effect. The false-negative rate will be high since all molecules not tested in the binding conformation will be filtered out, while molecules tested in the binding conformation are kept resulting in a relatively low false-positive rate.

Besides computing time issues, known protein–ligand complexes with the target protein may also help in selecting the most promising docking algorithm. First of all, the degree of conformational flexibility in the protein while binding different ligands should be analyzed. If there are significant changes, a docking tool considering these effects is necessary. Docking tools handling protein flexibility are very rare and often not fast enough to be used in screening scenarios. Some of these tools are mentioned in Section 7.2. As an alternative, a docking algorithm can be run several times with different structures of the target protein as input. Of course, this method can only be applied if the protein conformations fall into a low number of separate classes.

Besides analyzing protein flexibility, the shape of the active site and the orientation of different ligands is of importance. Often, the active site has a deep cavity as it is the case in the thrombin example shown in the introduction. This is a favorable situation for incremental construction algorithms since they achieve the best results if the can place a first anchoring group deep into the active site and then continue growing to the outside. The situation becomes more difficult if the active site is very shallow or tunnel-shaped (e.g., in HIV-protease). If, in addition, the key interactions to the protein are spread over the whole ligand structure, the chance of missing the global minimum when using an incremental construction algorithm increases. Here, a more generic search technique like a genetic algorithm or a monte carlo algorithm could be favorable.

7.5
Concluding remarks

Molecular docking tools are still far away from giving reliable results for all kinds of protein-ligand complexes. The complexity of the problem increases with the number of rotatable bonds in the ligand and the flexibility of the active site. In addition there is the lack of a reliable scoring function for the prediction of the free energy of binding.

The two main results of a docking calculation are the predicted complex structures for individual ligands as well as the ranking of binding molecules with respect to non-binding ones. Concerning structure prediction, docking software is able to generate a complex within a reasonable error bound (less than 2 Å RMSD) typically in about 70% of the test cases. In virtual screening, the current bottleneck is the scoring problem. Even if correct structures are generated, scoring functions often achieve only a weak correlation with experimental binding affinities. Nevertheless, molecular docking is already a quite useful tool in the process of structure-based drug design. It allows to search in large virtual chemical spaces in a time- and cost-efficient way. Docking tools are successfully applied to prioritize compound databases for screening, designing targeted combinatorial libraries, and improving the binding affinities of initial lead structures.

Within the next years, we can expect a further increase in computing speed and probably an improvement in the quality of scoring functions. Looking at whole molecular libraries in a closed rather than an enumerated form has a great potential of speeding up docking calculations. Algorithms for docking combinatorial libraries as discussed above have shown this already. It can be assumed that once methods are available which overcome the sequential screening paradigm by new search techniques, a speedup of several orders of magnitude can be achieved.

Scoring functions are of central importance for virtual screening applications. They have to give reliable estimates of differences in binding affinity and must be efficiently computable in order to be used on thousands of structures. Reviewing the literature on this topic, improvements can be seen in the quality of scoring functions over the past years. Very recently, the idea of consensus scoring emerged combining the individual strengths of several scoring functions. While, in the past, most scoring functions were developed on small sets of complexes with known structure and binding affinity, the focus is shifting now towards more realistic screening scenarios as the basis for developing and calibrating scoring functions. Although no breakthrough is in sight, further improvement in the next years can be expected.

The docking problem in its full generality is likely to remain an open problem even in the long term. Too many subproblems are still unsolved, the most urgent ones being the accurate prediction of binding affinity and the handling of protein conformational flexibility.

7.6
Software accessibility

Table 7.1 lists some available software packages for molecular docking discussed in this article. More information can be found on the corresponding Internet pages.

Acknowledgments

The author thanks the editor Thomas Lengauer and Martin Stahl (Fa.Hoffmann La-Roche, Basel) for carefully reading the first version of this manuscript and making lots of valuable comments. This work was partially funded by the BMBF (Bundesministerium für Bildung und Forschung) under grant 0311620 (Project Relimo).

Tab. 7.1
Software for automated molecular docking

Name	Refs.	WWW Homepage
AutoDock	[96, 98, 97, 83]	http://www.scripps.edu/pub/olson-web/doc/autodock/index.html
DOCK	[21, 26, 30, 25, 27, 28]	http://www.cmpharm.ucsf.edu/kuntz/dock.html
DockVision	[107]	http://www.dockvision.com
FlexX	[63, 64, 65, 72, 71, 149]	http://cartan.gmd.de/FlexX
GOLD	[79, 85]	http://www.ccdc.cam.ac.uk/prods/gold.html
ICM	[110, 168]	http://www.molsoft.com/icmpages/icmdock.htm
LIGIN	[93]	http://swift.embl-heidelberg.de/ligin
QXP	[108]	

References

1 J. M. BLANEY and J. S. DIXON. A good ligand is hard to find: Automated docking methods. *Perspectives in Drug Discovery and Design*, 1:301–319, **1993**.

2 H. KUBINYI. Structure-based design of enzyme inhibitors and receptor ligands. *Current Opinion in Drug Discovery and Development*, 1:4–15, **1998**.

3 J. D. HIRST. Predicting ligand binding energies. *Current Opinion in Drug Discovery and Development*, 1:28–33, **1998**.

4 R. M. A. KNEGTEL and P. D. J. GROOTENHUIS. Binding affinities and non-bonded interaction energies. *Perspectives in Drug Discovery and Design*, 9/10/11:99–114, **1998**.

5 J. R. H. TAME. Scoring functions: A view from the bench. *Journal of Computer-Aided Molecular Design*, 13:99–108, **1999**.

6 I. D. KUNTZ. Structure-based strategies for drug design and discovery. *Science*, 257:1078–1082, **1992**.

7 R. A. LEWIS and E. C. MENG. A discussion of various computational methods for drug design. **1993**.

8 W. C. GUIDA. Software for structure-based drug design. *Current Opinion in Structural Biology*, 4:777–781, **1994**.

9 P. M. COLMAN. Structure-based drug design. *Current Opinion in Structural Biology*, 4:868–874, **1994**.

10 T. P. LYBRAND. Ligand–protein docking and rational drug design. *Current Opinion in Structural Biology*, 5:224–228, **1995**.

11 R. ROSENFELD, S. VAJDA, and C. DELISI. Flexible docking and design. *Annual Reviews in Biophysics and Biomolecular Structure*, 24:677–700, **1995**.

12 H.-J. BÖHM. Current computational tools for de novo ligand design. *Current Opinion in Biotechnology*, 7:433–436, **1996**.

13 T. LENGAUER and M. RAREY. Computational methods for biomolecular docking. *Current Opinion in Structural Biology*, 6:402–406, **1996**.

14 M. J. E. STERNBERG, H. A. GABB, and R. M. JACKSON. Predictive docking of protein–protein and protein–DNA complexes. *Current Opinion in Structural Biology*, 8:250–256, **1999**.

15 P. PRABHAKAR and A. M. KAYASTHA. Mechanism of DNA–drug interactions. *Applied Biochemistry and Biotechnology*, 47:39–55, **1994**.

16 P. D. J. GROOTENHUIS, D. C. ROE, P. A. KOLLMAN, and I. D. KUNTZ. Finding potential DNA-binding compounds by using molecular shape. *Journal of Computer-Aided Molecular Design*, 8:731–750, **1994**.

17 M. ZACHARIAS and H. SKLENAR. Harmonic modes as variables to approximately account for receptor flexibility in ligand–receptor docking simulations: Application to DNA minor groove ligand complex. *Journal of Computational Chemistry*, 20:287–300, **1999**.

18 M. A. GALLOP, R. W. BARRETT, W. J. DOWER, P. A. FODOR, and E. M. GORDON. Applications of combinatorial technologies to drug discovery. 1. background and peptide combinatorial libraries. *Journal of Medicinal Chemistry*, 37:1233–1251, **1994**.

19 E. M. GORDON, R. W. BARRETT, W. J. DOWER, P. A. FODOR, and M. A. GALLOP. Applications of combinatorial technologies to drug discovery. 2. combinatorial organic synthesis, library screening strategies, and future directions. *Journal of Medicinal Chemistry*, 37:1386–1401, **1994**.

20 F. S. KUHL, G. M. CRIPPEN, and D. K. FRIESEN. A combinatorial algorithm for calculating ligand binding. *Journal of Computational Chemistry*, 5(1):24–34, **1984**.

21 I. D. KUNTZ, J. M. BLANEY, S. J. OATLEY, R. L. LANGRIDGE, and T. E. FERRIN. A geometric approach to macromolecule–ligand interactions. *Journal of Molecular Biology*, 161:269–288, **1982**.

22 F. M. RICHARDS. Areas, volumes, packing, and protein structure. *Annual Reviews in Biophysics and Bioengineering*, 6:151–176, **1977**.

23 M. L. CONNOLLY. Analytical molecular surface calculation. *Journal of Applied Crystallography*, 16:548–558, **1983**.

24 M. L. CONNOLLY. Molecular surface triangulation. *Journal of Applied Crystallography*, 18:499–505, **1985**.

25 B. K. SHOICHET and I. D. KUNTZ. Matching chemistry and shape in molecular docking. *Protein Engineering*, 6(7):723–732, **1993**.

26 B. K. SHOICHET, D. L. BODIAN, and I. D. KUNTZ. Molecular docking using shape descriptors. *Journal of Computational Chemistry*, 13(3):380–397, **1992**.

27 E. C. MENG, D. A. GSCHWEND, J. M. BLANEY, and I. D. KUNTZ. Orientational sampling and rigid-body minimization in molecular docking. *PROTEINS: Structure, Function and Genetics*, 17:266–278, **1993**.

28 T. J. A. EWING and I. D. KUNTZ. Critical evaluation of search algorithms for automated molecular docking and database screening. *Journal of Computational Chemistry*, 18:1175–1189, **1997**.

29 R. M. A. KNEGTEL, I. D. KUNTZ, and C. M. OSHIRO. Molecular docking to ensembles of protein structures. *Journal of Molecular Biology*, 266:424–440, **1997**.

30 E. C. MENG, B. K. SHOICHET, and I. D. KUNTZ. Automated docking with grid-based energy evaluation. *Journal of Computational Chemistry*, 13(4):505–524, **1992**.

31 E. C. MENG, I. D. KUNTZ, D. J. ABRAHAM, and G. E. KELLOGG. Evaluating docked complexes with the HINT exponential function and empirical atomic hydrophobicities. *Journal of Computer-Aided Molecular Design*, 8:299–306, **1994**.

32 D. A. GSCHWEND and I. D. KUNTZ. Orientational sampling and rigid-body minimization in molecular docking revisited: On-the-fly optimization and degeneracy removal. *Journal of Computer-Aided Molecular Design*, 10:123–132, **1996**.

33 B. K. SHOICHET, A. R. LEACH, and I. D. KUNTZ. Ligand solvation in molecular docking. *PROTEINS: Structure, Function and Genetics*, 34:4–16, **1999**.

34 M. Y. MIZUTANI, N. TOMIOKA, and A. ITAI. Rational automatic search method for stable docking models of protein and ligand. *Journal of Molecular Biology*, 243:310–326, **1994**.

35 H.-J. BÖHM. The computer program LUDI: A new method for the de novo design of enzyme inhibitors. *Journal of Computer-Aided Molecular Design*, 6:61–78, **1992**.

36 H.-J. BÖHM. LUDI: rule-based automatic design of new substituents for enzyme inhibitor leads. *Journal of Computer-Aided Molecular Design*, 6:593–606, **1992**.

37 M. C. LAWRENCE and P. C. DAVIS. CLIX: A search algorithm for finding novel ligands capable of binding proteins of known three-dimensional structure. *PROTEINS: Structure, Function and Genetics*, 12:31–41, **1992**.

38 P. J. GOODFORD. A computational procedure for determining energetically favorable binding sites on biologically important macromolecules. *Journal of Medicinal Chemistry*, 28:849–857, **1985**.

39 Y. LAMDAN and H. J. WOLFSON. Geometric hashing: A general and efficient model-based recognition scheme. *Proceedings of the IEEE International Conference on Computer Vision*, pages 238–249, **1988**.

40 D. FISCHER, R. NOREL, R. NUSSINOV, and H. J. WOLFSON. 3-D docking of protein molecules. In *Combinatorial Pattern Matching, Proceedings of the Fourth Annual Symposium (CPM'93)*, pages 20–34. Springer Lecture Notes in Computer Science No. 684, Heidelberg, Germany, **1993**.

41 D. FISCHER, S. L. LIN, H. L. WOLFSON, and R. NUSSINOV. A geometry-based suite of molecular docking processes. *Journal of Molecular Biology*, 248:459–477, **1995**.

42 S. LINNAINMAA, D. HARWOOD, and L. S. DAVIS. Pose determination of a three-dimensional object using triangle pairs. *IEEE Transactions on Pattern Analysis and Machine Intelligence*, 10(5):634–646, **1988**.

43 M. RAREY, S. WEFING, and T. LENGAUER. Placement of medium-sized molecular fragments into active sites of proteins. *Journal of Computer-Aided Molecular Design*, 10:41–54, **1996**.

44 D. R. Ferro and J. Hermans. A different best rigid-body molecular fit routine. *Acta Crystallographica*, A33:345–347, 1977.

45 M. J. Sippl and H. Stegebuchner. Superposition of three dimensional objects: A fast and numerically stable algorithm for the calculation of the matrix of optimal rotation. *Computers and Chemistry*, 15:73–78, 1991.

46 W. Kabsch. A solution for the best rotation to relate two sets of vectors. *Acta Crystallographica*, A32:922–923, 1976.

47 W. Kabsch. A discussion of the solution for the best rotation to relate two sets of vectors. *Acta Crystallographica*, A34:827–828, 1978.

48 M. Vieth, J. D. Hirst, and C. L. Brooks III. Do active site conformations of small ligands correspond to low free-energy solution structures? *Journal of Computer-Aided Molecular Design*, 12:563–572, 1998.

49 S. K. Kearsley, D. J. Underwood, R. P. Sheridan, and M. D. Miller. Flexibases: A way to enhance the use of molecular docking methods. *Journal of Computer-Aided Molecular Design*, 8:565–582, 1994.

50 T. F. Havel, I. D. Kuntz, and G. M. Crippen. The combinatorial distance geometry approach to the calculation of molecular conformation I. a new approach to an old problem. *Journal of Theoretical Biology*, 104:359–381, 1983.

51 T. F. Havel, I. D. Kuntz, and G. M. Crippen. The theory and practice of distance geometry. *Bulletin of Mathematical Biology*, 45(5):665–720, 1983.

52 G. M. Crippen and T. F. Havel. *Distance Geometry and Molecular Conformation*. Research Studies Press, Taunton, England, 1988.

53 M. D. Miller, S. K. Kearsley, D. J. Underwood, and R. P. Sheridan. FLOG: A system to select 'quasiexible' ligands complementary to a receptor of known three-dimensional structure. *Journal of Computer-Aided Molecular Design*, 8:153–174, 1994.

54 D. M. Lorber and B. K. Shoichet. Flexible ligand docking using conformational ensembles. *Protein Science*, 7:938–950, 1998.

55 R. L. DesJarlais, R. P. Sheridan, J. S. Dixon, I. D. Kuntz, and R. Venkataraghavan. Docking flexible ligands to macromolecular receptors by molecular shape. *Journal of Medicinal Chemistry*, 29:2149–2153, 1986.

56 B. Sandak, H. J. Wolfson, and R. Nussinov. Hinge-bending motion at molecular interfaces: Computerized docking of a dihydroxyethylene-containing inhibitor to the hiv-1 protease. *Journal of Biomolecular Structure and Dynamics*, 1:233–252, 1996.

57 B. Sandak, R. Nussinov, and H. J. Wolfson. 3-D flexible docking of molecules. *IEEE Workshop on Shape and Pattern Matching in Computational Biology*, pages 41–54, 1994.

58 B. Sandak, R. Nussinov, and H. J. Wolfson. An automated computer vision and robotics-based technique for 3-D flexible biomolecular docking and matching. *Computer Applications in Biological Sciences*, 11(1):87–99, 1995.

59 B. SANDAK, R. NUSSINOV, and H. J. WOLFSON. A method for biomolecular structural recognition and docking allowing conformational flexibility. *Journal of Compuational Biology*, 5:631–654, **1998.**

60 J. B. MOON and W. J. HOWE. Computer design of bioactive molecules: A method for receptor-based de novo ligand design. *PROTEINS: Structure, Function and Genetics*, 11:314–328, **1991.**

61 A. R. LEACH and I. D. KUNTZ. Conformational analysis of flexible ligands in macromolecular receptor sites. *Journal of Computational Chemistry*, 13:730–748, **1992.**

62 A. R. LEACH. Ligand docking to proteins with discrete side-chain flexibility. *Journal of Molecular Biology*, 235:345–356, **1994.**

63 M. RAREY, B. KRAMER, and T. LENGAUER. Time-efficient docking of flexible ligands into active sites of proteins. In C. Rawlings et al., editor, *Proceedings of the Third International Conference on Intelligent Systems in Molecular Biology*, pages 300–308. AAAI Press, Menlo Park, California, **1995.**

64 M. RAREY, B. KRAMER, T. LENGAUER, and G. KLEBE. A fast flexible docking method using an incremental construction algorithm. *Journal of Molecular Biology*, 261(3):470–489, **1996.**

65 M. RAREY, B. KRAMER, and T. LENGAUER. Multiple automatic base selection: Protein–ligand docking based on incremental construction without manual intervention. *Journal of Computer-Aided Molecular Design*, 11:369–384, **1997.**

66 G. KLEBE and T. MIETZNER. A fast and efficient method to generate biologically relevant conformations. *Journal of Computer-Aided Molecular Design*, 8:583–606, **1994.**

67 F. H. ALLEN, S. BELLARD, M. D. BRICE, B. A. CARTWRIGHT, A. DOUBLEDAY, H. HIGGS, T. HUMMELINK-PETERS, O. KENNARD, W. D. S. MOTHERWELL, J. R. RODGERS, and D. G. WATSON. The Cambridge Crystallographic Data Centre: computer-based search, retrieval, analysis and display of information. *Acta Crystallographica*, B35:2331–2339, **1979.**

68 J. GASTEIGER, C. RUDOLPH, and J. SADOWSKI. Automatic generation of 3D-atomic coordinates for organic molecules. *Tetrahedron Computer Methodology*, 3:537–547, **1990.**

69 J. SADOWSKI, J. GASTEIGER, and G. KLEBE. Comparison of automatic three-dimensional model builders using 639 x-ray structures. *Journal of Chemical Information and Computer Science*, 34:1000–1008, **1994.**

70 H.-J. BÖHM. The development of a simple empirical scoring function to estimate the binding constant for a protein–ligand complex of known three-dimensional structure. *Journal of Computer-Aided Molecular Design*, 8:243–256, **1994.**

71 M. RAREY, B. KRAMER, and T. LENGAUER. Docking of hydrophobic ligands with interaction-based matching algorithms. *Bioinformatics*, 15:243–250, **1999.**

72 M. RAREY, B. KRAMER, and T. LENGAUER. The particle concept: Placing discrete water molecules during protein–ligand docking predictions. *PROTEINS: Structure, Function and Genetics*, 34(1):17–28, **1999.**

73 H. CLAUßEN, C. BUNING, M. RAREY, and T. LENGAUER. Molecular docking into the flexible active site of aldose reductase using FlexE. In *Rational Approaches to Drug Design: Proceedings of 13th European Symposium on Quantitative Structure-Activity Relationships*. Prous Science, Barcelona, H.-D. HOLLJE, W. SIPPL (eds.), 324–333, **2001**.

74 H. CLAUßEN, C. BUNING, M. RAREY, and T. LENGAUER. FlexE: Efficient Molecular Docking into Flexible Protein Structures. *Journal of Molecular Biology*, 308:377–395, **2001**.

75 W. WELCH, J. RUPPERT, and A. N. JAIN. Hammerhead: fast, fully automated docking of flexible ligands to protein binding sites. *Chemistry & Biology*, 3:449–462, **1996**.

76 S. MAKINO and I. D. KUNTZ. Automated flexible ligand docking method and its application for database search. *Journal of Computational Chemistry*, 18(4):1812–1825, **1997**.

77 S. J. WEINER, P. A. KOLLMAN, D. A. CASE, U. C. SINGH, C. GHIO, G. ALAGONA, S. Jr. PROFETA, and P. WEINER. A new force field for molecular mechanical simulation of nucleic acids and proteins. *Journal of the American Chemical Society*, 106:765, **1984**.

78 D. A. PEARLMAN, D. A. CASE, J. W. CALDWELL, W. S. ROSS, T. E. III CHEATHAM, S. DEBOLT, D. FERGUSON, G. SEIBEL, and P. KOLLMAN. AMBER, a package of computer programs for applying molecular dynamics, normal mode analysis, molecular dynamics and free energy calculations to simulate the structural and energetic properties of molecules. *Computer Physics Communications*, 91:1–41, **1995**.

79 G. JONES, P. WILLETT, and R. C. GLEN. Molecular recognition of receptor sites using a genetic algorithm with a description of desolvation. *Journal of Molecular Biology*, 245:43–53, **1995**.

80 C. M. OSHIRO, I. D. KUNTZ, and J. S. DIXON. Flexible ligand docking using a genetic algorithm. *Journal of Computer-Aided Molecular Design*, 9:113–130, **1995**.

81 K. P. CLARK and AJAY. Flexible ligand docking without parameter adjustment across four ligand–receptor complexes. *Journal of Computational Chemistry*, 16(10):1210–1226, **1995**.

82 D. K. GELHAAR, G. VERKHIVKER, P. A. REJTO, D. B. FOGEL, L. J. FOGEL, and S. T. FREER. Docking conformationally flexible small molecules into a protein binding site through evolutionary programming. In J. R. McDonnell, R. G. Reynolds, and D. B. Fogel, editors, *Proceedings of the Fourth Annual Conference on Evolutionary Programming*, pages 615–627, **1995**.

83 G. M. MORRIS, D. S. GOODSELL, R. S. HALLIDAY, R. HUEY, W. E. HART, R. K. BELEW, and A. J. OLSON. Automated docking using a lamarckian genetic algorithm and an empirical binding free energy function. *Journal of Computational Chemistry*, 19:1639–1662, **1998**.

84 D. E. GOLDBERG. *Genetic Algorithms in Search Optimization and Machine Learning*. Addison–Wesley, Reading, MA, **1989**.

85 G. JONES, P. WILLETT, R. C. GLEN, A. R. LEACH, and R. TAYLOR. Development and validation of a genetic algorithm for flexible docking. *Journal of Molecular Biology*, 267:727–748, **1997**.

86 D. K. GEHLHAAR, G. M. VERKHIVKER, P. A. REJTO, C. J. SHERMAN, D. B. FOGEL, L. J. FOGEL, and S. T. FREER. Molecular recognition of the inhibitor ag-1343 by HIV-1 protease: conformationally flexible docking by evolutionary programming. *Chemistry & Biology*, 2:317–324, **1995**.

87 A. W. M. DRESS and T. F. HAVEL. Shortest path problems and molecular conformation. *Discrete Applied Mathematics*, 19:129–144, **1988**.

88 P. L. EASTHOPE and T. F. HAVEL. Computational experience with an algorithm for tetrangle inequality bound smoothing. *Bulletin of Mathematical Biology*, 51(1):173–194, **1989**.

89 A. K. GHOSE and G. M. CRIPPEN. Geometrically feasible binding modes of a flexible ligand molecule at the receptor site. *Journal of Computational Chemistry*, 6(5):350–359, **1985**.

90 A. S. SMELLIE, G. M. CRIPPEN, and W. G. RICHARDS. Fast drug-receptor mapping by site-directed distances: A novel method of predicting new pharmacological leads. *Journal of Chemical Information and Computer Science*, 31:386–392, **1991**.

91 M. BILLETER, T. F. HAVEL, and I. D. KUNTZ. A new approach to the problem of docking two molecules: The ellipsoid algorithm. *Biopolymers*, 26:777–793, **1987**.

92 V. SCHNECKE, C. A. SWANSON, E. D. GETZOFF, J. A. TAINER, and L. A. KUHN. Screening a peptidyl database for potential ligands to proteins with side-chain flexibility. *PROTEINS: Structure, Function and Genetics*, 33:74–87, **1998**.

93 V. SOBOLEV, R. C. WADE, G. VRIEND, and M. EDELMAN. Molecular docking using surface complementarity. *PROTEINS: Structure, Function and Genetics*, 25:120–129, **1996**.

94 C. A. BAXTER, C. W. MURRAY, D. E. CLARK, D. R. WESTHEAD, and M. D. ELDRIDGE. Flexible docking using tabu search and an empirical estimate of binding affinity. *PROTEINS: Structure, Function and Genetics*, 33:367–382, **1998**.

95 S. KIRKPATRIK, C. D. Jr. GELATT, and M. P. VECCHI. Optimization by simulated annealing. *Science*, 220:671–680, **1983**.

96 D. S. GOODSELL and A. J. OLSON. Automated docking of substrates to proteins by simulated annealing. *PROTEINS: Structure, Function and Genetics*, 8:195–202, **1990**.

97 G. M. MORRIS, D. S. GOODSELL, R. HUEY, and A. J. OLSON. Distributed automated docking of flexible ligands to proteins: Parallel applications of autodock 2.4. *Journal of Computer-Aided Molecular Design*, 10:293–304, **1996**.

98 D. S. GOODSELL, G. M. MORRIS, and A. J. OLSON. Automated docking of flexible ligands: Applications of AutoDock. *Journal of Molecular Recognition*, 9:1–5, **1996**.

99 S. YUE. Distance-constrained molecular docking by simulated annealing. *Protein Engineering*, 4(2):177–184, **1990**.

100 J. E. B. PLATZER, F. A. MOMANY, and H. A. SCHERAGA. Conformational energy calculations of enzyme–substrate interactions. *International Journal on Peptide and Protein Research*, 4:187–200, **1972**.

101 J. E. B. Platzer, F. A. Momany, and H. A. Scheraga. Conformational energy calculations of enzyme–substrate interactions. *International Journal on Peptide and Protein Research*, 4:201–219, **1972**.

102 A. Di Nola, D. Roccatano, and H. J. C. Berendsen. Molecular dynamics simulation of the docking of substrates to proteins. *PROTEINS: Structure, Function and Genetics*, 19:174–182, **1994**.

103 M. Mangoni, D. Roccatano, and A. Di Nola. Docking of flexible ligands to flexible receptors in solution by molecular dynamics simulation. *PROTEINS: Structure, Function and Genetics*, 35:153–162, **1999**.

104 J. A. Given and M. K. Gilson. A hierarchical method for generating low-energy conformers of a protein-ligand complex. *PROTEINS: Structure, Function and Genetics*, 33:475–495, **1998**.

105 B. A. Luty, Z. R. Wasserman, P. F. W. Stouten, C. N. Hodge, M. Zacharias, and J. A. McCammon. A molecular mechanics/grid method for evaluation of ligand–receptor interactions. *Journal of Computational Chemistry*, 16(4):454–464, **1995**.

106 Z. R. Wasserman and C. N. Hodge. Fitting an inhibitor into the active site of thermolysin: A molecular dynamics case study. *PROTEINS: Structure, Function and Genetics*, 24:227–237, **1996**.

107 T. N. Hart and R. J. Read. A multiple-start Monte Carlo docking method. *PROTEINS: Structure, Function and Genetics*, 13:206–222, **1992**.

108 C. McMartin and R. S. Bohacek. QXP: Powerful, rapid computer algorithms for structure-based drug design. *Journal of Computer-Aided Molecular Design*, 11:333–344, **1997**.

109 A. Wallqvist and D. G. Covell. Docking enzyme–inhibitor complexes using a preference-based free-energy surface. *PROTEINS: Structure, Function and Genetics*, 25:403–419, **1996**.

110 R. Abagyan, M. Totrov, and D. Kuznetsov. ICM – a new method for protein modeling and design: Applications to docking and structure prediction from the distorted native conformation. *Journal of Computational Chemistry*, 15(5):488–506, **1994**.

111 J. Apostolakis, A. Plückthun, and A. Caflisch. Docking small ligands in flexible binding sites. *Journal of Computational Chemistry*, 19:21–37, **1998**.

112 J. Y. Trosset and H. A. Scheraga. Reaching the global minimum in docking simulations: A Monte Carlo energy minimization approach using Bezier splines. *Proceedings of the National Academy of Sciences USA*, 95:8011–8015, **1998**.

113 J.-Y. Trosset and H. A. Scheraga. PRODOCK: Software package for protein modeling and docking. *Journal of Computational Chemistry*, 20:412–427, **1999**.

114 J.-Y. Trosset and H. A. Scheraga. Flexible docking simulations: Scaled collective variable Monte Carlo minimization approach using bezier splines, and comparison with a standard monte carlo algorithm. *Journal of Computational Chemistry*, 20:244–252, **1999**.

115 J. WANG, P. A. KOLLMAN, and I. D. KUNTZ. Flexible ligand docking: A multistep approach. *PROTEINS: Structure, Function and Genetics*, 36:1–19, **1999**.

116 DANIEL HOFFMANN, BERND KRAMER, TAKUMI WASHIO, TORSTEN STEINMETZER, MATTHIAS RAREY, and THOMAS LENGAUER. Two-stage method for protein–ligand docking. *Journal of Medicinal Chemistry*, 42:4422–4433, **1999**.

117 D. HOFFMANN, T. WASHIO, K. GESSLER, and J. JACOB. Tackling concrete problems in molecular biophysics using monte carlo and related methods: Glycosylation, folding, solvation. In P. Grassberger, G. Barkema, and W. Nadler, editors, *Proceedings of the workshop on: Monte Carlo approach to Biopolymers and Protein Folding*, pages 153–170, Singapore, **1998**. World Scientic. ISBN 981-02-3658-1.

118 H. KUBINYI. Combinatorial and computational approaches in structure-based drug design. *Current Opinion in Drug Discovery and Development*, 1:16–27, **1998**.

119 M. RAREY, B. KRAMER, C. BERND, and T. LENGAUER. Time-efficient docking of similar flexible ligands. In L. Hunter and T. Klein, editors, *Biocomputing: Proceedings of the 1996 Pacific Symposium (electronic version at* http://www.cgl.ucsf.edu/psb/psb96/proceedings/eproceedings.html*)*. World Scientific Publishing Co, Singapore, **1996**.

120 S. MAKINO and I. D. KUNTZ. ELECT++: Faster conformational search method for docking flexible molecules using molecular similarity. *Journal of Computational Chemistry*, 19:1834–1852, **1998**.

121 C. W. MURRAY, D. E. CLARK, T. R. AUTON, M. A. FIRTH, J. LI, R. A. SYKES, B. WASZKOWYCZ, D. R. WESTHEAD, and S. C. YOUNG. PRO_SELECT: Combining structure-based drug design and combinatorial chemistry for rapid lead discovery. 1. technology. *Journal of Computer-Aided Molecular Design*, 11:193–207, **1997**.

122 Y. SUN, T. J. A. EWING, A. G. SKILLMAN, and I. D. KUNTZ. CombiDOCK: Structure-based combinatorial docking and library design. *Journal of Computer-Aided Molecular Design*, 12:597–604, **1999**.

123 M. RAREY and T. LENGAUER. A recursive algorithm for efficient combinatorial library docking. *Perspectives in Drug Discovery and Design*, 20:63–81, **2000**.

124 E. K. KICK, D. C. ROE, A. G. SKILLMAN, L. GUANGCHENG, T. J. A. EWING, Y. SUN, I. D. KUNTZ, and J. A. ELLMAN. Structure-based design and combinatorial chemistry yield low nanomolar inhibitors of cathepsin d. *Chemistry & Biology*, 4:297–307, **1997**.

125 D. C. ROE and I. D. KUNTZ. BUILDER v.2: Improving the chemistry of a de novo design strategy. *Journal of Computer-Aided Molecular Design*, 9:269–282, **1995**.

126 H. J. BÖHM, D. W. BANNER, and L. WEBER. Combinatorial docking and combinatorial chemistry: Design of potent non-peptide thrombin inhibitors. *Journal of Computer-Aided Molecular Design*, 13:51–56, **1999**.

127 A. CAFLISCH. Computational combinatorial ligand design: Application to human α-thrombin. *Journal of Computer-Aided Molecular Design*, 10:372–396, **1996**.

128 S. MAKINO, T. J. A. EWING, and I. D. KUNTZ. DREAM++: Flexible docking program for virtual combinatorial libraries. *Journal of Computer-Aided Molecular Design*, 13:513–532, **1999**.

129 H.-J. BÖHM and M. STAHL. Rapid empirical scoring functions in virtual screening applications. *Medicinal Chemistry Research*, 9:445–462, **1999**.

130 AJAY, M. A. MURCKO, and P. F. W. STOUTEN. Recent advances in the prediction of binding free energy. In P. S. Charifson, editor, *Practical Application of Computer-Aided Drug Design*, pages 355–410. Marcel Dekker, **1997**.

131 AJAY and M. A. MURCKO. Computational methods to predict binding free energy in ligand–receptor complexes. *Journal of Medicinal Chemistry*, 38(26):4953–4967, **1995**.

132 T. A. HALGREN. Potential energy functions. *Current Opinion in Structural Biology*, 5:205–210, **1995**.

133 W. F. VAN GUNSTEREN and H. J. C. BERENDSEN. Computer simulation of molecular dynamics: Methodology, applications, and perspectives in chemistry. *Angewandte Chemie International Edition*, 29:992–1023, **1990**.

134 M. VIETH, J. D. HIRST, A. KOLINSKY, and C. L. BROOKS III. Assessing energy functions for flexible docking. *Journal of Computational Chemistry*, 19:1612–1622, **1998**.

135 M. STAHL and H. J. BÖHM. Development of filter functions for protein–ligand docking. *Journal of Molecular Graphics and Modelling*, 16:121–132, **1998**.

136 M. STAHL. Modifications of the scoring function in FlexX for virtual screening applications. To appear in Perspectives in Drug Discovery and Design.

137 A. N. JAIN. Scoring noncovalent protein–ligand interactions: A continuous differentiable function tuned to compute binding affinities. *Journal of Computer-Aided Molecular Design*, 10:427–440, **1996**.

138 R. D. HEAD, M. L. SMYTHE, T. I. OPREA, C. L. WALLER, S. M. GREEN, and G. R. MARSHALL. VALIDATE: A new method for the receptor-based prediction of binding affinities of novel ligands. *Journal of the American Chemical Society*, 118:3959–3969, **1996**.

139 M. D. ELDRIDGE, C. W. MURRAY, T. R. AUTON, G. V. PAOLINI, and R. P. MEE. Empirical scoring functions: I. the development of a fast empirical scoring function to estimate the binding affinity of ligands in receptor complexes. *Journal of Computer-Aided Molecular Design*, 11:425–445, **1997**.

140 C. W. MURRAY, T. R. AUTON, and M. D. ELDRIDGE. Empirical scoring functions. ii. the testing of an empirical scoring function for the prediction of ligand–receptor binding affinities and the use of bayesian regression to improve the quality of the model. *Journal of Computer-Aided Molecular Design*, 12:503–519, **1998**.

141 Y. TAKAMUTSU and A. ITAI. A new method for predicting binding free energy between receptor and ligand. *PROTEINS: Structure, Function and Genetics*, 33:62–73, **1998**.

142 R. WANG, L. LIU, L. LAI, and Y. TANG. SCORE: A new empirical method for estimating the binding affinity of a protein–ligand complex. *Journal of Molecular Modelling*, 4:379–394, **1998**.

143 J. B. O. Mitchell, R. A. Laskowski, A. Alex, and J. M. Thornton. BLEEP – a potential of mean force describing protein–ligand interactions: I. generating the potential. *Journal of Computational Chemistry*, 20:1165–1177, **1999**.

144 J. B. O. Mitchell, R. A. Laskowski, A. Alex, M. J. Forster, and J. M. Thornton. BLEEP – a potential of mean force describing protein–ligand interactions: II. calculation of binding energies and comparison with experimental data. *Journal of Computational Chemistry*, 20:1177–1185, **1999**.

145 I. Muegge and Y. C. Martin. A general and fast scoring function for protein–ligand interac tions: A simplified potential approach. *Journal of Medicinal Chemistry*, 42(5):791–804, **1999**.

146 H. Gohlke, M. Hendlich, and G. Klebe. Knowledge-based scoring function to predict protein–ligand interactions. *Journal of Molecular Biology*, 295:337–356, **2000**.

147 M. J. Sippl. Calculation of conformational ensembles from potentials of mean force – An approach to knowledge-based prediction of local structures in globular proteins. *Journal of Molecular Biology*, 213:859–883, **1990**.

148 M. J. Sippl. Boltzmann's principle, knowledge based mean fields and protein folding. An approach to the computational determination of protein structures. *Journal of Computer-Aided Molecular Design*, 7:474–501, **1993**.

149 B. Kramer, M. Rarey, and T. Lengauer. Evaluation of the FlexX incremental construction algorithm for protein-ligand docking. *PROTEINS: Structure, Function and Genetics*, 37:228–241, **1999**.

150 C. W. Murray, C. A. Baxter, and A. D. Frenkel. The sensitivity of the results of molecular docking to induced fit effects: Application to thrombin, thermolysin and neuraminidase. *Journal of Computer-Aided Molecular Design*, 13:547–562, **1999**.

151 M. Vieth, J. D. Hirst, B. N. Dominy, H. Daigler, and C. L. Brooks III. Assessing search strategies for flexible docking. *Journal of Computational Chemistry*, 19:1623–1631, **1998**.

152 D. R. Westhead, D. E. Clark, and C. W. Murray. A comparison of heuristic search algorithms for molecular docking. *Journal of Computer-Aided Molecular Design*, 11:209–228, **1997**.

153 N. C. J. Strynadka, M. Eisenstein, E. Katchalski-Katzir, B. K. Shoichet, I. D. Kuntz, R. Abagyan, M. Totrov, J. Janin, J. Cherfils, F. Zimmerman, A. Olson, B. Duncan, M. Rao, R. Jackson, M. Sternberg, and M. N. G. James. Molecular docking programs successfully predict the binding of a β-lactamase inhibitory protein to tem-1 β-lactamase. *Nature Structural Biology*, 3(3):233–239, **1996**.

154 J. S. Dixon. Evaluation of the casp2 docking section. *PROTEINS: Structure, Function and Genetics*, Suppl. 1:1(1):198–204, **1997**

155 T. N. Hart, S. R. Ness, and R. J. Read. Critical evaluation of the research docking program for the casp2 challenge. *PROTEINS: Structure, Function and Genetics*, Suppl. 1:1(1):205–209, **1997**.

156 V. SOBOLEV, T. M. MOALLEM, R. C. WADE, G. VRIEND, and M. EDELMAN. Casp2 molecular docking predictions with the ligin software. *PROTEINS: Structure, Function and Genetics*, Suppl. 1:1(1):210–214, **1997**.

157 M. TOTROV and R. ABAGYAN. Flexible protein–ligand docking by global energy optimization in internal coordinates. *PROTEINS: Structure, Function and Genetics*, Suppl. 1:1(1):210–214, **1997**.

158 B. KRAMER, M. RAREY, and T. LENGAUER. Casp-2 experiences with docking flexible ligands using flexx. *PROTEINS: Structure, Function and Genetics*, Suppl. 1:1(1):221–225, **1997**.

159 P. BURKHARD, P. TAYLOR, and M. D. WALKINSHAW. An example of a protein ligand found by database mining: Description of the docking method and its verification by a 2.3 Å x-ray structure of a thrombin-ligand complex. *Journal of Molecular Biology*, 277:449–466, **1998**.

160 R. M. A. KNEGTEL and M. WAGENER. Efficacy and selectivity in flexible database docking. *PROTEINS: Structure, Function and Genetics*, 37:334–345, **1999**.

161 P. S. CHARIFSON, J. J. CORKERY, M. A. MURCKO, and W. P. WALTERS. Consensus scoring: A method for obtaining improved hit rates from docking databases of three-dimensional structures into proteins. *Journal of Medicinal Chemistry*, 42:5100–5109, **1999**.

162 H.-J. BÖHM. Thrombin-Inhibitors, collected experimental data. Personal communication.

163 L. WEBER, S. WALLBAUM, C. BROGER, and K. GUBERNATOR. Optimization of the biological activity of combinatorial compound libraries by a genetic algorithm. *Angewandte Chemie International Edition*, 34:2280–2282, **1995**.

164 C. D. SELASSIE, Z. FANG, R. LI, C. HANSCH, G. DEBNATH, T. E. KLEIN, R. LANGRIDGE, and B. T. KAUFMAN. On the structure selectivity problem in drug design. A comperative study of benzylpyrimidine inhibition of vertebrate and bacterial dihydrofolate reductase via molecular graphics and quantitative structure-activity relationships. *Journal of Medicinal Chemistry*, 32:1895–1905, **1989**.

165 F. C. BERNSTEIN, T. F. KOETZLE, G. J. B. WILLIAMS, E. F. Jr. MEYER, M. D. BRICE, J. R. RODGERS, O. KENNARD, T. SHIMANOUCHI, and M. TASUMI. The protein data bank: a computer based archival file for macromolecular structures. *Journal of Molecular Biology*, 112:535–542, **1977**.

166 R. S. PEARLMAN, A. RUSINKO, J. M. SKELL, R. BALDUCCI, and C. M. McGARITY. *CONCORD 3D-structure generator*. Tripos Associates, Inc., St. Louis, Missouri, USA.

167 W. L. JORGENSEN. Free energy perturbations: A breakthrough for modeling organic chemistry in solution. *Accounts in Chemical Research*, 22:184–189, **1989**.

168 R. ABAGYAN and M. TOTROV. Biased probability monte carlo conformational searches and electrostatic calculations for peptides and proteins. *Journal of Molecular Biology*, 235:983–1002, **1994**.

8

Modelling Protein–Protein and Protein–DNA Docking

Michael J E Sternberg and Gidon Moont

8.1
Introduction

8.1.1
The need for protein–protein and protein–DNA docking

A description of many biological processes requires knowledge of the three-dimensional structure of macromolecular complexes. This structural information will provide insights into the specificity and can suggest lead compounds for the development of novel pharmaceutical agents. However, structural studies by crystallography and NMR often follow a reductionist approach so that the coordinates of the component molecules are available but the conformation of the complex is unknown. Indeed in the protein data bank [1], there is a large discrepancy between the number of different determined protein structures (c. 3000) and the number of protein–protein and protein–DNA complexes with proteins (c. 300). Thus computer algorithms are needed to predict the structure of macromolecular complexes starting from coordinates of the unbound components. This chapter describes computational strategies that can be employed to model the docking of two types of macromolecular complexes – protein–protein and protein–DNA.

The crystal structures of protein–protein and protein–DNA complexes have provided detailed descriptions of the interactions that lead to the specificity that is central to the biological activity of the system, such as the mechanism that leads to a disease. Currently we have details of a variety of protein–protein complexes including enzyme–inhibitor, antibody–antigen, hormones and their receptors and cell surface–cell surface proteins [2, 3, 4, 5]. The nature of the inter-protein recognition can be understood in terms of a general shape complementarity with specificity provided by particular spatial constraints from close packing and by hydrogen bonds and charge–charge ion pairings. From the structural information, one can start to alter one of the proteins to affect its recognition properties. For example, the details of a protein–protein complex can suggest that a particular loop is central to the interaction and this could lead to the design of lead compounds that could yield novel drugs. Indeed, there are several lead compounds that have been designed based on the structure of a protein receptor interacting with a small molecule ligand (e.g., see review by Colman [6]). With the increasing number of determined protein structures, the determination of protein–protein complexes by both experiment and modelling should also lead to suggestions of new pharmaceutical agents.

8.1.2
Overview of the computational approach

This chapter will describe computational strategies to predict the structure of protein–protein and protein–DNA complexes. The consensus approach to protein–protein docking is described in Figure 8.1. The precise order of implementing the steps can differ but essentially the key features are now overviewed.

 i. One starts with three-dimensional structures of the two unbound components and considers systems that are expected to have limited conformational change on macromolecular association.
 ii. The rigid-body approximation is then used which is that one can simulate the docking starting with the unbound components, given the limited conformational change.
iii. A search is performed over all possible associations. If there is no biological information about which parts of the molecules interact then all possible associations must be considered. If however there are some constraints, these can be used either to limit the initial search or as a post search filter. The search will sample the space of possible associations and consequently there will be a lower limit on the difference in conformations between two docked complexes that determines the resolution of the search procedure.
 iv. A function is developed to score the quality of the docked complex often using simplified terms to evaluate shape and electrostatic complementarity. There are two reasons for a simple function at this stage. First, as a large number of different complexes are generated in global scan, the energy term must be computationally fast to evaluate. Second, one requires a soft potential that is not unduly sensitive to the conformational differences between the docked structures formed from unbound components and the true complex with the bound components.
 v. Ideally, the docking algorithm would thereby identify a single complex that is a close approximation to the true docked complex based on this complex having the best score (i.e., the lowest energy complex). In

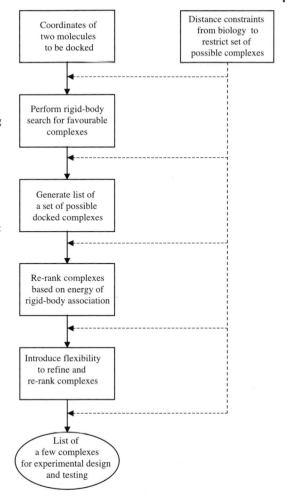

Fig. 8.1

Overview of a typical docking procedure. This diagram summarizes the consensus approach based on rigid body docking followed by re-ranking and refinement. Approaches vary according to whether the steps are combined or performed independently in a particular program. Many strategies only perform some stages, particularly the initial docking and scoring or just subsequent re-ranking and refinement. Biological distance constraints can be included at different stages in the procedure.

practice, the current status of the algorithms is that they generate a limited list of possible complexes ranked on the score and the objective is that one member of the list should be close to the true complex.

vi. At this stage, a re-ranking of the resultant complexes can be undertaken possibly using computationally more intensive calculations.

vii. Conformational flexibility is generally introduced into the algorithm when there is only a limited number of complexes to consider. Perturbations to the structure of the docked complex are made and the energies of the resultant conformation is evaluated. The aims are both to improve the structural quality of the predicted complex and to improve the power of identifying the true solution from the set of false structures that have been generated.

A summary of the current status of protein-protein docking is that when this strategy has been applied to around ten test systems, one can generate a close model (2.5 Å rmsd for the Cα atoms) to the true complex at a rank of ten or less for many of the systems studied.

8.1.3
Scope of this chapter

Given the reliance of most methods on starting by a rigid-body docking, the next section in this chapter will describe the extent of conformational change on protein–protein association. It will be shown that for many systems, the change is sufficiently limited to suggest that the rigid-body approach is viable.

Next, a computational strategy developed in our Laboratory for protein–protein will be described. The results will show that protein–protein docking can often yield a limited list of possible complexes one of which is close to the native. The chapter will then describe the application of this procedure to protein–DNA docking. Protein–DNA docking is harder than protein–protein docking given the more uniform shape of double-stranded DNA and the more extensive conformational change that tends to occur on association. The results will show that despite these problems, modelling protein–DNA complexes is viable. The description of our approach will serve to illustrate many aspects of the general strategy outlined above (Section 8.1.2) and thereby report the current status of protein docking.

We will then survey several alternative approaches for protein–protein docking. A true test of any predictive method is its success in blind trials since this prevents bias for over optimisation during the development of the approach. There have been two such blind trials and the results will be described. In addition to this chapter, the reader is referred to other reviews of the field [7–9].

8.2
Structural studies of protein complexes

The first step in the development of a protein docking algorithm is to examine the known complexes to establish the principles of molecular recognition. Following earlier work [10], there are reviews that examined the interactions between hetero-protein complexes [4, 7]. These analyses have focussed on the static structure of the complexes. A major problem in protein docking is to cope with the conformational flexibility that occurs on complex formation. Accordingly, we have analyzed the conformational changes on complex formation for 39 pairs of structures of proteins in complexes and in their unbound states [2]. The data set mainly consisted of enzyme–inhibitor and antibody–antigen complexes but also included other systems such as human growth hormone and its receptor.

The conformational differences were evaluated in terms of root mean square deviations (rmsd) of Cα positions and of side-chain positions. Residues were identified as exposed when their total relative side-chain accessible area (main-chain for Gly) was greater than 15%. Interface residues were defined as having at least one atom within 4 Å of the other component in the complex. To assess the significance of the differences between bound and unbound structures, it is necessary to identify the differences in coordinates that can occur simply from the crystallographic determination of the structure. Accordingly, 12 pairs of independently solved crystal structures of identical proteins were analysed. 95% of this data set (11 out of 12) had an rmsd for exposed Cα atoms of <0.6 Å and for exposed side-chain atoms of <1.7 Å. These values were taken as the control cut-offs. Four measures were adopted for overall conformational change – rmsd of Cα and of side-chain atoms for just the interface atoms and for exposed non-interface atoms. 19 out of the 39 proteins that were involved in complex formation do not have any of their four measures for conformational change above the cut off values. Many of the other proteins showed changes just above the control cut offs. Thus for many systems, there is limited overall conformational change on protein–protein association which suggests that the rigid-body approach should be widely applicable.

In addition to overall conformational changes, one or a few individual residues might show particularly large conformational changes and this could markedly affect the viability of the rigid-body approach. The control cut-offs for the movement of an individual residue are 3.0 Å for a Cα and 5.6 Å for side-chain displacements. Examination of the complexes showed that all large movements of exposed residues that were not in the interface can be explained by either their close proximity to the interface or by structural disorder. For a few of the systems, there are movements of individual or a set of residues in the interface that are above the control cut-offs. These shifts are intimately involved in the complex formation. Thus there are several complexes with substantial conformational change for one or a few residues whilst still there is overall a limited structural perturbation.

The general conclusion from this analysis is that protein–protein interactions are described by the induced fit model. However for many systems, the extent of conformational change is limited and the lock-and-key model is a valid first approximation. Accordingly, for many systems it is appropriate to develop a protein–protein docking algorithms that starts with the docking of the components as rigid bodies and then as a refinement considers limited conformational changes. The major caveat is that the systems analyzed were dominated by enzyme–inhibitor and antibody–antigen complexes; consequently other biological systems may exhibit a greater degree of conformational change on complex formation.

8.3
Methodology of a protein–protein docking strategy

In this section we describe a computational strategy developed in our Laboratory for docking. Our suite of programs is available from http:// www.bmm.icnet.uk and models the docking two medium-sized proteins (50–500 residues). In outline (see Figure 8.2) the method is:

i. a rigid-body search for docked complexes that are favourable in terms of shape complementarity and electrostatics,
ii. the application of an empirical scoring function to re-rank the docked structures generated by (i),
iii. the use of biological distance constraints, particularly details of the binding sites in one or both proteins, to screen the putative complexes from (ii),
iv. the final refinement and further re-ranking of the rigid body structure by a consideration of side-chain conformational change.

8.3.1
Rigid body docking by Fourier correlation theory

The first step in most approaches including ours is the rigid-body docking of the two molecules to generate a set of complexes. There are two major requirements:

i. The method must be able within a realistic computation time to generate and evaluate trial complexes that cover all relative orientations at sufficient resolution to yield at least one solution that is close to the true structure.
ii. The approach must use scoring functions that are sufficiently soft to cope with a limited conformational change on association yet be able to identify a solution that is close to the true complex.

The Fourier correlation approach introduced by Katchalski-Katzir et al. [11] was developed to meet these requirements. We have augmented their method to include electrostatic effects that often are required to obtain results starting from unbound complexes. Our initial work (FTDOCK1) was described in Gabb et al. [12], but the approach has been re-implemented with minor changes and this algorithm (FTDOCK2) is now described (see Figure 8.2).

8.3.1.1 The generation of the grid representation
The Katchalski-Katzir method is a grid based approach to evaluate the shape complementarity between two proteins (see Figure 8.3). Consider two mol-

Fig. 8.2
Schematic of our strategy
for generating and
screening docked protein
complexes.

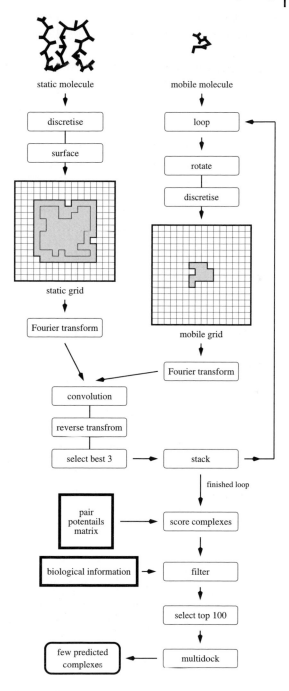

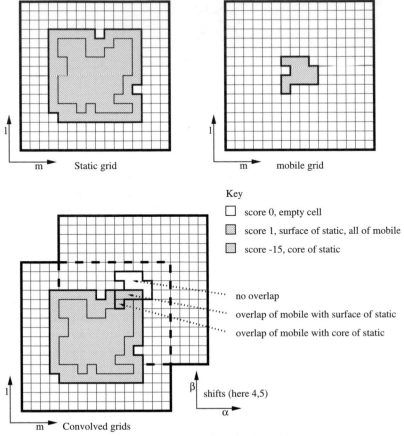

Fig. 8.3
Scoring shape complementarity using a grid-based scoring approach. The Figure illustrates the approach used in FTDOCK. The top two grids show the individual discretisation of the static and mobile molecules. The lower grid shows the convolved grids with the calculation of shape complementarity.

ecules *A* and *B*. Let *A* be the larger of the two and taken to be the static molecule. B will be the smaller mobile component to be docked against *A*.

Each molecule is placed onto a 3-dimensional grid and the algorithm requires that both grids are the same dimension ($N \times N \times N$). The size of grid cell must be sufficiently small to model the atomic structure of the molecules. However the computational time increases as the cell size decreases. We currently use a value of 0.7 Å that is sufficiently small that generally only one atom (excluding hydrogen) is in an individual cell. To evaluate the required *N*, for each protein the maximum distance from any atom to the geometric centroid is calculated ($DMAX_A$ and $DMAX_B$). The value *XN* is calculated from:

$$XN = (2 \times (DMAX_A + DMAX_B) + 1.0)/0.7$$

N is set to the nearest integer to XN. For ease of computation in the Fourier method, N should be even and so if N is odd, the next highest even integer is used. The result of this calculation of N ensures that the mobile molecule can always be placed to touch the static molecule and there will remain grid cells at the boundary unoccupied by the protein.

The discretization is done with each molecule at the centre of each grid. The empty space not filled by the molecule is necessary for the algorithm. To perform the discretization, each grid cell within which an atomic position is found is turned 'on'. Grid cells whose centre is within 1.8 Å of any atomic position are also turned 'on'. This value of 1.8 Å is chosen to approximate an effective van der Waals radius for an atom combined with any hydrogen atoms that are bound to it. Thus the surface of the resulting grids will represent the atomic surface of the molecules.

Next, the larger static molecule (A) is assigned a surface thickness below its atomic surface in order to be able to calculate the quality of the fit. For algorithmic speed, this surfacing procedure is done to the static molecule, and accordingly it needs to be done only once by the program. The depth of this surface is 1.4 Å, so that the surface is never more than two cells thick. After the surface has thereby been assigned, the core cells still cover all the actual atomic positions, but the van der Waals radius has been effectively removed, so allowing closer interactions between the two components of the complex. This representation of the surface provides a model for complex formation in which there is some side-chain rearrangement followed by side-chain interdigitation.

8.3.1.2 Evaluation of shape complementarity

The grid values $a_{l,m,n}$ for the static molecule at node l, m, n are given by

$$a_{l,m,n} = \begin{cases} 1 & \text{for grid points on the surface of the molecule} \\ \rho & \text{for the core of the molecule} \\ 0 & \text{for outside the molecule} \end{cases}$$

where ρ is negative (we use -15, see below).

For the second molecule (B) the grid values

$$b_{l,m,n} = \begin{cases} 1 & \text{for the molecule} \\ 0 & \text{for outside the molecule.} \end{cases}$$

The two grids can then be superimposed and the movable grid (B) translated by shifts α, β, γ (see Figure 8.3). The value of

$$a_{l,m,n} \cdot b_{l-\alpha, m-\beta, n-\gamma}$$

gives the extent of shape complementarity for grid point l, m, n of the grid of molecule A. A value of 1 for the product indicates that the cell from molecule B is superimposed on the surface of A which is favorable shape complementarity. A value of ρ indicates a steric clash with the cell from B superimposed on the core of A. The value of ρ is chosen so as to penalise but not totally prevent steric clashes. A value of zero means that there is no overlap between the two molecules. Thus the total shape complementarity for the two superimposed grids $c_{\alpha, \beta, \gamma}$ is evaluated from

$$c_{\alpha, \beta, \gamma} = \sum_{l=1}^{N} \sum_{m=1}^{N} \sum_{n=1}^{N} a_{l, m, n} \cdot b_{l-\alpha, m-\beta, n-\gamma} \tag{1}$$

8.3.1.3 Use of discrete Fourier transforms

The value of c is a convolution and its calculation requires approximately N^3 multiplications and a summation for every N^3 α, β, γ shifts. Katchalski-Katzir et al. introduced the use of discrete Fourier transforms to speed up the process of calculating c. First, calculate the discrete Fourier transforms of the discrete functions for the cell values a and b and let these transforms be denoted as $DFT(a)$ and $DFT(b)$. Then, calculate the complex conjugate of $DFT(a)$ which is denoted as $DFT^*(a)$. Then by Fourier theory, the discrete Fourier transform of c, $DFT(c)$, is given by

$$DFT(c) = DFT^*(a) \cdot DFT(b)$$

Hence, $c = IFT(DFT^*(a) \cdot DFT(b))$ where IFT denotes the inverse Fourier transform.

The discrete Fourier transform, if performed efficiently, takes $N^3 \log N$. This must be performed three times, but this does not increase the order of magnitude of the calculation. The pointwise multiplication takes N^3. Thus the total runtime takes $N^3 \log N + N^3$, which is of the same order as $N^3 \log N$. In contrast without the transform in real space the calculation would be of order N^6. As N is of the order of 100, the discrete Fourier transform decreases the time required by more than a factor of 10 000.

8.3.1.4 The global search

The movable molecule is rotated to sample all possible rotations in as fair a way as possible. In order to describe the orientation of a three-dimensional object in a three-dimensional space, three rotational angles are required. Therefore, in order to orient a molecule it is necessary to perform three rotations (see Figure 8.4). For FTDOCK2 the rotations were in order; around the z-axis (z twist), around the y axis (theta), and around the z axis again (phi). Since only integer values are used to describe the rotations, this limits

Fig. 8.4
A schematic diagram of the rotational
search.

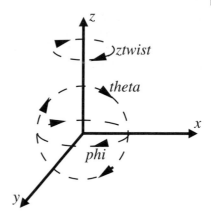

the angle step used to being an integer factor of 180. Once the angle step
had been chosen for the sampling rate, in order to create an even sampling,
the angle step used for phi was calculated dependent on the value of theta.
This results in a fair sampling of rotational space. For angle step of 20°, this
method gives 1980 rotations. For an angle step of 15° it gives 4656, and for
12° it gives 9240.

8.3.1.5 Electrostatic effects

Both shape complementarity and electrostatic effects are important in the
recognition process in protein complex formation. Accordingly a treatment
of electrostatics was introduced into the Fourier correlation approach. The
charge–charge interaction is evaluated from point charges of the mobile
molecule B interacting with the potential from static molecule A. This
choice results in having to perform the potential calculation only once (for
the static molecule) whilst the charge calculation (see below) is performed
for every rotation of the mobile molecule. In the above treatment of rigid
body docking based on shape complementarity, it is possible to place two
charges closer together than would be allowed by van der Waals packing.
Since the potential energy of two interacting charges depends inversely on
their separation, this would result in an artificially highly favorable or very
unfavourable interaction. These artificial terms are prevented by the method
now described to calculate the potential.

Charges (see Gabb et al. [12]) are assigned to the atoms of molecule A
and the electrostatic potential evaluated from

$$\phi_{l,m,n} = \sum_j \frac{q_j}{\varepsilon(r_{ij})r_{ij}}$$

where $\phi_{l,m,n}$ is the potential at node l, m, n (position i), q_j is the charge on atom j, $r_{i,j}$ is the distance between i and j (with a minimum value of 2 Å to avoid artificially large values of the potential) and $\varepsilon(r_{ij})$ is a distance dependent dielectric function. The sigmoidal function of Hingerty et al. [72] is used:

$$\varepsilon(r_{ij}) = \begin{cases} 4 & \text{for } r_{ij} \leq 6\,\text{Å} \\ 38r_{ij} - 224 & \text{for } 6\,\text{Å} < r_{ij} < 8\,\text{Å} \\ 80 & \text{for } r_{ij} \geq 8\,\text{Å}. \end{cases}$$

This function was introduced for modelling the effective dielectric between atoms in proteins. The rationale for this function is that at close separation ($r_{ij} \leq 6\,\text{Å}$) when there is no intervening water molecules, the effective dielectric is that of protein atoms and a value of 4 is appropriate. For separations of 8 Å or more then the dielectric is dominated by the screening effect of the intervening water and the value for bulk water (80) is used. Between these two separations a linear interpolation is used. In FTDOCK, we do not use precise atomic positions, but still need to model the complex dielectric behavior of proteins in solvent and this function was used and found to perform well.

The potential $\phi_{l,m,n}$ is only assigned to nodes outside and on the surface region of molecule A. Inside the core of molecule A, $\phi_{l,m,n}$ is zero. For the mobile molecule B, the charges on the charged atoms are distributed amongst the closest 8 grid cells, see Figure 8.5. Given the atomic position of the charged atom normalised onto the centre of the 8 neighbouring cells as (x, y, z), and the normalised centre of one of those closest 8 cells as

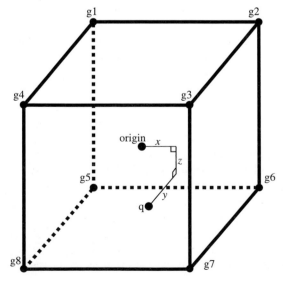

Fig. 8.5
Schematic diagram of the approach to transfer an electronic charge to its nearest eight grid nodes.

(X, Y, Z), then the charge given to that cell is

$$(q_{l, m, n}/8) * ((x + X)/X) * ((y + Y)/Y) * ((z + Z)/Z)$$

The electrostatic interaction $e_{\alpha, \beta, \gamma}$ for a shift of α, β, γ is calculated from

$$e_{\alpha, \beta, \gamma} = \sum_{l=1}^{N} \sum_{m=1}^{N} \sum_{n=1}^{N} \phi_{l, m, n} \cdot q_{l-\alpha, m-\beta, n-\gamma} \tag{2}$$

Eq. (2) is analogous to Eq. (1) for shape complementarity and accordingly the Fourier correlation approach can be also used to speed up the calculation.

8.3.1.6 Generation of putative complexes

In trials with FTDOCK, it was found that the electrostatic term worked best as a binary filter. Complexes with unfavourable electrostatics are discarded and the remaining docked structures ranked by shape complementarity. For a given orientation of the movable molecule, the shape complementarity correlation function c is examined and the three most favourable solutions stored. After all orientations are sampled, the top set (typically 10 000) of complexes are kept for subsequent examination.

8.3.2
Use of residue pair potentials to re-rank docked complexes

After the global search for shape complementarity and favourable electrostatic interaction, a list of several putative complexes is generated. A model close to native (see below for quantitative measures) is rarely at the top of the list of complexes ranked by shape complementarity. It is necessary, therefore, to develop further scoring functions to identify good solutions from the list of alternatives. The strategy we have developed is to use a residue-residue scoring scheme to evaluate the docked complexes. Such an approach has an energy landscape that is smoother than an approach based on all (heavy) atoms but incorporates information that cannot be included in the grid-based first step. The terms to quantify the residue-residue interactions are derived empirically from known protein structures and these terms are often known as empirical pair potentials.

The theory of empirical pair potentials is that since they are derived from observations they will incorporate the dominant thermodynamic effects without explicitly having to partition and quantify the effects. Empirical pair potentials have been widely used in protein structure prediction, particularly in fold recognition following their introduction by Sippl [13] and their use

Tab. 8.1

Complexes used to generate the empirical scoring function. The protein data bank codes [1] of the complex are given.

1a14	1a1y	1a22	1a2x	1a2y	1a4y	1agr	1ahw	1aip	1ak4	1akj	1am4
1avw	1avz	1ay7	1azs	1b6c	1bd2	1bdj	1bgx	1bh9	1bi8	1bjl	1bkd
1blx	1bqq	1bvk	1d0g	1dan	1dfj	1dvf	1eai	1eay	1efn	1efu	1fdl
1fin	1fle	1gc1	1got	1gua	1hia	1hlu	1igc	1itb	1jrh	1kb5	1ld9
1lfd	1lpb	1mct	1mda	1mel	1mlc	1nfd	1nfi	1nmc	1noc	1oak	1osp
1qbk	1raa	1rrp	1seb	1sgp	1slu	1spb	1tab	1taw	1tx4	1uea	1ugh
1viw	1vpp	1wej	1ycs	2jel	2pcc	2seb	2trc	3sic	4cpa	4htc	

by Jones et al. [14]. For a review, see Vajda et al. [15]. In addition, empirical functions have been used to evaluate protein-low molecular weight ligand complexes [16, 17, 18, 19]. Here we present the derivation and application of pair potentials specifically designed to re-rank docked protein–protein complexes.

The natural source to derive empirical potentials for protein docking is the complexes themselves. However at the time we first developed the potentials there was insufficient data to obtain the potentials and the statistics were taken from observations of residue-residue pairing within protein structures [20]. Recently, with the increase in the number of determined complexes, we have been able to generate the pair potential from these data and this approach is now described. By use of the SCOP (Structural Classification of Proteins) database [21] version 1.48, a set of 83 non-homologous complexes was obtained (see Table 8.1). In this dataset homology between two complexes (A–B) and (C–D) is defined as when both A is homologous to C and B is homologous to D, taking homology as being in the same SCOP superfamily.

Two residues were taken as pairing across the interface if any pair of atoms between them was closer than a cut off d_{cut}. A value of d_{cut} of 4.5 Å is used to represent two typical van der Waals radii (1.5 Å each) plus an additional 1 Å. Let the frequencies of observed pairings between residues of type a and b be F_{ab}.

These observed frequencies are compared to those for a random state. The model for the random state is based on random associations of exposed residues on one protein with exposed residues on another protein. To evaluate this one needs the frequencies of exposed residues in the database of proteins used to generate the observed pairings. From the coordinates of a protein, the program DSSP [22] estimates the extent to which each protein residue is accessible to solvent contact. This is represented by an accessibility score and a value of 10 units is equivalent to one water molecule in contact with the residue. We find that a value of an accessible score > 5 units identifies residues that are not primarily buried within the protein core.

The expected number of random pairings of residue types a and b, E_{ab}, is based on frequencies of the exposed residues. Let T be the total number of all pairings observed in calculating the observed residue-residue pairings,

$$T = \sum_{a=1}^{20} \sum_{b=1}^{20} F_{ab}$$

Let n_a and n_b be the total occurrences of exposed residues of types a and b. The fraction of residues that are of type a and type b are respectively

$$\frac{n_a}{\sum_{a=1}^{20} n_a}$$

and

$$\frac{n_b}{\sum_{b=1}^{20} n_b}$$

Thus the expected frequency for the a-b pairing is given by

$$E_{ab} = T \cdot \frac{n_a}{\sum_{a=1}^{20} n_a} \cdot \frac{n_b}{\sum_{b=1}^{20} n_b}$$

A log-odds score for a pairing of residue types a and b is derived:

$$S_{ab} = \log_{10}\left(\frac{F_{ab}}{E_{ab}}\right)$$

These scores are shown schematically in Figure 8.6. As expected, pairings between hydrophobic residues and between residues with opposite charges are favourable. The total score for a complex is obtained by summing the S_{ab} values for all residue pairings between the two molecules with the distance less than d_{cut}. High scoring complexes are evaluated by this function to be more favourable. In many other applications of empirical pair potentials the log odds scores are converted to a potential of mean force by the application of Boltzmann's principle. However certain groups have questioned the validity of this approach [23, 24] and accordingly we simply treat the scores as a statistical measure of relative likelihood. This screening procedure is implemented in the program RPSCORE.

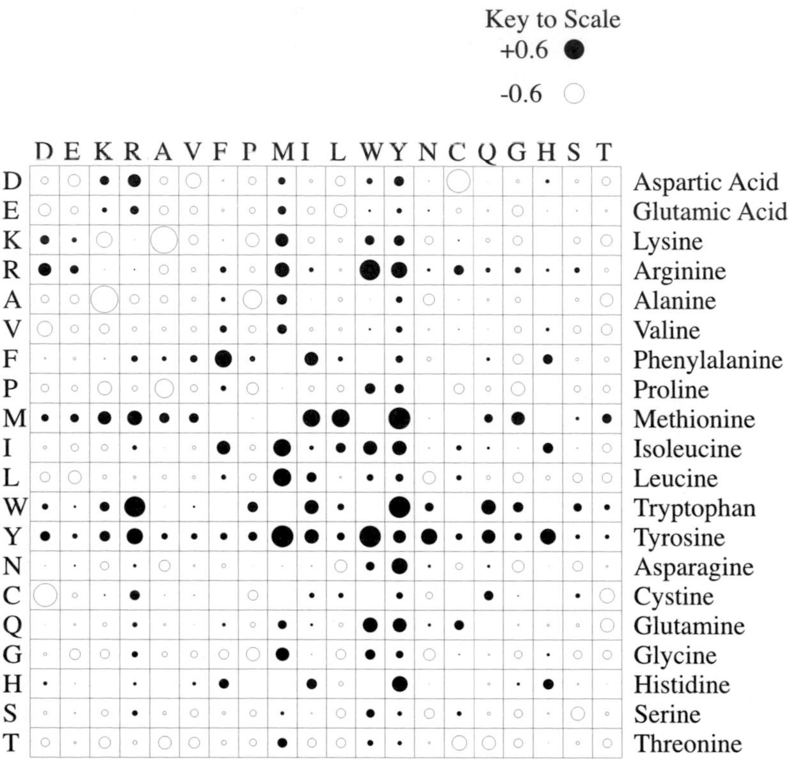

Key to Scale
+0.6 ●
-0.6 ○

D E K R A V F P M I L W Y N C Q G H S T

D		Aspartic Acid
E		Glutamic Acid
K		Lysine
R		Arginine
A		Alanine
V		Valine
F		Phenylalanine
P		Proline
M		Methionine
I		Isoleucine
L		Leucine
W		Tryptophan
Y		Tyrosine
N		Asparagine
C		Cystine
Q		Glutamine
G		Glycine
H		Histidine
S		Serine
T		Threonine

Fig. 8.6
Empirical residue–residue docking potentials. The empirically derived residue–residue contact scores are illustrated. The largest solid sphere represents a favourable score of +0.6 whilst the largest open sphere the most unfavourable score of −0.6.

8.3.3
Use of distance constraints

The location of the binding site can yield distance constraints to filter possible solutions and this is implemented in our procedure. An intermolecular residue–residue interaction is defined if any pair of atoms is closer than a 4.5 Å cut off. This cut off considers the probable error in conformation of any predicted complex. One feature of the Fourier correlation method implemented in FTDOCK, is that biological constraints on one molecule cannot be used to reduced the initial search. For any orientation of the movable molecule, it is placed at every translational position with respect to the static molecule. However, if constraints were available for both molecules, the range of orientations sampled by the movable molecule could be restricted.

8.3.4
Refinement and additional screening of complexes

The above strategy explores and ranks rigid-body associations of the molecules with the pair potentials operating at the level of residue-residue interaction. The next step is to allow for conformational change for side-chains
and to consider the interactions between the proteins in the complex at
the atomic level. We [25] have developed a procedure (MULTIDOCK) that
models both side-chain conformational changes in a mean field approach
together with limited rigid-body shifts between the component molecules.
The energy of interaction is evaluated from a molecular mechanics function.

8.3.4.1 Potential energy function
The proteins are represented at the atomic level by multiple copies of
side-chains on a fixed peptide backbone modelled according to a rotamer
library [26] that gives the commonly occurring side chain conformations.
The model to be described considers protein-protein interactions. The terms
for van der Waals interactions are taken from the AMBER force field [27]
and atomic charges from the PARSE parameters [28]. A cut off of 10 Å is
used in the calculation of the non-bonded interactions (van der Waals and
electrostatic) between atoms. The dielectric screening between charges is
represented by the widely-used distance dependent dielectric (dielectric
ε = distance of separation in Å). The effect of this dielectric model is that for
close separations the value is low and represent the dielectric due to protein
atoms. Thus at a separation of 4 Å the dielectric is 4. For larger separation
the dielectric increases representing the greater electrostatic screening due
to water molecules (not included explicitly).

This approach uses a rotamer library so side chains adopt a limited
number of conformations rather than being able to sample every value for
bond rotation. In addition, the starting model generated by FTDOCK (or
another rigid body docking approach) may be a few Ångstrom rmsd (Cα
atoms) from the true structure. The consequence is that there can be an
unrealistic close approach of atoms that could distort the modelling due to
high repulsive van der Waals interactions or electrostatic effects of large
magnitude. To reduce this effect, van der Waals interactions are truncated to
a maximum value of 2.5 kcal/mol. Similarly, an electrostatic interaction
scheme was introduced in which a minimum allowed distance separation
between two interacting charges q_i and q_j is set so that atom pairs that come
closer than allowed are re-scaled to realistic values which is no greater than
the approximate sum of their van der Waals radii. The minimum allowed
distances for two charges are 3 Å for two heavy atoms, 2 Å for one heavy
atom with hydrogen and 1 Å for two hydrogen atoms. Note that the treatment of the screening effect of water for larger separations and the preven-
tion of anomalous electrostatic effects are also included in FTDOCK.

8.3.4.2 Refinement procedure

The object of the refinement procedure is to move from a completely rigid-body docking scheme to include both flexibility of the side-chains together with a limited re-orientation of the two interacting molecules. Thus the refinement procedure is an iterative two step approach repeated until convergence involving:

i. optimisation of the protein side chain conformations by a self-consistent mean field approach [29, 30]. The resultant protein conformations then undergo

ii. rigid-body energy minimisation to relax the protein interface.

The mean-field approach is now described. The side chain degrees of freedom are defined by a conformational matrix, CM, where each rotamer, k, has a probability of $CM(i, k)$, where the sum of the probabilities for a given residue, i, must be equal to 1. The potential of mean force, $E(i, k)$, on the k-th rotamer of residue, i, is given by;

$$E(i, k) = V(\chi_{ik}) + V(\chi_{ik}, \chi_{mc}) + \sum_{\substack{j=1 \\ j \neq i}}^{N} \sum_{l=1}^{K_j} CM(j, l) V(\chi_{ik}, \chi_{jl})$$

where V is the potential energy, χ_{ik} are the coordinates of atoms in rotamer k of residue i and χ_{mc} are the coordinates of atoms in the protein main chain. N is the number of residues in the protein and K_j is the number of rotamers for residue j. The first term models the internal energy of the rotamer whilst the second represents the interaction energy between the rotamer and all the main chain atoms. These two values are constant for a given rotamer on a given main chain. The third term models the interaction energy between the rotamer and all the rotamers of other residues weighted by their respective probabilities.

Given the effective potentials acting on all K_i possible rotamers of residue, i, the Boltzmann principle can be used to calculate the probability of a particular rotamer:

$$CM(i, k) = \frac{e^{-E(i, k)/RT}}{\sum_{k=1}^{Ki} e^{-E(i, k)/RT}}$$

where R is the Boltzmann constant and T the temperature. The values of $CM(i, k)$ are substituted back into the equation describing $E(i, k)$ and its new value recalculated. This process is repeated until values of $CM(i, k)$ converge. The predicted structure corresponds to the rotamer of each residue with the highest probability. Trials showed that the procedure converged to the same side-chain rotamers using a number of different schemes for initiating the starting probabilities in the *CM* matrix.

After a complete cycle of mean field optimization of side-chain conformation, a rigid-body minimization is performed on the resultant coordinates of the interacting protein molecules. Only interface residues whose $C\beta$ atoms ($C\alpha$ for Gly) are within 15 Å of the $C\beta$ atom of the other molecule are included in the minimization. The larger molecule is kept stationary while the three rotational and three translational degrees of freedom of the smaller molecule are moved according to the path determined by the derivatives to minimize the intermolecular interaction energy. The steepest descents approach for minimization is used.

8.3.5
Implementation of the docking suite

The software for the strategy is available from our Web site under the home page http://www.bmm.icnet.uk and is free to academic users. The original distributed version of FTDOCK (FTDOCK1) had two implementations of the Fast Fourier Transforms (FFT) to search the translational binding space of two rigid molecules. One used the FFT routines from Numerical Recipes Software [31] and can be implemented on a variety of hardware platforms. The other implementation used the more efficient Silicon Graphics library functions.

Our recent implementation (FTDOCK2), is written in ANSI C and uses the "The Fastest Fourier Transform in the West," developed by M. Frigo and S. G. Johnson (see http://www.fftw.org/). As this is a fast public domain Fourier Transform Routine, we can distribute the entire source code for FTDOCK (as FTDOCK2). FTDOCK2 has not been parallelised which facilitates its implementation on a wide variety of platforms, including Linux. The program requires at least 128 Mbytes of RAM to prevent excessive paging. On a single R10000 Silicon Graphics processor a typical docking took between 9 and 36 hours, depending on the size of the molecules. In addition, PERL 5.003 or later is required for pre-processing the protein data bank coordinates.

To run FTDOCK2, the user first runs a supplied program to pre-process the coordinate files. Then the user need simply to identify the coordinate files for the static and the mobile molecules although other parameters can be adjusted.

The RPSCORE program that re-ranks the generated solution using the empirical pairing score is written in ANSI C and the source code is available. The program will automatically use our chosen matrix but the user can input a matrix of their choice.

The FILTER program implements the distance constraints. The program takes a list of inter-molecular constraints which can either be residue to residue, chain to residue, or chain to chain. Only one from a given list of constraints, needs to be satisfied for the program to accept a complex as passing the filter.

The code for MULTIDOCK is available as an executable for a Silicon Graphics computer. In our development of MULTIDOCK we included a version that explicitly introduced solvent molecules. The typical run times for MULTIDOCK to examine a single complex such as that of a serine protease – inhibitor system are 10 minutes with solvent and 3 minutes without solvent. For a larger system, an antibody docking to hen lysozyme, the corresponding run times are 40 minutes with solvent and 7 minutes without solvent. In an application of the procedure, it is essential to include a large number of trial solutions to ensure not missing the predicted complex that is close to the native. This can mean that up to 200 solutions may need to be screened (see Results section below). Thus for larger systems a solvent based screening could take several days. However, trials showed that inclusion of solvent does not turn a native-like solution with a poor *in vacuo* energy into a high ranking solution. At present we therefore distribute MULTIDOCK for use without explicit solvent molecules.

The total run time to examine a single protein-protein complex is on one processor about 2 days. Many users would be interested in how a particular pair of molecules will dock and then this run time is acceptable. In addition, one can evaluate the strategy for a number of test systems (typically 10–20) over about one month. However this length of run time prevents one exploring the use of the procedure to study problems such as all pairwise associations within a set of (say) 100 proteins.

8.4
Results from the protein–protein docking strategy

We are currently benchmarking the protein docking procedure on a set of 15 complexes with coordinates available for both components in their unbound state. The results will be placed on our Web page (www.bmm.icnet.uk). Preliminary inspection suggests the results of the new benchmark will be similar to our original benchmark. Here we report the results of the original benchmark. The approach used to obtain these results was (i) FTDOCK (as reported in Gabb et al. [12]), (ii) filtering on biological distance constraints [12], (iii) RPSCORE [20] and (iv) MULTIDOCK [25].

The benchmark consists of six enzyme/inhibitor and four antibody/antigen complexes. With the exception of the coordinates of two antibodies (HyHEL5 and HyHEL10), all the coordinates were from the unbound state. Table 8.2 presents the results. In each study, 4000 solutions were generated by FTDOCK1. The total number remaining after the application of the biological filter is shown in column 2 of Table 8.2 and forms set 1. All the en-

Tab. 8.2

Results of protein–protein docking. αCHYN–α-chymotrypsinogen, α-CHY–α-chymotrypsin, HPTI–human pancreatic trypsin inhibitor, BPTI–bovine pancreatic trypsin inhibitor, CHYI–chymotrypsin inhibitor, Subtilisin I–subtilisin inhibitor D1.3, D44.1, HyHEL5 and HyHEL10 are monoclonal antibodies. For further details of coordinates see Gabb et al. [12]. In the table some degenerate identical complexes included our earlier studies have been excluded.

System	Total no after FTDOCK & FILTR (set 1)	N ≤ 2.5 Å in FTDOCK list	Rank FTDOCK in set 1	Rank RPSCORE in set 1 (making set 2)	Rank MULTI-DOCK in set 1	Rank MULTI-DOCK in set 2 (i.e., after RPSCORE)	rmsd Cα atoms (Å)
αCHYN–HPTI	93	1	3	2	2	2	2.0
αCHY–ovomucoid	85	5	11	6	1	1	1.1
Kallikrein–BPTI	349	16	128	13	2	2	1.0
Subtilisin–CHY I	26	2	8	1	12	2	2.0
Subtilisin–subtilisin I	–	–	–	–	–	–	–
Trypsin–BPTI	205	7	12	14	23	3	1.7
D1.3–lysozyme	636	2	149	75	211	38	1.8
D44.1–lysozyme	539	4	34	24	101	21	2.5
HyHEL5–lysozyme	498	2	218	36	29	29	1.5
HyHEL10–lysozyme	700	4	48	6	9	2	1.1

zymes were serine proteases and the distance filter was that at least one residue in the inhibitor must be in contact with the one of the catalytic triad (i.e., His, Ser or Asp). For the antibodies, the constraint was that the antigen must contact either the third complementarity determining region of the light or the heavy chain (CDR-L3 or CDR-H3). In this list there are at least one, but often several, complexes that are considered a good prediction. In this study a good prediction is defined as within 2.5 Å of the correct structure superposing Cα all atoms of the predicted complex generated from unbound coordinates onto the X-ray bound complex. If just FTDOCK is used, the rank of the first good solution is given in column 4. In four out of the six enzyme/inhibitor complexes a good solution is obtained within 16 solutions. However no suitable complex is obtained for subtilisin docking to its inhibitor. The results for docking antibodies with antigens are less successful.

The results from FTDOCK can then be re-ranked using empirical scores (RPSCORE) to yield set 2. In general the rank at which one finds a good solution is decreased. A good solution is always found in the top 12% of the list of structures generated by FTDOCK (apart for subtilisin with its inhibitor). Set 2 can be further re-ranked using MULTIDOCK. The strategy was to take the top 25% of structures generated by RPSCORE, which is over double the fraction required in this benchmark. These are solutions are then ranked using MULTIDOCK *in vacuo* (see column 7). For the five serine proteases (i.e., excluding subtilisin and its inhibitor), a good solution is at rank 3 or better. The results for the antibody–antigen complexes are poorer with ranks as large as 38 that must be examined to ensure a good solution. Possibly the lower affinity for antibody–antigen complex formation compared to the enzyme–inhibitors studied [32] is reflected in the poorer discrimination in modelling their docking. Table 8.2 also presents the results of re-ranking the results of FTDOCK directly with MULTIDOCK (column 6). The approach of first using RPSCORE and then MULTIDOCK is always as good as, and generally superior, to using just MULTIDOCK for screening. Table 8.2 also gives the rmsd of the first ranked good solution. For about half of the simulations, the predicted structure is better than 2.0 Å rmsd of the bound coordinates.

Figure 8.7 illustrates some of the predicted complexes. Figures 8.7a and b show the Cα trace of the predicted complex generated by FTDOCK superposed on the X-ray complex for trypsin–BPTI (bovine pancreatic trypsin inhibitor) and Fab D1.3–hen lysozyme. Figure 8.7c illustrates the results of side-chain refinement using MULTIDOCK. Three side-chain in BPTI that mediate the interaction with trypsin are shown. After MULTIDOCK two of the three move closer to the correct position than their original location following rigid-body docking using FTDOCK. In this benchmark, the agreement between predicted and native structure is sufficient to suggest further experiments, such as mutagenesis, to probe the interaction. Furthermore, the identification of the loop and key side-chains in an inhibitor

Fig. 8.7
Comparison of native and
modelled protein–protein
complexes. In the figures
the black line denotes the
X-ray structure of the
complex and the grey lines
the predicted complex. *Fig.
a.* Superposition of the
predicted and native
complexes for bovine
pancreatic trypsin inhibitor
(BPTI) docking to trypsin.
A Cα trace is shown. The
predicted structure is
generated by FTDOCK1
and the rmsd of the
superposition is 1.7 Å. *Fig.
b.* Superposition of the
predicted and native
complexes of Fab D1.3 and
hen lysozyme. A Cα trace
is shown. The predicted
structure is generated by
FTDOCK1 and the rmsd of
the superposition is 1.8 Å.

(a)

(b)

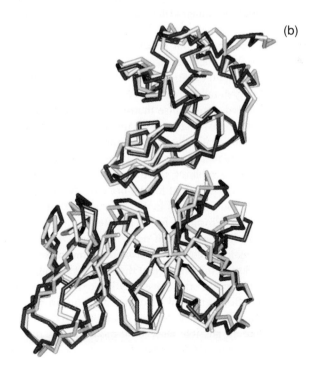

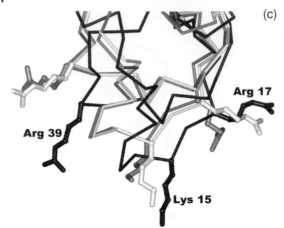

(C)

Fig. 8.7 (continued)
Fig. c. Refinement of the docking of BPTI to trypsin. Only BPTI is shown for clarity. The figure shows a Cα trace with three side-chain that are central fore the interaction of BPTI with trypsin. The grey lines are the initial results from FTDOCK1, the light lines, the prediction after refinement by MULTIDOCK. The black lines are the X-ray structure of the complex.

could be a valuable guide for the design of novel compounds of pharmaceutical use.

8.5
Modelling protein–DNA complexes

Knowledge of the structure of protein-DNA complexes provides insights into the structural basis of many fundamental biological processes [33]. For example an experimental structure of a transcriptional repressor with DNA will (generally) provide an understanding of why certain base sequences are required to recognise the repressor and of which residues in the repressor are responsible for interacting with DNA. This information provides both fundamental understanding and in the future opens the route for the design of drugs that specifically recognise certain DNA sequences. As for protein–protein complexes, there are often coordinates for an unbound protein that binds DNA but not for the protein-DNA complex. Thus predictive protein–DNA could be used to derive structural information, and the consequential biological insights, for such systems.

In the formation of many protein–DNA complexes there are gross conformational changes that would prevent employing a rigid-body approximation as the first step of the docking strategy. Nevertheless, there are some systems that often have only limited conformation change and to evaluate the effectiveness of predictive docking on these targets we have modelled transcriptional repressors recognising DNA [34]. The objective was to explore all possible association modes of the macromolecules. Previous work on protein–DNA docking focussed on exploring specificity once the protein has been oriented in a correct recognition mode, such as the recognition α-helix along the major groove [35, 36].

8.5.1
Method

The approach employs the first three stages of the protein-protein docking strategy amended to tackle protein–DNA complexes [34]. The steps are:

  i. rigid-body docking;
 ii. filtering with a protein–DNA scoring function and
iii. the use of biological distance filters.

Eight repressor–DNA complexes were examined (see Table 8.3). They show a diversity of recognition modes. Five (CRO, GAL, LAC, LAM and PUR) involved the α-helix/turn/α-helix recognising the major DNA groove, two (ARC and MET) involved a two stranded antiparallel β-sheet recognising bases in the major groove and in TRP there is both major and minor groove recognition. The unbound DNA was constructed according to standard geometry. For seven of the eight complexes, the unbound repressor structure was used but for one (LAM), bound coordinates had to be employed.

In the study FTDOCK1 was used with a $128 \times 128 \times 128$ grid that yielded a resolution of 0.51–0.87 Å. The DNA was kept as the fixed molecule. Rotational sampling was at $12°$. A specific charge set was developed for DNA that exaggerates the partial charges on the chemical groups in the DNA helix groove and dampens the phosphate charges. The electrostatic potential is only calculated for grid nodes within 2 Å of a charged atom (on the DNA). For the protein, charges are assigned to the main-chain atoms and to Asp, Glu, Arg, Lys Asn, Gln and His. The top 4000 structures generated by FTDOCK1 were stored.

Empirical potentials were derived to score amino-acid–nucleotide interactions. The parameters were derived from the structures of 20 protein–DNA complexes, none of which were homologous to the eight systems being modelled. An amino acid of type a was in contact with a nucleotide of type b if the distance between the $C\beta$ atom and the base glycosidic N was less than a cut off d_{cut}. As for the protein–protein pairing scores, the random model was based on composition (i.e., a molar fraction). In a cross-validated trial on the eight complexes, the best parameters for screening were found to be d_{cut} of 12 Å and using a sparse matrix that only scored interactions with charged or polar amino-acids (C, D, E, H, K, N, Q, R, S, and T).

Two types of distance filtering were considered. In the first (filter 1), it was assumed that there is knowledge of the central two base pairs involved in recognition. Such information is commonly derived from experimental

Tab. 8.3

Results of modelling repressor docking to DNA. The complexes modelled with the Protein Data Bank codes of the bound and unbound components are: ARC (arc repressor, 1par,1arr) CRO (cro repressor, 3cro, 2cro); GAL (CD2-GAL4 DNA binding domain, 1d66, 125d); LAC (lactose repressor,1lbg,1lqc); LAM (lamda phage operator repressor; 1lmb, 1lrp); MET (met repressor, 1cma, 1cmc); PUR (purR repressor, 1pnr, 1pru); TRP (trp repressor, 1tro, 2wrp). For coordinate details see Aloy et al. [34]. Filter 1 employs knowledge of the central two base pairs. Filter 2 employs filter 1 plus knowledge of some critical residues on the repressor. %CC is the percentage of correct contacts. rmsd is the deviation of the superimposed predicted and native complexes.

Complex	Ranking by shape complementarity				Ranking by empirical score			
	Rank after filter 1	Rank after filter 2	%CC	rmsd (Å)	Rank after filter 1	Rank after filter 2	%CC	rmsd (Å)
ARC	91	22	69	4.1	1	1	69	4.0
CRO	12	3	85	3.0	121	9	80	3.8
GAL	37	26	75	3.6	2	2	75	3.6
LAC	30	13	72	4.0	133	117	77	3.4
LAM	22	3	84	3.0	4	4	98	1.6
MET	–	–	–	–	–	–	–	–
PUR	9	2	68	4.3	28	28	92	2.3
TRP	4	2	65	4.2	1	1	67	3.2

work, particularly DNA footprinting, and historically this information was often available prior to determination of the structure of the complex. The second more restrictive distance constraint (filter 2) was that in addition to the base sequence, there was knowledge about one or a few repressor residues that contacted the DNA. Such information could be derived from mutagenesis experiments or can be predicted from identification of spatial clusters of conserved residues (Aloy and Sternberg, unpublished). To establish if there was a protein–DNA contact, each side-chain type was assigned an effective side-chain length (L) ranging from 0.5 Å (Gly) to 6.0 Å (Arg) as detailed in Gabb et al. [12]. The distances (D) between protein Cα atoms and the base glycosidic N atoms were calculated. There was a contact if any $D < L + 4.5$ Å.

8.5.2
Results

The agreement between the predicted and native complex was quantified using the standard rmsd superposing representing the protein by Cα atoms and the DNA by C1′ nucleotide atoms. However inspection of the results suggested that this did not provide a good guide to the quality of the prediction. Instead the percentage of correct contacts in the interface was used. First, the protein–nucleotide pairs in the native interface were defined as those pairs that had at least one non-hydrogen protein–DNA atom–atom contact of <5 Å. For these pairs, the Cα-C1′ distances were measured in the native and the predicted complexes and if the difference in distance was <4 Å, then a correct contact was assigned to the predicted structure. A "good" prediction was taken if the percentage correct contacts compared to the total number of interface contacts (%CC) was >65%.

Table 8.3 gives the rank of the first "good" prediction in lists ordered by the scoring function after the two types of distance filters ranking by shape complementarity and by the empirical scoring scheme. Out of the eight systems only for MET did the approach fail to generate a "good" solution. After filter 2, shape complementarity gave 4 solutions with a rank in the top 5. When ranking by the empirical score, four complexes were in the top 5 rank after filter 1 or after filter 2. However other complexes were below the top 100 solutions. At present, the empirical score would be useful to generate a limited list (say 5) of suggestions which has about a 50% chance of including a "good" solution. In contrast, shape complementarity would be more suitable to generate a list for subsequent refinement. The predicted complexes superimposed on the native structures are shown in Figure 8.8. The first "good" solution ranked on shape complementarity is shown.

To our knowledge, this work is the only systematic study of a simulation that performed a global search of docking proteins to DNA. The results show a somewhat poorer level of discrimination that we and other groups

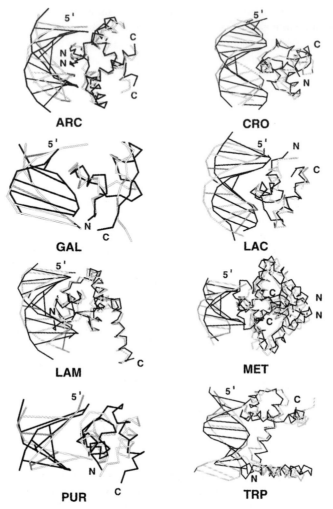

Fig. 8.8

Comparison of native and modelled repressor–DNA complexes. Superposition of native (black) and modelled (grey) complexes. The DNA is shown as the phosphate backbone with lines for the base pairs. The repressor is shown as a Cα trace. The first correct structure from Table 8.3 is shown. For MET the best model is described. Reproduced with permission from Figure 8.2 of Aloy et al. [34].

are obtaining for protein–protein docking. The method therefore would have to be used cautiously, for example in systems where there is substantial experimental data or when experiments would be designed to evaluate the prediction. In addition, the repressor–DNA system studied has limited conformational change on complex formation and many systems exhibit far larger changes in structure. In general, one does not know of the extent of conformational change on complex formation in advance of knowing the structure.

8.6
Strategies for protein–protein docking

Different groups have explored a variety of methods for protein–protein docking and we now review some of the major strategies that have been used, also see reviews [7–9]. We will emphasise the recent developments.

8.6.1
Evaluation of the results of docking simulations

There are several difficulties in trying to compare the results of the different approaches for protein–protein docking. Firstly there are a variety of methods to quantify the rmsd between the predicted and native complexes. Normally this is expressed as the rmsd between superimposed Cα coordinates of the predicted and a reference set of coordinates. There are however many approaches to calculate this rmsd. The reference molecule can be taken from the true (experimental) complex. However other groups first optimally superpose the unbound receptor onto the bound receptor in the complex and then the unbound ligand onto the bound ligand. The coordinates of the superposed unbound components is then taken as a reference as this represents the best structure one could achieve in rigid body docking. (This latter approach will be referred to here as rmsd from best attainable prediction.) Sometimes the optimal superposition of predicted receptor-ligand onto true receptor–ligand is evaluated. Other groups optimally superpose one of the molecules (typically the larger receptor) and then calculate the rmsd for the resultant predicted and true inhibitor positions. This second measure generally yields a substantially larger rmsd than the first. In addition, some groups will superimpose the entire molecules whilst others, particularly for larger molecules, will consider only residues in the interface.

One then needs also to define what rmsd constitutes a "good" prediction and groups use different cut offs. Generally with present day algorithms, the good structure rarely is consistently at top rank. Thus results cite the rank in a list at which a good model occurs. Comparison of ranks will depend on the fineness of sampling alternate docked arrangements and these vary between algorithms.

In evaluating algorithms, it is important to distinguish between results on docking starting with the components taken simply from the bound complex and studies that docking of both components starting from their unbound states. Although docking bound complexes might be helpful to develop an algorithm and to understand how it operates, there is a danger in over-optimising for stereochemical complementarity that will not occur in rigid body docking starting from unbound complexes. We therefore focus on results from unbound docking in this review. In addition, we note that in early work on docking, many of the algorithms were time consuming and the developers sometimes presented results on a very few systems. Results on limited data sets can suffer from the problem of over optimising the approach to provide success on a few specific test cases.

8.6.2
Fourier correlation methods

The Fourier correlation method that is the basis of our program FTDOCK described above was originally introduced by Katchalski-Katzir et al. [11] and the original paper emphasised the application to docking bound complexes. One of the authors of this work was Vakser who has continued with this approach and explored its application to low resolution docking [36, 37]. In this approach, a low resolution grid (typically several Ångstroms) is used and this leads to representing only the major spatial features of the two molecules. The energy surface is therefore smoother, i.e., not at the atomic level, and the search far faster as the number of cells used to represent the two molecules far fewer. Recently Vakser [39] explored the use of low resolution docking on a database of 475 co-crystallized protein–protein complexes. Most of the database were multimeric proteins but it also included complexes of the type conventionally studied in docking. He considered the average separation between the mass centres of the receptor and the ligand in his database (48 Å). The study considered the binding region of the receptor which is defined as within a 10 Å arc traced out by the centre of mass of the ligand at this separation. Note that the orientation of the ligand is not considered, only the location of the binding site of the receptor. The docking program GRAMM was run at grid spacing of 6.8 Å and a rotational step of 20°. The receptor binding site was recognised In 52% of the complexes using this low resolution docking. For 113 complexes with a large interface area ($>4000 \text{ Å}^2$), the success rate rose to 76%. Thus low resolution docking provides a valuable tool to delineate a putative binding site in a protein. This could then be used to provide a filter for results of higher resolution procedures.

A new approach for rapid protein docking using Fourier correlation, HEX, has been introduced by Ritchie and Kemp [40]. They employ spherical polar Fourier correlations which removes the time consuming requirement of FTDOCK and GRAMM for explicitly generating different orientations for the mobile molecule. They consider both shape complementarity and electrostatic effects. Note that their approach uses a Fourier correlation but does not use the discrete Fourier transform employed by GRAMM and FTDOCK. A full search on a single workstation takes around two hours. However, constraints on the location of the receptor binding site can readily be incorporated and this reduces the calculation to the order of minutes. In the conventional Fourier approach (FTDOCK and GRAMM), knowledge of the binding site in one of the two molecules (say the static) does not provide a constraint. Every orientation of the mobile molecule still needs to be sampled. Their program has been extensively benchmarked on unbound docking and yields results that suggest their program is a valuable tool for protein docking.

8.6.3
Other rigid-body docking approaches

Janin and coworkers [41] used spheres of different radii to replace residues and docked the resultant structures so as to maximise the buried surface area. These initial docked complexes were refined using a Monte Carlo search in which trial perturbations are generated randomly. A perturbation is always accepted if it reduces the energy and may also be accepted if the energy increases but is below a randomly chosen value determined by Boltzmann's principle. Finally a fully atomic representation (apart from hydrogen) is restored and an energy minimisation performed that allows the side-chains in the predicted interface to move. Note that the initial selection of associations to examine including flexibility is constrained by the rigid-body approximation. Cherfils et al. [41] tested the approach on two unbound docking systems and for one obtained a native-like solution at rank three.

The DOCK algorithm, developed by Kuntz and his group, is widely used in the community for the docking of low molecular weight ligands to protein receptor. It can however also be applied to protein-protein docking [42]. The DOCK approach fills the binding site of one protein (the receptor) with a cluster of overlapping spheres. Then the algorithm matches the sphere centres of this cluster with similar clusters from the ligand protein. Predicted complexes are then ranked in terms of a score for residue–residue contact. In the early study on docking starting with unbound complexes [42], individual atoms from the molecules had to be manually truncated to obtain good predictions. Although there were suggestions of which atoms may be mobile from the crystallographic thermal parameters present in the coordinate file, in other simulations one would not know in advance which atoms need to be truncated. Recently, the approach has been developed further for macromolecular docking [43]. The approach consists of three steps: defining the shape-based sites that define putative docking locations; docking using these site points; and scoring the docked complex. Complexes were scored using van der Waals and electrostatic interactions calculated from the AMBER program with united-atom parameters [27]. The method was benchmarked on re-docking several complexes starting from the *bound* coordinates. The approach was then used to suggest a model for the docking of human growth hormone to its receptor.

The groups of Wolfson and Nussinov have developed an approach for docking approach based on matching critical points [44–46]. These points define the knobs and holes on the two interacting surfaces. Both surface points and surface normals are matched and the approach can be implemented speedily. A global search takes of the order of minutes. After the search, putative solutions are checked to penalise overlap. In addition, the extent of hydrophobic packing across the interface is assessed. The method was recently tested on four different protease–inhibitor systems starting with unbound components and two bound antibody – unbound antigens complexes. The rmsd was taken between the true and predicted ligand after optimally superposing the receptor. Rankings of the first structure better than 5 Å rmsd were between 1 and 600, with several systems having rank 10 or better. For each system studied, several sets of coordinate were used in different trials. This showed that the method proved sensitive to the precise starting set of coordinates for a given biological complex. This highlights that for other algorithms (including FTDOCK) it is important to assess the dependency of the results on the precise starting coordinates.

Lenhof [47] has developed an approach for docking based on the identification of points on the surface of each molecule that could be equivalenced in a close-packed association. The search for possible rigid-body associations is then speeded up by considering which sets of three points on one molecule could be equivalenced to three points on the other molecule. Suitable docked complexes are then scored in terms of the geometric match between atoms followed by consideration of the chemical complementarity of the match. Trials with docking starting from unbound components showed that for several, but not all, systems the method yielded lists of a few complexes (<10) one of which was close (<4 Å rmsd) to the native. This method has been extended to include a treatment of side-chain flexibility in a subsequent screening that is described in 6.6 below [48].

Recently, Ausiello et al. [49] have developed a new docking procedure, ESCHER. The method starts with shape complementarity based on a slices along the Z axis of the protein surface mapped onto sets of polygons. Complementarity is assessed from the close approach of polygon vertices between the two docked molecules. Steric clashes and charge complementarity are then evaluated. The quality of the results is assessed from the rmsd of the predicted complex from the best attainable prediction. Only one true unbound docking system was studied (chymotrypsin–ovomucoid inhibitor) and good structures (rmsd for the complex <2.0 Å) was obtained at rank 3. In addition similar results were obtained for two complexes in modelling the docking of bound antibody to unbound antigen.

8.6.4
Flexible protein–protein docking

Abagyan and Tortov [50] have developed a method that introduces side-chain flexibility early in the search procedure. The approach was applied to unbound docking hen lysozyme to the bound conformation of the combining site of antibody HyHEL5. A trial set of starting conformations are generated to sample space. Then a Monte Carlo search is performed starting with random rigid-body shifts (see 8.6.3 for a brief description of the Monte Carlo procedure). Following each random step, side chain torsion angles were allowed to vary and the energy of the resultant complex minimised. This procedure was run to identify the set of conformations close to (within 20 kcal/mol) to the best structure. In this system, this leads to 30 trial structures that were then subject to extensive energy minimisation including both rigid-body shifts and side-chain rotation. The lowest energy structure was close the native complex – when the antibody coordinates were superimposed the rmsd for the Cα, C and N backbone lysozyme atoms was 1.6 Å. The procedure was reported in 1994 and was time consuming taking 500 hours of computing time on a state-of-the-art workstation (AXP3000/400).

8.6.5
Rigid-body treatment to re-rank putative docked complexes

Evaluation of a scoring function to assess the stereochemistry of trial complexes is a central feature of all docking algorithms. However additional methods can be applied to re-rank putative complexes. An important distinction must be made between whether the re-evaluation treats the complexes as rigid bodies or if the procedure first includes conformational flexibility to refine the complex before re-ranking. This distinction is illustrated by the distinction in our protein-protein docking procedure described in sections 3 and 4. The scoring potentials implemented in RPSCORE are used to re-rank rigid-body complexes. In contrast MULTIDOCK refines side-chain conformation and then re-ranks the complexes.

Prior to the development of RPSCORE and MULTIDOCK, in our Laboratory we introduced a rigid-body re-ranking procedure referred to as the continuum model [51]. This approach evaluated the electrostatic and hydrophobic contribution to relative energies of a series of putative docked complexes treating the solvent macroscopically (i.e., as a continuum) rather than explicitly including the solvent atoms. The total electrostatic energy of binding involved the loss of interaction between the solvent and each of the protein components independently followed by the interaction between the two proteins. The system treated each protein as a low dielectric surrounded by a high dielectric solvent. The electrostatic contributions were evaluated using the program DelPhi [52]. This program maps the protein–solvent system onto a grid and then calculates the resultant electrostatic effects considering both the local dielectric (protein or solvent) and the local charge distribution. In addition, in the continuum model we optimized the position of polar hydrogens prior to the calculation of the electrostatic effects. The hydrophobic effect was quantified in terms of the change to the molecular surface (MS) when the two proteins dock [53]. Generally the hydrophobic effect is quantified in continuum modelling as being proportional to the change in solvent accessible surface (SAS) area [54], where SAS is the surface traced by the centroid of a hypothetical probe water molecule as it rolls along the surface of the protein [53]. Molecular surface can be considered as the surface representing the protein/solvent-probe interface. Previously, Jackson and Sternberg [55] suggested that MS provides a better model for the hydrophobic effect than SAS. The continuum method was applied to re-rank the results on three enzyme-inhibitor systems generated in unbound docking using DOCK by [42]. The continuum model was able to identify a near native solution as having a particularly low energy. We are planning to evaluate the use of this continuum model to screen the results of complexes generated by our docking suite after their refinement by MULTIDOCK.

Recently an approach to re-rank complexes treated as rigid bodies was developed by [56]. They developed a new soft mean-field potential derived from analysis of protein–protein contacts in crystal structures. The hydrophobic–hydrophobic atom potential was applied to screen docked complexes. The series of docked complexes were generated by the approach of Cherfils et al. [41], see above. Four systems were studied: a reconstitution of the bound components of barnase and barstar, two docking starting with unbound coordinates (β-lactamase–β-lactamase inhibitor and chymotrypsin –ovomucoid) and one docking bound antibody coordinates to lysozyme. For the first three of these systems studied, the lowest free energy trial complex was a "good" prediction being within 2.5 Å of the true complex. In the fourth system, such a "good" solution was found at rank 2 on ordering by increasing energy. Thus the potentials are highly effective in screening for a "good" docked structure.

8.6.6
Introduction of flexibility to re-rank putative docked complexes

A method to include side-chain flexibility into the refinement and re-ranking of docked complexes has been developed by [57]. They considered the three enzyme–inhibitor complexes generated using unbound docking by [42] and subsequently studied by the continuum model by [51], see Section 8.6.5 above. The conformation of inhibitor side chains buried in the docked complex with the enzyme were examined by ehaustive conformational search for energetically more stable positions using CONGEN [58]. The resultant complexes were then scored by a measure of their relative energetic stability. The function considering the electrostatic interaction between the molecules, desolvation and side-chain conformational entropy. Desolvation was evaluated using the rapid approach of being proportionality to the change in accessible surface area where the constant of proportionality depends on the nature of the atom [59]. Side-chain conformational entropy was assumed to be proportional to the change in solvent accessible surface [60, 61]. For each complex the procedure lead to the identification of a native-like structure as having the lowest energy of association. In addition, in general there was an improvement in the agreement between the native and predicted geometry of the side-chains whose conformation was adjusted.

A method that provides an extensive sampling of side-chain conformations has been recently reported [48]. The set of rigid-body docked complexes for screening were generated by the method of [47] described in Section 8.6.4. All side chains with rotatable bonds to non-hydrogen atoms that are part of the putative interface are considered. A combinatorial search for favourable orientations in undertaken using computational methods (dead-end elimination and branch-and-bound) to prune the search space and thereby speed up the calculation. The docked structures are evaluated in a method similar to the continuum model of Jackson and Sternberg [51]. They workers considered three enzyme-inhibitor complexes and for each the lowest energy conformation was close to the native complex.

8.7
Blind trials of protein–protein docking

There are several problems in comparing the success of docking approaches from different groups on a set of previously known structures. We have above (8.6.1) described the variety of different measures used to report the agreement between a predicted and the true complex and the problems of the fineness of sampling the space of docked conformation. In addition, working on reproducing the docked structures of known complexes can lead to optimising an approach until it is successful. A developer might be aware of specific features of the stereochemistry of the known complexes and include these features in the algorithm leading to a bias towards known rather than unknown targets. In recognition of these problems, there have been two blind tests of protein–protein docking. In these tests, docking groups were supplied coordinates of the components and challenged to predict the structure of the complex prior to its structure being reported.

The first test, the Alberta challenge, was organised by James and Strynadka [62] and involved docking the coordinates of unbound β-lactamase to those of its unbound inhibitor. Six groups submitted entries (see Table 8.4). The results were impressive because all entrants identified as their favored suggestion a model that had an rmsd for superimposed Cα atoms of no more than 2.5 Å. However some groups submitted other entries that were far from the correct structure. Many of the entries used a biological filter requiring that the inhibitor docked to the known active site of β-lactase. The closest prediction to the true structure, submitted by Eisenstein and Katchalski-Katzir, had an rmsd of all superimposed Cα atoms of 1.1 Å which corresponds to an rmsd for the inhibitor Cα atoms of 4.6 Å when the enzyme was optimally superposed. Their successful approach employed the Fourier correlation method developed by Katchalski-Katzir et al. [11]. Their version only considered rigid-body shape complementarity which clearly was sufficient in this system for a successful prediction. However, in trials with

Tab. 8.4

Results of the Alberta Docking Challenge. The complex was the docking of unbound
β-lactamase coordinates with unbound inhibitor coordinates. The value of rmsd-all is
the rmsd of superposed main-chain atoms of the complex. The value of rmsd-inhibitor
is the rmsd for the main-chain atoms of the inhibitor after optimally superposing the
enzyme. Note that the best rank was often decided by the user rather than
automatically by the algorithm.

Group	No of models	rmsd-all of best ranked (Å)	rmsd-all of other models (Å)	rmsd inhibitor of best ranked (Å)
Abagyan & Totrov	3	1.9	11.3–16.2	6.6
Duncan, Rao and Olson	14	1.9	2.0–17.7	4.5
Eisenstein & Katchalski-Katzir	3	1.1	13.4–14.1	3.4
Jackson & Sternberg	1	1.9	N/A	4.0
Janin, Cherfils & Zimmerman	4	2.5	2.5–16.0	6.1
Shoichet & Kuntz	15	1.8	2.3–18.7	3.8

FTDOCK, we found that for many other systems electrostatic effects need to
be included.

In the Alberta challenge, four other groups all performed a global search
and submitted a model between 1.9 Å and 2.5 Å from the true complex corresponding to an rmsd for the inhibitor of between 4.0 Å and 6.6 Å. The
methods were diverse. Two methods were totally rigid-body dockings – the
DOCK approach from Shoichet and Kuntz (see 8.6.3 above and Shoichet
and Kuntz [42]) and the comparison of protein surfaces using a smoothed
representation (by Duncan Rao and Olson [63]). The model submitted by
Janin, Cherfils and Zimmerman (see Section 8.6.3 above and Cherfils et al.
[41]) started with a rigid-body docking but then includes side-chain optimisation. Only one approach, from Abagyan and Totrov, used a procedure
(see Section 8.6.4 and Totrov and Abagyan [64]) that incorporated flexibility
at an early stage of the search procedure. These results suggest that for this
system, the rigid-body approximation is appropriate for docking simulation.
In addition, the different approach to match surfaces yield broadly similar
results. In contrast to the other five submissions that performed a global
search, the use of the continuum model (see Section 8.6.5 and Jackson and
Sternberg [51]) for screening was evaluated by Jackson and Sternberg. They
used the results generated from DOCK and were able to identify a single
preferred complex that was close to the native.

The second test was as part of the second comparative assessment of
protein structure prediction (CASP2) [65]. The target was an antibody–
hemagglutinin complex and coordinates of unbound hemagglutinin and
bound antibody were supplied. This was a difficult target given the size of
the complex and only four groups entered. Multiple entries were allowed
and confidences then had to be assigned so that the total was 100%. No

Tab. 8.5
Results of the CASP2 Docking Challenge. The complex was the docking of unbound hemagglutinin coordinates with bound antibody coordinates. The rmsd refers to rmsd for the interface Fab Cα atoms after optimal superposition of the hemagglutinins. The interface atoms of the antibody are those Cα atoms within 8 Å of the hemagglutinin–Fab interface. Correct residue–residue contacts are defined as inter-protein atom-atom distances less than the sum of the van der Waals radii plus 1 Å. There were a total of 59 contacts. The mean values refer to the weighted scores from all the predictions. Min and max are the minimum and maximum values.

Group	No of models	rmsd (Å)		No of correct contacts	
		mean	min	mean	max
DeLisi	2	18.3	15.1	4.5	5
Rees	2	32.3	30.6	0	0
Sternberg	8	20.2	8.5	1.8	8
Vakser	1	9.5	9.5	0	0

group submitted any entry that was close to the true complex (Table 8.4). The averaged best prediction, from Vakser, used the Fourier correlation approach. Only one model was submitted and this yielded an rmsd for the interface Cα atoms of the antibody of 9.5 Å, calculated after optimally superposing the hemagglutinin. However this prediction did not have any correct contacts (defined as atoms that are separated by their van der Waals radii plus 1 Å). This submission was based on the Katchalski-Katzir Fourier method implemented for low-resolution search in Vakser's program GRAMM [37, 66]. The single prediction that was closest to the native (an rmsd of 8.5 Å for the interface antibody Cα atoms) was also based on the Fourier correlation method as applied by the ICRF group of Sternberg, Jackson and Gabb using preliminary versions of FTDOCK and MULTIDOCK. However the approach when applied did not provide a single confident prediction and accordingly the group submitted eight entries. There were entries using two other approaches – a matching of surfaces using graph theory from Rees and coworkers [67] and the approach of DeLisi and coworkers (see 8.6.6 and Weng et al. [57]).

From only two blind trials, one cannot draw definitive conclusions. However, in both challenges the Fourier correlation approach of Katchalski-Katzir yielded the best submission which suggests that it must be considered as a valuable strategy for macromolecular docking.

8.8
Energy landscape for protein docking

This chapter has considered predictive protein–protein and protein–DNA docking. However there remains a related question of how two molecules can associate within the time observed biologically. The problem is that association rates for protein–protein docking would be $\sim 10^3$ M^{-1} s^{-1} if it were just governed by diffusion and a correction for orientational constraints. However observed rates are typically far faster being of the order of 10^5–10^7 M^{-1} s^{-1}. This increase in speed is often attributed to long range effects, particularly long range electrostatic steering. Janin [68] has shown that long-range electrostatic steering can enhance association rates by up to 10^5 fold.

Recently, the energy surface near the native docked complex has been modelled in terms of empirical atomic contact energies and Coulomb electrostatic interactions [69]. The Coulomb interaction has a distant-dependent dielectric and a cut off of 20 Å. Thus the energy surface does not consider long-range effects. The study showed that the energy gradient provided by the surface provides a funnel towards the docked structure that increases the probability that an encounter will evolve into the stable complex by about 400-fold. Given the simplified treatment of the interaction energy, in real system energy funnels could provide even greater enhancements for the rates of association. Thus even without long-range electrostatic effects, energy funnels provide a possible explanation for the observed relatively rapid association rates. More generally, the role of funnels in directing protein folding and function has been reviewed by Tsai et al. [70].

8.9
Conclusions

This review presents the current status of protein-protein and protein–DNA docking. The field will advance and new approaches will be proposed. To identify useful approaches, it is essential that results are presented on docking starting with unbound components. In addition, our experience is that good results can readily be obtained for a few systems and it therefore important to benchmark an algorithm on a large test set. Indeed part of the design principle for an algorithm should be that it sufficiently fast to be tested on a large data set. However, the best approach to compare approaches is via blind tests.

How useful is predictive protein docking in current work in structural bioinformatics? In general, the status of the better algorithms is that for systems that do not involve substantial conformational change, one can generate a solution close to the native structure ($<3.0 \text{ Å}$ rmsd for superposed interface $C\alpha$ atoms) and this solution can be ranked at 10 or better. However results are this type are not guaranteed – typically it will occur for only half the systems studied.

We will soon have the genome sequences for man and many pathogens. Structural genomics initiatives [71] will exploit this sequence information

and over the next few years the three-dimensional structures for many proteins in these genomes will be determined. There will remain the difficulty
of obtaining structural information of how the proteins interact. Thus the
application are for predictive docking will increase drastically over the next
few years. The predicted structures of docked complexes will suggest further experimentation such as mutagenesis to probe activity and alter binding affinities between molecules. For example, the systematic modification
of the sequence of the combining site of antibodies based on a predicted
structure could lead to molecules that recognise antigens with improved
or altered specificity. These engineered antibodies should prove to have a
major role in both medical and commercial applications, for example in
recognising tumors. However in many application, high molecular weight
proteins are not suitable as pharmaceutical agents due to problems such as
being metabolised rapidly or generating an allergic response. However predictive models of a protein–protein complex could suggest the design of low
molecular weight compounds that mimic the recognition region these could
provide the lead for drug discovery.

The development of the algorithms for protein-protein docking has been
highly successful over the 1990s and now is a viable approach for modelling. Over the next decade, we envisage that will there be further advances
in the methodology together with many application of predictive docking in
both fundamental and applied research.

Acknowledgements

We thank the following colleagues with whom we worked on developing
the docking programs: Dr Henry Gabb (FTDOCK); Dr Richard Jackson
(MULTIDOCK); Mr Patrick Aloy, Professor Francesc Aviles and Professor
Enrique Querol (DNA docking) and Dr Mathew Betts (conformational
changes on docking). Mr Aloy and Professors Aviles and Querol are members of the Institut de Biologia Fondamental, Universitat Autonoma de
Barcelona, Spain.

References

1 BERMAN, H. M., WESTBROOK, J., FENG, Z., GILLILAND, G.,
BHAT, T. N., WEISSIG, H., SHINDYALOV, I. N. & BOURNE, P. E.
(2000). The Protein Data Bank. *Nucleic Acid Res.* 28, 235–242.
2 BETTS, M. J. & STERNBERG, M. J. E. (1999). An analysis of
conformational changes on protein–protein docking:
implications for predictive docking. *Prot. Eng.* 12, 271–283.

3 CONTE, L. L., CHOTHIA, C. & JANIN, J. (1999). The atomic structure of protein–protein recognition sites. *J. Mol. Biol.* 285, 2177–2198.

4 JONES, S. & THORNTON, J. M. (1996). Principles of protein–protein interactions. *Proc. Natl. Acad. Sci. USA* 93, 13–20.

5 SUNDBERG, E. J. & MARIUZZA, R. A. (2000). Luxury accommodations: the expanding role of structural plasticity in protein–protein interactions [In Process Citation]. *Structure Fold Des* 8, 137–142.

6 COLMAN, P. M. (1994). Structure-based drug design. *Current Opinion in Structural Biology* 4, 868–874.

7 JANIN, J. (1995). Protein–protein recognition. *Prog. Biophys. Molec. Biol.* 64, 145–166.

8 SHOICHET, B. K. & KUNTZ, I. D. (1996). Predicting the structure of protein complexes: a step in the right direction. *Chemistry & Biology* 3, 151–156.

9 STERNBERG, M. J. E., GABB, H. A. & JACKSON, R. M. (1998). Predictive docking of protein–protein and protein–DNA complexes. *Curr Opin Structural Biology* 8, 250–256.

10 CHOTHIA, C. & JANIN, J. (1975). Principles of protein–protein recognition. *Nature (London)* 256, 705–708.

11 KATCHALSKI-KATZIR, E., SHARIV, I., EISENSTEIN, M., FRIESEM, A. A., AFLALO, C. & VAKSER, I. A. (1992). Molecular surface recognition: determination of geometric fit between proteins and their ligands by correlation techniques. *Proc. Natl. Acad. Sci. USA* 89, 2195–2199.

12 GABB, H. A., JACKSON, R. M. & STERNBERG, M. J. E. (1997). Modelling protein docking using shape complementarity, electrostatics and biochemical information. *J. Mol. Biol.* 272, 106–120.

13 HINGERTY, B. E., RITCHIE, R. H., FERRELL, T. L. & TURNER, J. E. (1985). *Biopolymers* 24, 427.

13 SIPPL, M. J. (1990). Calculation of conformational ensembles from potentials of mean force. An approach to the knowledge-based prediction of local structures in globular proteins. *J. Mol. Biol.* 213, 859–883.

14 JONES, D. T., TAYLOR, W. R. & THORNTON, J. M. (1992). A new approach to protein fold recognition. *Nature* 358, 86–89.

15 VAJDA, S., SIPPL, M. & NOVOTNY, J. (1997). Empirical potentials and functions for protein folding and binding. *Curr. Opin. in Struct. Biol.* 7, 222–228.

16 MITCHELL, J. B. O., LASKOWSKI, R. A., ALEX, A. & THORNTON, J. M. (1999a). BLEEP – Potential of mean force describing protein–ligand interactions: I. Generating potential. *J. Comp. Chem.* 20, 1165–1176.

17 MITCHELL, J. B. O., LASKOWSKI, R. A., ALEX, A. & THORNTON, J. M. (1999b). BLEEP – Potential of mean force describing protein–ligand interactions: II. Calculation of binding energies and comparison with experimental data. *J. Comp. Chem.* 20, 1177–1185.

18 MUEGGE, I. & MARTIN, Y. C. A general and fast scoring function for protein–ligand interactions: a simplified potential approach. *J. Med. Chem.* 42, 791–804.

19 WALLQVIST, A., JERNIGAN, R. L. & COVELL, D. G. **(1995)**. A preference-based free-energy parameterization of enzyme-inhibitor binding. Applications to HIV-1–protease inhibitor design. *Protein Science* 4, 1881–1903.

20 MOONT, G., GABB, H. A. & STERNBERG, M. J. E. **(1999)**. Use of pair potentials across protein interfaces in screening predicted docked complexes. *Proteins* 35, 364–373.

21 HUBBARD, T. J. P., AILEY, B., BRENNER, S. E., MURZIN, A. G. & CHOTHIA, C. **(1999)**. SCOP: A structural classification of proteins database. *Nucleic Acids Research* 27, 254–256.

22 KABSCH, W. & SANDER, C. **(1983)**. Dictionary of protein secondary structure: pattern recognition of hydrogen-bonded and geometrical features. *Biopolymers* 22, 2577–2637.

23 SKOLNICK, J., JAROSZEWSKI, L., KOLINSKI, A. & GODZIK, A. **(1997)**. Derivation and testing of pair potentials for protein folding. When is the quasichemical approximation correct? *Prot. Sci.* 6, 676–688.

24 THOMAS, P. D. & DILL, K. A. **(1996)**. Statistical potentials extracted from protein structures: how accurate are they? *J. Mol. Biol.* 257, 457–469.

25 JACKSON, R. M., GABB, H. A. & STERNBERG, M. J. E. **(1998)**. Rapid refinement of protein interfaces incorporating solvation: application to the docking problem. *J. Mol. Biol.* 276, 265–285.

26 TUFFERY, P., ETCHEBEST, C., HAZOUT, S. & LAVERY, R. **(1991)**. A new approach to the rapid determination of protein side-chain conformations. *J. Biomol. Struct. Dynam.* 8, 1267–1289.

27 WEINER, S. J., KOLLMAN, P. A., CASE, D. A., SINGH, U. C., GHIO, C., ALAGONA, G., PROFETA, S. & WEINER, P. **(1984)**. A new force field for molecular mechanical simulation of nucleic acids and proteins. *J. Am. Chem. Soc.* 106, 765–784.

28 SITKOFF, D., SHARP, K. A. & HONIG, B. **(1994)**. Accurate calculation of hydration free energies using macromolecular solvent models. *J. Phys. Chem.* 98, 1978–1988.

29 KOEHL, P. & DELARUE, M. **(1994)**. Application of a self-consistent mean field theory to predict protein side-chains conformation and estimate their conformational entropy. *J. Mol. Biol.* 239, 249–275.

30 LEE, C. **(1994)**. Predicting protein mutant energetics by self-consistent ensemble optimization. *J. Mol. Biol.* 236, 918–939.

31 PRESS, W. H., TEUKOLSKY, S. A., VETTERLING, W. T. & FLANNERY B. P. **(1992)**. *Numerical recipes in FORTRAN – The art of scientific computing.* 2nd edn, Cambridge University Press, Cambridge.

32 LAWRENCE, M. C. & COLMAN, P. M. **(1993)**. Shape complementarity at protein/protein interfaces. *J. Mol. Biol.* 234, 946–950.

33 JONES, S., HEYNINGEN, P. V., BERMAN, H. M. & THORNTON, J. M. **(1999)**. Protein-DNA Interactions: A Structural Analysis. *J. Mol. Biol.* 287, 877–896.

34 ALOY, P., MOONT, G., GABB, H. A., QUEROL, E., AVILES, F. X. & STERNBERG, M. J. E. **(1998)**. Modeling repressor proteins binding to DNA. *Proteins* 33, 535–549.

35 KNEGTEL, R. M. A., ANTOON, J., RULLMANN, C., BOELENS, R. & KAPTEIN, R. (1994a). MONTY: a Monte Carlo approach to protein–DNA recognition. *J. Mol. Biol.* 235, 318–324.

36 KNEGTEL, R. M. A., BOELENS, R. & KAPTEIN, R. (1994b). Monte Carlo docking of protein–DNA complexes: incorporation of DNA flexibility and experimental data. *Prot. Eng.* 7, 761–767.

37 VAKSER, I. A. (1995). Protein docking for low-resolution structures. *Protein Engineering* 8, 371–377.

38 VAKSER, I. A. (1996). Long-distance potentials: an approach to the multiple-minima problem in ligand–receptor interaction. *Protein Engineering* 9, 37–41.

39 VAKSER, I. A., MATAR, O. G. & LAM, C. F. (1999). A systematic study of low-resolution recognition in protein–protein complexes. *Proc. Natl. Acad. Sci. U.S.A.* 96, 8477–8482.

40 RITCHIE, D. W. & KEMP, G. J. L. (2000). Protein docking using spherical polar Fourier correlations. *Proteins* 39, 178–194.

41 CHERFILS, J., DUQUERROY, S. & JANIN, J. (1991). Protein–protein recognition analyzed by docking simulation. *Proteins* 11, 271–280.

42 SHOICHET, B. K. & KUNTZ, I. D. (1991). Protein docking and complementarity. *J. Mol. Biol.* 221, 327–346.

43 HENDRIX, D. K., KLIEN, T. E. & KUNTZ, I. D. (1999). Macromolecular docking of a three-body system: The recognition of human growth hormone by its receptor. *Protein Science.* 8, 1010–1022.

44 FISCHER, D., LIN, S. L., WOLFSON, H. J. & NUSSINOV, R. (1995). A suite of molecular docking processes. *J. Mol. Biol.* 248, 459–477.

45 FISCHER, D., NOREL, R., WOLFSON, H. & NUSSINOV, R. (1993). Surface motifs by a computer vision technique: searches, detection, and implications for protein-ligand recognition. *Proteins: Structure, Function, and Genetics* 16, 278–292.

46 NOREL, R., PETREY, D., WOLFSON, H. J. & NUSSINOV, R. (1999). Examination of shape complementarity in docking of *unbound* proteins. *Proteins* 36, 307–317.

47 LENHOF, H. P. (1997). New contact measures for the protein docking problem. In *RECOMB97 – Proceedings of the first annual international conference on computational molecular biology.* ACM.

48 ALTHAUS, E., KOHLBACHER, O., LENHOF, H. P. & MULLER, P. (2000). A combinatorial approach to protein docking with flexible side-chains. In *RECOMB 2000 – Proceedings of the fourth international conference on computational molecular biology.* ACM.

49 AUSIELLO, G., CESARENI, G. & HELMER-CITTERICH, M. (1997). ESCHER: A new docking procedure applied to the reconstruction of protein tertiary structure. *Proteins* 28, 556–567.

50 TOTROV, M. & ABAGYAN, R. (1994). Detailed ab initio prediction of lysozyme-antibody complex with 1.6 Å accuracy. *Nature Structural Biology* 1, 259–263.

51 JACKSON, R. M. & STERNBERG, M. J. E. (1995). A continuum model for protein–protein interactions: Application to the docking problem. *J. Mol. Biol.* 250, 258–275.

52 GILSON, M. K. & HONIG, B. **(1988)**. Calculation of the total electrostatic energy of a macromolecular system: solvation energies, binding energies, and conformational analysis. *Proteins: Structure, Function, and Genetics* 4, 7–18.

53 RICHARDS, F. M. **(1977)**. Areas, volumes, packing, and protein structure. *Annu. Rev. Biophys. Bioeng.* 6, 151–176.

54 CHOTHIA, C. **(1974)**. Hydrophobic bonding and accessible surface area in proteins. *Nature (London)* 248, 338–339.

55 JACKSON, R. M. & STERNBERG, M. J. E. **(1994)**. Application of scaled particle theory to model the hydrophobic effect: implications for molecular association and protein stability. *Prot. Eng.* 7, 371–383.

56 ROBERT, C. H. & JANIN, J. **(1998)**. A soft, mean-field potential derived from crystal contacts for predicting protein-protein interactions. *J. Mol. Biol.* 283, 1037–1047.

57 WENG, Z., VAJDA, S. & DELISI, C. **(1996)**. Prediction of protein complexes using empirical free energy functions. *Prot. Sci.* 5, 614–626.

58 BRUCCOLERI, R. E. & NOVOTNY, J. **(1992)**. Antibody modeling using the conformational search program CONGEN. *Immunomethods* 1, 96–106.

59 EISENBERG, D. & McLACHLAN, A. D. **(1986)**. Solvation energy in protein folding and binding. *Nature* 319, 199–203.

60 PICKETT, S. D. & STERNBERG, M. J. E. **(1993)**. Empirical scale of side-chain conformational entropy in protein folding. *J. Mol. Biol.* 231, 825–839.

61 VAJDA, S., WENG, Z., ROSENFELD, R. & DELISI, C. **(1994)**. Effect of conformational flexibility and solvation on receptor–ligand binding free energies. *Biochemistry* 33, 13977–13988.

62 STRYNADKA, N. C. J., EISENSTEIN, M., KATCHALSKI-KATZIR, E., SHOICHET, B., KUNTZ, I., ABAGYAN, R., TOTROV, M., JANIN, J., CHERFILS, J., ZIMMERMAN, F., OLSON, A., DUNCAN, B., RAO, M., JACKSON, R., STERNBERG, M. & JAMES, M. N. G. **(1996)**. Molecular docking programs successfully determine the binding of a beta-lactamase inhibitory protein to TEM-1 beta-lactamase. *Nature Structural Biology* 3, 233–238.

63 DUNCAN, B. S. & OLSON, A. J. **(1993)**. Approximation and characterization of molecular surfaces. *Biopolymers* 33, 219–229.

64 TOTROV, M. & ABAGYAN, R. **(1994)**. Detailed ab initio prediction of lysozyme–antibody complex with 1.6 Å accuracy. *Nature Structural Biology* 1, 259–263.

65 DIXON, J. S. **(1997)**. Evaluation of the CASP2 docking section. *Proteins Supplement* 1, 198–204.

66 VAKSER, I. A. **(1997)**. Evaluation of GRAMM low-resolution docking methodology on the hemagglutinin–antibody complex. *Proteins* Supplement 1, 226–230.

67 WEBSTER, D. M. & REES, A. R. **(1993)**. Macromolecular recognition: antibody–antigen complexes. *Prot Eng* 65, 94.

68 JANIN, J. **(1997)**. The kinetics of protein–protein recognition *Proteins* 28, 153–161.

69 ZHANG, C., CHEN, J. & DELISI, C. **(1999)**. Protein–protein recognition: exploring the energy funnels near the binding sites. *Proteins* 34, 255–267.

70 Tsai, C. J., Kumar, S., Ma, B. Y. & Nussinov, R. **(1999).** Folding funnels, binding funnels, and protein function. *Protein Science* 8, 1181–1190.

71 Burley, S. K., Almo, S. C., Bonanno, J. B., Capel, M., Chance, M. R., Gaasterland, T., Lin, D. W., Sali, A., Studier, F. W. & Swaminathan, S. **(1999).** Structural genomics: beyond the human genome project. *Nature Genetics* 23, 151–157.

72 Hingerty, B. E. & Broyde, S. **(1985).** Carcinogen-base stacking and base-base stacking in depdg modified by (+) and (−) anti-bpde. *Biopolymers* 24, 2279–2299.

Appendix

Thomas Lengauer

Glossary of algorithmic terms in bioinformatics

A* algorithm

A heuristic procedure for navigating in large search spaces that are formally described by graphs. The algorithm originated in the field of Artificial Intelligence in the 1960s (Hart et al., 1968). The algorithm finds a cheapest path in a graph from a start vertex to one among a selected set of goal vertices. As usual with path search algorithms the path is extended away from the start vertex step by step. At each step the next most promising vertex v to proceed to is selected. The selection is based on a scoring function that encompasses two terms

- a term amounting to the cost of the initial path segment from the start vertex to v
- a term estimating the cost of the optimal remaining path segment from v to the closest goal vertex.

If we require that the estimate underestimates the actual costs of the remaining path segment then we can prove that the A* algorithm always finds a cheapest path. If the estimator is always zero, then the path search amounts to a normal breadth-first traversal of the graph (Russell & Norvig, 1995).

In bioinformatics, the A* algorithm has been used in conformational search in docking a (rigid) ligand into a flexible protein structure ((Leach, 1994), see Chapter 7 of Volume I). The path assigns rotameric states to side chains in the protein binding pocket, one after the other. A goal vertex is one that corresponds to a complete assignment of rotameric states. The start vertex is the one where no rotameric states have been assigned.

The estimation of the cost of the optimal remaining path segment is performed by generalizing a process called *dead-end elimination* (Desmet et al., 1992). This process precomputes for each rotamer j of a side chain whether there is another rotamer k of the same side chain that leads to an energetically better conformation, independent of rotameric choices for other side chains. If so, rotamer j is excluded from the search. Leach (Leach, 1994) generalizes this process from just excluding alternatives to computing a lower bound on path cost by considering optimum rotamers independently for each side chain yet to be placed and adding up the corresponding cost terms. This extension, which Leach also adapted to the placement of side-chains in homology-based modeling (Leach & Lemon, 1998), is described in detail in Chapter 5 of Volume I.

Branch & bound

Search method in tree-structured search spaces (Cormen et al., 1990). The original problem instance corresponds to the root of the tree. Internal nodes of the tree are generated iteratively, beginning with the root, by decomposing a problem instance into alternatives. Each alternative becomes a child of the node representing the subdivided problem. The solution of any of the alternatives (children) solves the original problem. Problem instances that are not decomposed further correspond to leaf nodes of the tree. They can be solved directly without any further subdivision. The solution of the original problem then amounts to completing any path from the root to a leaf node. Each interior node represents a partial problem solution that amounts to the set of alternatives already decided upon that are represented by the nodes along the path.

In general the tree is very large but not very deep. The depth of the tree (the length of the longest path from a root to a leaf) corresponds to the number of decisions to be taken to assemble the solution, which is limited in general (linear or a small-degree polynomial in the problem size). In contrast, the number of leaves of the tree, i.e. the number of alternative solutions, is usually huge, often exponential in the problem size. Thus traversing the tree efficiently becomes a major problem.

Often, problem solutions (complete or partial) are labeled with costs. Then, an optimal solution is represented by a path from the root to a leaf with optimal cost. In general, in order to find an optimal solution, the whole tree has to be enumerated. Complete enumeration can be avoided if we can provide for each node v inside the tree a lower bound (for minimization, upper bound for maximization) on the cost of all leaves that are contained in the subtree rooted at v. If this bound exceeds the cost of any solution that we have found before in the tree traversal, the whole subtree rooted at v can be disregarded. The sharper the lower bound the larger the regions of the tree that can be eliminated in this branch & bound procedure.

When traversing the whole tree, a branch & bound algorithm is certain to produce an optimal solution. Even when stopped early, the algorithm generally has found some solution already that it can return. In addition, the lower bounds computed so far can be used to generate a lower bound for the optimal solution of the problem and thus a quality certificate for the solution returned. The longer the algorithm runs the better the solution it returns and the closer the lower bound to the cost of the optimal solution. In typical cases, a branch & bound algorithm that runs to completion will spend about 10% to 20% of its time finding the optimal solution. The remaining 80% to 90% are spent on completing the tree traversal in order to make sure that there is no better solution.

A branch & bound algorithm for protein threading has been introduced by Lathrop and Smith (Lathrop & Smith, 1996). Here, the alternatives amount to different alignments of secondary structure blocks in proteins, and the cost function estimates the energy of the resulting protein structure. The algorithm is described in detail in Chapter 6 of Volume I. A branch & bound algorithm for docking ligands into protein structures with flexible side chains has been described in (Althaus et al., 2000), see also Chapter 8 of Volume I. The → A* algorithm with dead-end elimination is a variant of a branch & bound algorithm, as described in Chapter 5 of Volume I (for side chain placement).

Cluster analysis

A basic technique for data classification. Cluster analysis is aimed at grouping items in a set into clusters. The item in each cluster should share some commonalities that justify to group them. The similarity or difference between different items is usually measured by a function of pairs of items. If we measure difference (similarity), the values decrease (increase) with increasing similarity between a pair of items.

Clustering procedures can be distinguished with respect to the structure of the item set. In bioinformatics, mostly, the items are points in Euclidean space of some dimension or else they are nodes in a graph or hypergraph.

Hierarchical clustering procedures iteratively partition the item set into disjointed subsets. There are top-down and bottom-up techniques. The top-down techniques partition can be into two or more subsets, and the number of subsets can be fixed or variable. The aim is to maximize the similarity of the items within the subset or to maximize the difference of the items between subsets. The bottom-up techniques work the other way around and build a hierarchy by assembling iteratively larger clusters from smaller clusters until the whole item set is contained in a single cluster. A popular hierarchical technique is nearest-neighbor clustering, a technique that works bottom up by iteratively joining two most similar clusters to a new cluster.

Algorithms for computing phylogenies are closely related to hierarchical clustering methods (Gusfield, 1997).

Nonhierarchical techniques do not build up a tree structure of clusters.

Probably the most popular nonhierarchical clustering technique groups the items – in this case points in Euclidean space – around a fixed number k of cluster centers. The clusters are first determined randomly and then the following two steps are iterated. First the center of each cluster is computed, then the items are reassigned to the cluster whose center is closest. The iteration proceeds until the clusters do not change any more. If we extend this method such as to recalculate cluster centers everytime that we reassign an item to a new cluster then we obtain the widely used k-means clustering algorithm (Kaufman & Rousseeuw, 1990).

In bioinformatics, clustering is used in the analysis of protein families – sequence families and multiple alignment (Chapter 2 of Volume I), structure families (Chapter 6 of Volume I), or families of proteins with the same function. Furthermore gene expression data are clustered with respect to similarity in the expression profiles (Chapter 5 of Volume II), clustering is used in grouping similar molecular conformations in protein modeling (Chapter 5 of Volume I) or molecular structures or interaction geometries in docking (Chapters 7 and 8 of Volume I). In proteomics, gel images are clustered (Chapter 4 of Volume II). Clustering is also a major approach to analyzing molecular similarity (Chapter 6 of Volume II).

Divide & conquer

One of the basic algorithmic techniques, which solves a problem by subdividing it into a usually small set of subproblems, solving those problems and then assembling their solution to a solution of the original problem (Cormen et al., 1990). The subproblems are solved in the same fashion, unless they can be solved directly without further subdivision. In contrast to the tree search described above under → branch & bound, in divide & conquer algorithms we need to solve not just one but all subproblems of a problem to solve the problem. The procedure is most effective, if the subdivision is balanced in the sense that all subproblems of the same problem have about equal size.

Divide & conquer algorithms for sequence alignment are described in Chapters 2 of Volume I and Volume II, applications to the analysis of molecular similarity are described in Chapter 6 of Volume II.

Dynamic programming

Tabular method of discrete optimization. The basis of dynamic programming is the *principle of prefix optimality*. This principle states that the optimal solution to the optimization problem can be composed of optimal solutions of a limited number of smaller instances of the same type of problem. The tabular method incrementally solves subinstances of increasing size. The solutions of larger problem instances are assembled from previously computed solutions of smaller problem instances. The method stops when the original problem instance has been solved. The runtime for solving each subinstance tends to be constant or linear in the problem size. The number of subinstances can often be limited to a low-degree polynomial in the size of the problem instance (Cormen et al., 1990). For hard optimization problems this number may grow exponentially, however.

Sequence alignment (Chapter 2 of Volume I) is a prime example of a problem in bioinformatics that can be solved efficiently with dynamic programming. Dynamic programming for sequence alignment is also discussed extensively in (Gusfield, 1997; Waterman, 1995). Other examples of dynamic programming algorithms occur in Chapters 3 and 6 of Volume I and Chapter 2 of Volume II.

Energy minimization

Optimization method that aims at minimizing the energy of a molecular system that is usually described in the coordinates of the participating atoms. Since we usually deal with large molecules, the dimension of the space over which the energy is to be minimized is large (goes into the many thousands). Usually the energy function has a large number of local minima whose energy is close to or exactly the global minimum. In this context, finding the global minimum is very difficult (multi-minima problem). Depending on the type of energy function, different optimization methods are used. If we use local optimization methods that descend to the next local minimum we have to judiciously choose starting points for the optimization. Still, in the presence of many local minima, with local optimization we cannot hope to sample the solution space adequately to find the global minimum. Thus Monte Carlo methods such as → simulated annealing are employed that manage to escape from local minima.

Often the energy minimization is hampered by the fact that the energy function used is inaccurate. In addition, nature actually minimizes the *free energy* as opposed to the internal energy. Accurate computational models for free energy scoring require sampling of many conformations and are hence burdened by either substantial statistical error or extensive computing resource requirements. A particular problem is the incorporation of water, which makes large entropic contributions to the free energy. If water is modeled as a large set of discrete molecules large computing requirements result. The alternative is to model water as a continuum which, however, means another approximation.

Energy minimization is a voluminous field with dominating physico-chemical and thermodynamic aspects (Christensen & Jorgensen, 1997; Church & Shalloway, 2001; Trosset & Scheraga, 1998).

Fourier transform

A very important mathematical operation that transforms functions (continuous Fourier transform) or vectors (discrete Fourier transform) into the frequency domain (Bracewell, 1986). This means that the function is described in terms of the contributions of basic (sine) waves with different frequencies that compose the function. Besides its very fundamental physical relevance, the Fourier transform, especially its discrete version, has computational significance (Cormen et al., 1990). The reason is twofold. First, the Fourier transform of a vector of length n can be computed quite quickly, namely in time $O(n \log n)$ by the so-called fast Fourier transform (FFT) algorithm (Cooley & Tukey, 1965). Second, the important but quite complex operation on vectors that is called convolution, can be reduced to the much easier pointwise multiplication, when transformed into the frequency domain. The convolution of two vectors of length n requires time $O(n^2)$ when performed in a straightforward fashion. If we instead first Fourier transform the vectors, then pointwise multiply the result and then inverse Fourier transform, we spend $O(n \log n)$ time doing the transforms and $O(n)$ time for the multiplications, which amounts to a total of $O(n \log n)$ time. The convolution operation is related to polynomial and integer multiplication and therefore these are operations that can be sped up by using the Fourier transform. In protein-protein docking (Chapter 8 of Volume I), the convolution operation comes up in the analysis of the steric complementarity of two rigid protein structures. Thus Fourier transform helps there, as well.

Genetic algorithm

A class of algorithms that mimics nature's way of optimization by variation and selection (Goldberg, 1989). Genetic algorithms work on populations of solutions to an optimization problem. Each solution (phenotype) is encoded in a string form called the chromosome (genotype). Chromosomes are varied by mutation only (evolutionary algorithm (Fogel et al., 1966)) or by mutation and cross-over, i.e. merging parts of two chromosomes in analogy to the recombination process taking place during meiosis (Holland, 1973). A selection function is applied to the resulting phenotypes. This process is iterated through a number of generations. Typical population sizes rank from 10 to a few hundred. Generally, a few dozen generations are computed. The key ingredients of the development of an evolutionary algorithm are the suitable choice of the chromosomal representation of the solutions, the mutation and cross-over operators and the diverse parameters governing the execution of the algorithm.

In bioinformatics, genetic algorithms are popular as a tool for fast prototyping. For this reason genetic algorithms can be found in practically all facets of bioinformatics. In contrast, exploiting all of the power that lies in the approach of genetic algorithms requires an expert, and is an involved task.

This book mentions genetic algorithms in Chapters 6 and 7 of Volume I. An overview of genetic algorithms in molecular recognition is given in (Willett, 1995). In addition, we mention here recent application of genetic algorithms to protein structure superposition (Szustakowski & Weng, 2000), protein folding (Schulze-Kremer, 2000), protein-protein docking (Gardiner et al., 2001), and drug selection from combinatorial libraries (Illgen et al., 2000).

Greedy algorithm

A heuristic optimization strategy that operates on the same kind of tree-structured optimization problem as the $\rightarrow$ branch&bound algorithm. Whereas branch&bound algorithms aim at covering the whole tree, greedy algorithms generate only one path from the root of the tree to a (hopefully) close to optimal leaf node. The path is extended from the root towards the leaf by selecting one alternative to proceed at each step. This selection is usually made heuristically. Therefore, we also speak of greedy heuristics. The notion "greedy" originates from the fact, that each decision for an alternative, when taken, is never reconsidered or taken back.

Since greedy algorithms generate only one path in the tree they are generally much faster than branch&bound algorithms, Therefore greedy algorithms are an alternative to branch&bound algorithms, whenever runtime efficiency is of prime importance or if no effective lower bounding procedures are available.

In contrast, greedy algorithms usually provide solutions of significantly lower quality than branch&bound algorithms. Also no quality certificates can be given for solutions computed by greedy algorithms Generally, greedy algorithms are used in an early stage of algorithm development. There is a subclass of optimization problem for which greedy algorithms exist that are guaranteed provide the optimal solution. These problems are called matroid problems (Cormen et al., 1990).

Greedy algorithms are explicitly or implicitly mentioned in many chapters of this book.

Gibbs sampler

A specific, in some sense the simplest Markov Chain Monte Carlo algorithm ($\rightarrow$ Monte Carlo methods). In Gibbs sampling one samples the variables of a solution space one after the other. Each variable is drawn from a conditional distribution on the values of the other variables that were previously chosen. The conditional distributions must be accessible. The order in which variables are sampled one by one or in groups differs among different Gibbs sampling methods (Geman et al., 1984; Lawrence et al., 1993).

Hidden Markov models

Hidden Markov models (HMMs) are a popular variant of stochastic models for analyzing biomolecular sequences. HMMs are used for characterizing protein families (Baldi & Chauvin, 1994), (multiply) aligning protein sequences (Krogh et al., 1994), detecting remote protein homologies (Karplus et al., 1998), protein threading (Karplus et al., 1997), predicting signal peptides (Nielsen & Krogh, 1998) and transmembrane domains (Sonnhammer et al., 1998) in proteins, finding secondary structures in proteins (Asai et al., 1993), finding and characterizing promoters in DNA sequences (Pedersen et al., 1996), and gene finding (Henderson et al., 1997; Kulp et al., 1996).

An HMM is essentially a Markov chain ($\rightarrow$ Monte Carlo methods). Each state inside the Markov chain can produce a letter, and it does so by a stochastic Poisson process that chooses one of a finite number of letters in an alphabet. Each letter has a characteristic probability of being chosen. That probability depends on the state and on the letter. After the state produces a letter, the Markov chain moves to another state. This happens according to a transition probability that depends on the prior and succeeding state.

We exemplify HMMs with Figure 1, which is taken from (Eddy, 1998). Figure 1 depicts a Markov model for protein sequence alignment. Sometimes this specific topology of a hidden Markov model is called a *profile hidden Markov model* There is a specific *start* state b and a specific *final* state e. The *Match* states (m1 to m3) represent sequence positions. The *Insert*

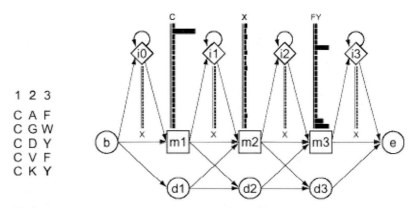

Fig. 1
A profile hidden Markov model (with permission of Oxford University Press).

(i0 to i3) and *Delete* (d1 to d3) states allow for inserting or jumping over sequence positions. Above each match state and below each insert state is a histogram of the probabilities with which each letter from the alphabet of amino acids is generated by the model. The model has been trained on the five three-letter sequences shown on the left.

The length of the hidden Markov model is the number of match states. A profile HMM stands for a family of – in this case protein – sequences.

Basically there are three algorithmic problems associated with an HMM.

1. **Training:** Set the probabilities of an HMM such that the protein sequences you want the protein family modeled by the HMM to belong to are generated with highest probability possible (Baum et al., 1970; Rabiner, 1989). Special measures are taken to avoid overfitting (see *Regularization* below). Training is by far the most complex of the three problems. It amounts to a minimization in a high-dimensional space with an irregular cost function with many minima. General solution methods boil down to local optimization procedures combined with a judicious choice of a limited set of starting solutions. Much progress can still be made on this problem.

2. **Classification:** Given a trained model and a protein sequence, evaluate the probability with which the model generates the sequence. If this probability is above a suitable threshold predict that the sequence belongs to the family modeled by the HMM. Otherwise predict the opposite. For profile HMMs and other special cases of HMMs this problem can be solved in linear time in the length of the model with → dynamic programming methods. In general the algorithm requires time $O(n^2 t)$ where n is the number of states in the HMM and t is the length of the sequence to be generated by the HMM (Rabiner, 1989).

3. **Alignment:** Given a trained model and a protein sequence, determine, which path through the model is the most probable to generate the sequence. In the case of profile HMMs this path represents an alignment of the sequence to the model. The alignment can also be done in linear time in the length of the model with → dynamic programming methods (Forney, 1973; Rabiner, 1989). Thus, by aligning many sequences to the model one after the other we can generate a multiple alignment in linear time in the length of the model times the number of sequences. This is in contrast to many other methods of multiple sequence alignment whose runtime grows exponentially in the number of sequences. Of course, HMMs transfer much of this complexity into the training phase.

A very nice general tutorial of HMMs is given in (Rabiner, 1989). An extension of HMMs from finite automata (which is the deterministic version of a Markov chain) to context-free languages leads to the concept of a *stochastic context-free grammar*. Such grammars have been used in an analogous way to predict RNA secondary structures (Grate et al., 1994).

Molecular dynamics

In molecular dynamics methods the equations of motion of a molecular assembly are integrated numerically (Sprik, 1996). Thus we obtain not only a molecular structure but a trajectory of molecular motion. Due to high frequency atomic oscillations the time step for the numerical integration must be very small (about 10^{-15} sec). This results in extraordinarily high computing requirements even for computing trajectories that span a few nanoseconds. In contrast, proteins fold in a few seconds. This difference in time scale cannot be overcome by sheer computing power. Thus methods are under investigation that can smooth out the high frequency oscillations without invalidating the final outcome (Elber et al., 1999).

Molecular dynamics methods are primarily used for the refinement of structural models (Li et al., 1997) or the analysis of molecular interactions (Cappelli et al., 1996; Kothekar et al., 1998). In both cases the time scales to be simulated are within range of current computing technology. Another application area is the study of allosteric movements of proteins (Tanner et al., 1993). Molecular Dynamics approaches to protein-ligand docking are described in Chapter 7 of Volume I.

Monte Carlo methods

General class of algorithmic methods that involve a stochastic element, i.e., that let the computer make random decisions (Binder & Landau, 2000). An important subclass, the so-called *Markov Chain Monte Carlo (MCMC) methods* can be understood as acting on Markov chains. A Markov chain is a stochastic finite automaton that consists of states and transitions between states (Feller, 1968). At each time we consider ourselves as resident in a certain state of the Markov chain. At discrete time steps we leave this state and move to another state of the chain. This transition is taken with a certain probability characteristic for the given Markov chain. The probability depends only on the two states involved in the transition. Often one can only move to a state in a small set of states from a given state. This set of states is called the *neighborhood* of the state from which the move originates.

A Markov chain is *ergodic* if it eventually reaches every state. If, in addition, a certain symmetry condition – the so-called criterion of *detailed balance* or *microscopic reversibility* – is fulfilled, the chain converges to the same *stationary probability distribution* of states, as we throw dice to decide which state transitions to take one after the other, no matter in which state we start. Thus, traversing the Markov chain affords us with an effective way of approximating its stationary probability distribution (Baldi & Brunak, 1998).

Monte Carlo algorithms pervade bioinformatics. Especially the chapters dealing with molecular structures mention this class of algorithms fre-

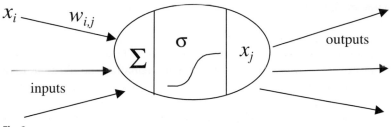

Fig. 2
Neuron j in a neural network

quently, for instance, to generate diverse ensembles of molecular structures or to perform energy minimization. → Simulated annealing is a special instance of a Monte Carlo algorithm. Chapter 7 of Volume I contains a detailed overview of Monte Carlo approaches to protein-ligand docking.

Neural network

A computational connection pattern that is formed in (simplified) analogy to biological (cortical) neural networks and used successfully for the classification of unstructured data. Neural networks have a broad range of applications in biology and chemistry (Schneider & Wrede, 1998).

In general, a neural network is a directed graph with nodes called neurons (see Figure 2). There may be a directed edge (i, j) from each neuron i to each neuron j. The directed edges (i, j) of the network are labeled with numbers $w_{i,j}$ called *weights*. Each neuron is equipped with an activation function σ. Neural networks are distinguished via the type of activation function they use. The more powerful class of neural networks uses nonlinear, e.g., sigmoid activation functions like $\sigma(x) = 1/(1 + e^{-x})$.

The neural network operates as follows. It is presented numbers $x_1, \ldots, x_n$ at distinguished input nodes. Each neuron applies its activation function to combine the numbers presented on its incoming edges and propagate the result along its outgoing edges. Specifically, the neuron j with, say, k incoming edges from neurons $i_1, \ldots, i_k$ computes the value $x_j = \sigma(w_{i_1,j} x_{i_1} + \cdots + w_{i_k,j} x_{i_k})$. The numbers produced at the outputs of the network are the result of the computation of the neural network. Data classification can be performed, e.g., by thresholding the single output of a neural network or presenting a separate output for each class and comparing the numbers produced at each of a set of distinguished output nodes.

Neural networks can be distinguished by their topology. *Layered feedforward* networks, the most popular type, are acyclic and organized in layers of neurons. They always produce outputs. The class of functions they can compute increases with the number of layers of neurons. In contrast, neural networks with cycles are more complicated. They can become unstable or oscillate or exhibit chaotic behavior.

The training procedure for neural networks optimizes the weights along the arcs of the network in order to optimally reproduce the function to be learned. This function is implicitly represented by the training data. Here we have to take care to avoid overfitting ($\rightarrow$ *Regularization*). The training problem is hard ($\rightarrow$ *hidden Markov models*) and the training procedure approaches the problem heuristically, usually by performing some type of local optimization (Russell & Norvig, 1995).

The application of neural feed-forward networks has proved effective in protein 2D structure prediction (see Chapter 6 of Volume I, (Rost et al., 1994)). Other applications of neural networks in bioinformatics include the recognition of protein sequence features such as transmembrane domains (Jacoboni et al., 2001) and recognition of features in DNA sequences (promoters (Matis et al., 1996; Nair et al., 1995), coding regions (Cai & Chen, 1995) etc.) as well as the interpretation of NMR spectra for protein structure resolution (Hare & Prestegard, 1994) and the discrimination between drug-like compounds and those that are not drug-like (Sadowski & Kubinyi, 1998).

A particularly interesting kind of neural network for bioinformatics is the *Kohonen map* (Kohonen, 1990). This is a two-dimensional field of artificial neurons that affords the projection of adjacencies in high-dimensional Euclidean spaces down to two dimensions. Among other things, Kohonen maps are used to classify sets of organic compounds (Anzali et al., 1996; Cai et al., 1998; Kirew et al., 1998), to distinguish between different protein binding sites (Stahl et al., 2000), to investigate protein families (Ferran et al., 1994) or aspects of protein structure (Ding & Dubchak, 2001; Schuchhardt et al., 1996), to analyze the codon usage in different bacterial genomes (Wang et al., 2001) and to cluster gene expression data (Tamayo et al., 1999).

Regularization

There is a general problem with methods that learn structure from limited amounts of training data, such as $\rightarrow$ hidden Markov models, $\rightarrow$ neural networks or $\rightarrow$ support vector machines mentioned in this glossary. The problem is called *overfitting*. If we carry the training procedures mentioned in the respective sections to their extreme, we will memorize all that we have seen in the training data, including the regularities that generalize to other data for which we want to make predictions in the future, and the irregularities that are specific to the training data and carry no information on data seen in the future. In order to avoid overfitting, the training data have to be supplemented with information that accounts for the lack of our knowledge on the data yet to be seen. In some sense, the classification function to be learned is smoothed out. It takes the training data into account, but also the lack of information on what is yet to come. Regularization is an involved concept that can be approached with theory (Vapnik, 2000). In practice, regularization is often performed heuristically, sometimes even in an *ad hoc* fashion.

The regularization procedure takes a different form for each method of statistical learning. When training hidden Markov models, the derived probabilities do not only take the observed sequences into account but also use so-called *prior* distributions that formulate some hypothesis on the occurrence of output characters in the case that we have no additional information (such as the observed sequences). An expected background distribution of amino acids or nucleotides is a natural starting point for such a prior distribution.

For → neural networks, a common implementation of regularization is to not carry the training procedure to the end but stop sometime before. → Support Vector Machines are automatically regularized by maximizing the margin of separation of the two classes of training data. The amount of regularization can be directly controlled by the parameter that guides the tradeoff between correct classification of the training data and the size of the margin.

Simulated annealing

Another Markov Chain Monte Carlo (MCMC) algorithm. Here each state of the Markov chain is labeled with a "cost" or "energy". Moves to neighboring states are accepted if they involve a decrease in cost. If they involve an increase in cost, they are only accepted with a certain probability. This probability is determined by the Metropolis criterion $e^{-c/kT}$, where c is the increase in cost, T is a *temperature* parameter and k is the Boltzmann constant (Metropolis et al., 1953). T starts out at a finite initial temperature and is then decreased according to a so-called *cooling schedule* until it approaches zero. This means that, as the algorithm proceeds, increases in cost become less and less likely. Partly because of its paradigmatic relationship with statistical physics, simulated annealing is a popular version of an MCMC algorithm. The algorithm has the facility of escaping from local minima reached during execution of the algorithm. It is frequently used in global optimization (Kirkpatrick, 1984). Simulated annealing cooling schedules are also very popular with Molecular Dynamics, e.g. for structural refinements using experimental constraints (Brunger et al., 1997) or for generating and refining molecular conformations in homology-based protein modeling (Chapter 5 of Volume I). Simulated annealing approaches for protein-ligand docking are described in Chapter 7 of Volume I.

Support vector machine

Support vector machines are a technology developed by Vapnik and others for classifying unstructured data (Burges, 1998; Christianini & Shawe-Taylor, 2000) that has gained much attention in bioinformatics lately. Support vector machines classify points in multi-dimensional Euclidean space into two classes. They extend linear classifiers (see Figure 3 a) that classify the points by placing a hyperplane into the space that separates the space into two half spaces. Training the classifier amounts to selecting the hyperplane such that it best fits a subset of preclassified points. This means that the smallest number of preclassified points comes to lie in the wrong half space. In fact, this is not quite the case, since we want to avoid overfitting to the training data (→ *Regularization*).

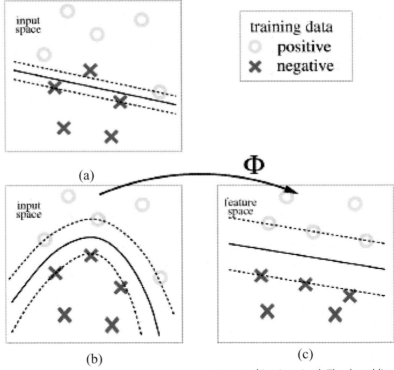

(a)

(b)

(c)

Fig. 3
Support Vector Machines: (a) linear classifier, (b) nonlinear classifier, (c) nonlinear classifier mapped to a high-dimensional space, where the classification becomes linear. Crosses and circles denote pre-classified points on which the support vector machine is trained. The dotted lines denote the margin between the separating surface and the closest preclassified points. The mapping from (b) to (c) is performed implicitly by a so-called *kernel function* Φ

Often hyperplanes are not sufficient for the classification of point sets. Support vector machines allow nonlinear separation between the two point classes (Figure 3 b). They do so by mapping the points nonlinearly into a higher-dimensional feature space and performing linear separation there (Figure 3 c). The mapping is done implicitly by a so-called *kernel function*, thereby making very high-dimensional feature spaces computationally affordable. The choice of the kernel function is the essential ingredient in the development of the support vector machine.

Support vector machines receive their name through their second feature. Rather than taking all preclassified points into account, they only consider those points that come to lie close to the separating surface. These points are called the support vectors. The surface is placed such that the distance to the closest support vectors is maximized (modulo preventing overfitting, → Regularization). Here we allow a limited number of misplacements.

Support vector machines have been applied to a wide variety of classification problems in bioinformatics including finding translation initiation sites in DNA sequences (Zien et al., 2000), classifying genes via their promoter regions (Pavlidis et al., 2001a), analyzing gene expression data (Brown et al., 2000; Furey et al., 2000), prediction of gene function (Pavlidis et al., 2001b), determining secondary structures of proteins (Hua & Sun, 2001), protein fold recognition (Ding & Dubchak, 2001), prediction of protein-protein interactions (Bock & Gough, 2001), and protein sorting (Cai et al., 2000).

Tabu search

Meta-heuristic for local search strategies. The goal is to avoid getting caught in cycles while traversing the search space. Tabu search starts out by looking for local minima. To avoid retracing the moves just made the method protocols characteristic features of solutions that were recently visited in one or several tabu lists, each list for a certain class of solution attributes. Tabu search methods vary in the details of the management of the tabu lists (Glover, 1989; Glover, 1990). Tabu search is mentioned in Chapter 7 of Volume I. Additonal material on tabu search in docking can be found in (Hou et al., 1999; Westhead et al., 1997).

Acknowledgements

I am grateful to Holger Claußen, Daniel Hoffmann, Jochen Selbig, Alexander Zien, and Ralf Zimmer for helpful comments on this glossary. Alexander Zien provided Figure 3.

References

ALTHAUS, E., KOHLBACHER, O., LENHOF, H. P. & MÜLLER, P. (2000). A combinatorial approach to protein docking with flexible side-chains. In *Annual International Conference on Computational Molecular Biology (RECOMB)*, pp. 8–14. ACM Press.

ANZALI, S., BARNICKEL, G., KRUG, M., SADOWSKI, J., WAGENER, M., GASTEIGER, J. & POLANSKI, J. (1996). The comparison of geometric and electronic properties of molecular surfaces by neural networks: application to the analysis of corticosteroid-binding globulin activity of steroids. *J Comput Aided Mol Des* 10(6), 521–34.

ASAI, K., HAYAMIZU, S. & HANDA, K. (1993). Prediction of protein secondary structure by the hidden Markov model. *Comput Appl Biosci* 9(2), 141–6.

BALDI, P. & BRUNAK, S. (1998). *Bioinformatics: the machine learning approach*. Adaptive computation and machine learning, MIT Press, Cambridge, Mass.

BALDI, P. & CHAUVIN, Y. (1994). Hidden Markov Models of the G-protein-coupled receptor family. *J Comput Biol* 1(4), 311–36.

BAUM, L. E., PETRIE, T., SOULES, G. & WEISS, N. (1970). A maximization technique occurring in the statistical analysis of probabilistic functions of Markov chains. *Ann Math Stat* 41(1), 164–171.

BINDER, K. & LANDAU, D. P. (2000). *A Guide to Monte Carlo Simulations in Statistical Physics*, Cambridge University Press.

BOCK, J. R. & GOUGH, D. A. (2001). Predicting protein-protein interactions from primary structure. *Bioinformatics* 17(5), 455–60.

BRACEWELL, R. N. (1986). *The Fourier transform and its applications*. 2nd, rev. edit. McGraw-Hill series in electrical engineering. Circuits and systems, McGraw-Hill, New York.

BROWN, M. P., GRUNDY, W. N., LIN, D., CRISTIANINI, N., SUGNET, C. W., FUREY, T. S., ARES, M., Jr. & HAUSSLER, D. (2000). Knowledge-based analysis of microarray gene expression data by using support vector machines. *Proc Natl Acad Sci U S A* 97(1), 262–7.

BRUNGER, A. T., ADAMS, P. D. & RICE, L. M. (1997). New applications of simulated annealing in X-ray crystallography and solution NMR. *Structure* 5(3), 325–36.

BURGES, C. (1998). A Tutorial on Support Vector Machines for Pattern Recognition. *Data Mining and Knowledge Discovery* 2(2), 121–167.

CAI, Y. & CHEN, C. (1995). Artificial neural network method for discriminating coding regions of eukaryotic genes. *Comput Appl Biosci* 11(5), 497–501.

CAI, Y. D., LIU, X. J., XU, X. & CHOU, K. C. (2000). Support vector machines for prediction of protein subcellular location. *Mol Cell Biol Res Commun* 4(4), 230–3.

CAI, Y. D., YU, H. & CHOU, K. C. (1998). Artificial neural network method for predicting HIV protease cleavage sites in protein. *J Protein Chem* 17(7), 607–15.

CAPPELLI, A., DONATI, A., ANZINI, M., VOMERO, S., DE BENEDETTI, P. G., MENZIANI, M. C. & LANGER, T. (1996). Molecular structure and dynamics of some potent 5-HT3 receptor antagonists. Insight into the interaction with the receptor. *Bioorg Med Chem* 4(8), 1255–69.

CHRISTENSEN, I. T. & JORGENSEN, F. S. (1997). Molecular mechanics calculations of proteins. Comparison of different energy minimization strategies. *J Biomol Struct Dyn* 15(3), 473–88.

CHRISTIANINI, N. & SHAWE-TAYLOR, J. **(2000)**. *An Introduction to Support Vector Machines*, University Press, Cambridge, UK.

CHURCH, B. W. & SHALLOWAY, D. **(2001)**. Top-down free-energy minimization on protein potential energy landscapes. *Proc Natl Acad Sci U S A* 98(11), 6098–103.

COOLEY, J. W. & TUKEY, J. W. **(1965)**. An algorithm for the nachine calculation of complex Fourier series. *Mathematics of Computation* 19(90), 297–301.

CORMEN, T. H., LEISERSON, C. E. & RIVEST, R. L. **(1990)**. *Introduction to algorithms*, MIT Press, Cambridge, MA, USA.

DESMET, J., DE MAEYER, M., HAZES, B. & LASTERS, I. **(1992)**. The dead-end elimination theorem and its use in protein side-chain positioning. *Nature* 356, 539–542.

DING, C. H. & DUBCHAK, I. **(2001)**. Multi-class protein fold recognition using support vector machines and neural networks. *Bioinformatics* 17(4), 349–58.

EDDY, S. R. **(1998)**. Profile Hidden Markov Models. *Bioinformatics* 14(9), 755–63.

ELBER, R., MELLER, J. & OLENDER, R. **(1999)**. Stochastic path approach to compute atomically detailed trajectories: application to the folding of C peptide. *Journal of Physical Chemistry B* 103(6), 899–911.

FELLER, W. **(1968)** *An Introduction to Probability Theory & Its Applications*, Vol 1. Probability & Mathematical Statistics.

FERRAN, E. A., PFLUGFELDER, B. & FERRARA, P. **(1994)**. Self-organized neural maps of human protein sequences. *Protein Sci* 3(3), 507–21.

FOGEL, L. J., OWENS, A. J. & WALSH, M. J. **(1966)**. Intelligent decision making through a simulation of evolution. *Behav Sci* 11(4), 253–72.

FORNEY, G. D., Jr. **(1973)**. The Viterbi algorithm. *Proceedings of the IEEE* 61(3), 268–78.

FUREY, T. S., CRISTIANINI, N., DUFFY, N., BEDNARSKI, D. W., SCHUMMER, M. & HAUSSLER, D. **(2000)**. Support vector machine classification and validation of cancer tissue samples using microarray expression data. *Bioinformatics* 16(10), 906–14.

GARDINER, E. J., WILLETT, P. & ARTYMIUK, P. J. **(2001)**. Protein docking using a genetic algorithm. *Proteins* 44(1), 44–56.

GEMAN, S. G. D., GEMAN, S. & GEMAN, D. **(1984)**. Stochastic relaxation, Gibbs distributions, and Bayesian restoration of images. *IEEE Transactions on Pattern Analysis & Machine Intelligence* PAMI-6(6), 721–41.

GLOVER, F. **(1989)**. Tabu search. 1. *ORSA Journal on Computing* 1(3), 190–206.

GLOVER, F. **(1990)**. Tabu search. II. *ORSA Journal on Computing* 2(1), 4–32.

GOLDBERG, D. E. **(1989)**. *Genetic Algorithms in Search, Optimization and Machine Learning*, Addison-Wesley.

GRATE, L., HERBSTER, M., HUGHEY, R., HAUSSLER, D., MIAN, I. S. & NOLLER, H. **(1994)**. RNA modeling using Gibbs sampling and stochastic context free grammars. *Proc Int Conf Intell Syst Mol Biol* 2, 138–46.

GUSFIELD, D. **(1997)**. *Algorithms on strings, trees, and sequences: Computer science and computational biology*, Cambridge University Press, Cambridge [England]; New York.

SCHNEIDER, G. & WREDE, P. **(1998).** Artificial neural networks for computer-based molecular design. *Progress in Biophysics & Molecular Biology* 70(3), 175–222.

SCHUCHHARDT, J., SCHNEIDER, G., REICHELT, J., SCHOMBURG, D. & WREDE, P. **(1996).** Local structural motifs of protein backbones are classified by self-organizing neural networks. *Protein Eng* 9(10), 833–42.

SCHULZE-KREMER, S. **(2000).** Genetic algorithms and protein folding. *Methods Mol Biol* 143, 175–222.

SONNHAMMER, E. L., von HEIJNE, G. & KROGH, A. **(1998).** A hidden Markov model for predicting transmembrane helices in protein sequences. *Proc Int Conf Intell Syst Mol Biol* 6, 175–82.

SPRIK, M. **(1996).** Introduction to molecular dynamics methods. *Proceedings of the Conference on Monte Carlo and Molecular Dynamics of Condensed Matter Systems* 49, 43–88.

STAHL, M., TARONI, C. & SCHNEIDER, G. **(2000).** Mapping of protein surface cavities and prediction of enzyme class by a self-organizing neural network. *Protein Eng* 13(2), 83–8.

SZUSTAKOWSKI, J. D. & WENG, Z. **(2000).** Protein structure alignment using a genetic algorithm. *Proteins* 38(4), 428–40.

TAMAYO, P., SLONIM, D., MESIROV, J., ZHU, Q., KITAREEWAN, S., DMITROVSKY, E., LANDER, E. S. & GOLUB, T. R. **(1999).** Interpreting patterns of gene expression with self-organizing maps: methods and application to hematopoietic differentiation. *Proc Natl Acad Sci U S A* 96(6), 2907–12.

TANNER, J. J., SMITH, P. E. & KRAUSE, K. L. **(1993).** Molecular dynamics simulations and rigid body (TLS) analysis of aspartate carbamoyltransferase: evidence for an uncoupled R state. *Protein Sci* 2(6), 927–35.

TROSSET, J. Y. & SCHERAGA, H. A. **(1998).** Reaching the global minimum in docking simulations: a Monte Carlo energy minimization approach using Bezier splines. *Proc Natl Acad Sci U S A* 95(14), 8011–5.

VAPNIK, V. N. **(2000).** *The nature of statistical learning theory.* 2nd edit. Statistics for engineering and information science, Springer, New York.

WANG, H. C., BADGER, J., KEARNEY, P. & LI, M. **(2001).** Analysis of codon usage patterns of bacterial genomes using the self-organizing map. *Mol Biol Evol* 18(5), 792–800.

WATERMAN, M. S. **(1995).** *Introduction to computational biology: Maps, sequences and genomes.* 1st edit, Chapman & Hall, London; New York, NY.

WESTHEAD, D. R., CLARK, D. E. & MURRAY, C. W. **(1997).** A comparison of heuristic search algorithms for molecular docking. *J Comput Aided Mol Des* 11(3), 209–28.

WILLETT, P. **(1995).** Genetic algorithms in molecular recognition and design. *Trends Biotechnol* 13(12), 516–21.

ZIEN, A., RATSCH, G., MIKA, S., SCHOLKOPF, B., LENGAUER, T. & MULLER, K. R. **(2000).** Engineering support vector machine kernels that recognize translation initiation sites. *Bioinformatics* 16(9), 799–807.

Subject Index

Index entries give volume number first, then page number.
Boldface numbers refer to term definitions.

Name Index

Index entries give volume number first, then page number.